The Handbook of Environmental Chemistry

Volume 3 Anthropogenic Compounds Part T

O. Hutzinger
Editor-in-Chief

Springer

Berlin
Heidelberg
New York
Hong Kong
London
Milan
Paris
Tokyo

Traffic and Environment

Volume Editor: Dušan Gruden

With contributions by
W. Berg · K. Borgmann · D. Gruden · O. Hiemesch
H.-P. Lenz · S. Prüller

Springer

Volume Editor

Prof. Dr. Dušan Gruden
Porsche Aktiengesellschaft
Porschestraße
71287 Weissach
Germany
E-mail: dusan.gruden@porsche.de

ISSN 1433-6847
ISBN 3-540-00050-x
DOI 10.1007/b13434
Springer-Verlag Berlin Heidelberg New York

Library of Congress Cataloging-in-Publication Data
The Natural environment and the biogeochemical cycles / with contributions by P. Craig ... [et al.].
v. <A–F > : ill. ; 25 cm. – (The Handbook of environmental chemistry : v. 1) Includes bibliographical references and indexes.
ISBN 0-387-09688-4 (U.S). – ISBN 3-540-55255-3 (pt. F : Berlin). –
ISBN 0-387-55255-3 (pt. F : New York)
1. Biogeochemical cycles. 2. Environmental chemistry.
I. Craig, P. J., 1944– . II. Series.
QD31. H335 vol. 1 [QH344] 628.5 s

Springer-Verlag Berlin Heidelberg New York
a member of BertelsmannSpringer Science+Business Media GmbH
http//www.springer.de

Printed in Germany

Production Editor: Christiane Messerschmidt, Rheinau
Cover Design: E. Kirchner, Springer-Verlag
Typesetting: Fotosatz-Service Köhler GmbH, Würzburg
Printed on acid-free paper 52/3020 – 5 4 3 2 1 0

The Handbook of Environmental Chemistry Also Available Electronically

Environmental chemistry is a rather young and interdisciplinary field of science. Its aim is a complete description of the environment and of transformations occurring on a local or global scale. Environmental chemistry also gives an account of the impact of man's activities on the natural environment by describing observed changes.

"The Handbook of Environmental Chemistry" provides the compilation of today's knowledge. Contributions are written by leading experts with practical experience in their fields. The Handbook will grow with the increase in our scientific understanding and should provide a valuable source not only for scientists, but also for environmental managers and decision makers.

As a rule, contributions are specially commissioned. The editors and publishers will, however always be pleased to receive suggestions and supplements information. Papers for *The Handbook of Environmental Chemistry* are accepted in English.

In reference The Handbook of Environmental Chemistry is abbreviated Handb. Environ. Chem. and is cited as a journal.

Springer WWW home page: http://www.springer.de
Visit the MS home page at www.springerlink.com/series/hec/
or http://link.springer.ny.com/series/hec/

For all customers with a standing order for The Handbook of Environmental Chemistry we offer the electronic form via SpringerLink free of charge. Please contact your librarian who can receive a password for free access to the full articles. By registration at:

http://link.springer.de/orders/index.htm

However, if you do not have a standing order, you can browse through the table of contents of the volumes and the abstracts of each article at:

http://www.springerlink.com/series/hec

There you will also find information about the

- Editorial Board
- Aims and Scope
- Instructions for Authors

Preface

Environmental Chemistry is a relatively young science. Interest in this subject, however, is growing very rapidly and, although no agreement has been reached as yet about the exact content and limits of this interdisciplinary discipline, there appears to be increasing interest in seeing environmental topics which are based on chemistry embodied in this subject. One of the first objectives of Environmental Chemistry must be the study of the environment and of natural chemical processes which occur in the environment. A major purpose of this series on Environmental Chemistry, therefore, is to present a reasonably uniform view of various aspects of the chemistry of the environment and chemical reactions occurring in the environment.

The industrial activities of man have given a new dimension to Environmental Chemistry. We have now synthesized and described over five million chemical compounds and chemical industry produces about hundred and fifty million tons of synthetic chemicals annually. We ship billions of tons of oil per year and through mining operations and other geophysical modifications, large quantities of inorganic and organic materials are released from their natural deposits. Cities and metropolitan areas of up to 15 million inhabitants produce large quantities of waste in relatively small and confined areas. Much of the chemical products and waste products of modern society are released into the environment either during production, storage, transport, use or ultimate disposal. These released materials participate in natural cycles and reactions and frequently lead to interference and disturbance of natural systems.

Environmental Chemistry is concerned with reactions in the environment. It is about distribution and equilibria between environmental compartments. It is about reactions, pathways, thermodynamics and kinetics. An important purpose of this Handbook, is to aid understanding of the basic distribution and chemical reaction processes which occur in the environment.

Laws regulating toxic substances in various countries are designed to assess and control risk of chemicals to man and his environment. Science can contribute in two areas to this assessment; firstly in the area of toxicology and secondly in the area of chemical exposure. The available concentration ("environmental exposure concentration") depends on the fate of chemical compounds in the environment and thus their distribution and reaction behaviour in the environment. One very important contribution of Environmental Chemistry to the above mentioned toxic substances laws is to develop laboratory test methods, or mathematical correlations and models that predict the environmental fate of new chemical compounds. The third purpose of this Handbook is to help in the basic understanding and development of such test methods and models.

The last explicit purpose of the Handbook is to present, in concise form, the most important properties relating to environmental chemistry and hazard assessment for the most important series of chemical compounds.

At the moment three volumes of the Handbook are planned. Volume 1 deals with the natural environment and the biogeochemical cycles therein, including some background information such as energetics and ecology. Volume 2 is concerned with reactions and processes in the environment and deals with physical factors such as transport and adsorption, and chemical, photochemical and biochemical reactions in the environment, as well as some aspects of pharmacokinetics and metabolism within organisms. Volume 3 deals with anthropogenic compounds, their chemical backgrounds, production methods and information about their use, their environmental behaviour, analytical methodology and some important aspects of their toxic effects. The material for volume 1, 2 and 3 was each more than could easily be fitted into a single volume, and for this reason, as well as for the purpose of rapid publication of available manuscripts, all three volumes were divided in the parts A and B. Part A of all three volumes is now being published and the second part of each of these volumes should appear about six months thereafter. Publisher and editor hope to keep materials of the volumes one to three up to date and to extend coverage in the subject areas by publishing further parts in the future. Plans also exist for volumes dealing with different subject matter such as analysis, chemical technology and toxicology, and readers are encouraged to offer suggestions and advice as to future editions of "The Handbook of Environmental Chemistry".

Most chapters in the Handbook are written to a fairly advanced level and should be of interest to the graduate student and practising scientist. I also hope that the subject matter treated will be of interest to people outside chemistry and to scientists in industry as well as government and regulatory bodies. It would be very satisfying for me to see the books used as a basis for developing graduate courses in Environmental Chemistry.

Due to the breadth of the subject matter, it was not easy to edit this Handbook. Specialists had to be found in quite different areas of science who were willing to contribute a chapter within the prescribed schedule. It is with great satisfaction that I thank all 52 authors from 8 countries for their understanding and for devoting their time to this effort. Special thanks are due to Dr. F. Boschke of Springer for his advice and discussions throughout all stages of preparation of the Handbook. Mrs. A. Heinrich of Springer has significantly contributed to the technical development of the book through her conscientious and efficient work. Finally I like to thank my family, students and colleagues for being so patient with me during several critical phases of preparation for the Handbook, and to some colleagues and the secretaries for technical help.

I consider it a privilege to see my chosen subject grow. My interest in Environmental Chemistry dates back to my early college days in Vienna. I received significant impulses during my postdoctoral period at the University of California and my interest slowly developed during my time with the National Research

Council of Canada, before I could devote my full time of Environmental Chemistry, here in Amsterdam. I hope this Handbook may help deepen the interest of other scientists in this subject.

Amsterdam, May 1980 *O. Hutzinger*

Twentyone years have now passed since the appearance of the first volumes of the Handbook. Although the basic concept has remained the same changes and adjustments were necessary.

Some years ago publishers and editors agreed to expand the Handbook by two new open-end volume series: Air Pollution and Water Pollution. These broad topics could not be fitted easily into the headings of the first three volumes. All five volume series are integrated through the choice of topics and by a system of cross referencing.

The outline of the Handbook is thus as follows:

1. The Natural Environment and the Biogeochemical Cycles,
2. Reaction and Processes,
3. Anthropogenic Compounds,
4. Air Pollution,
5. Water Pollution.

Rapid developments in Environmental Chemistry and the increasing breadth of the subject matter covered made it necessary to establish volume-editors. Each subject is now supervised by specialists in their respective fields.

A recent development is the accessibility of all new volumes of the Handbook from 1990 onwards, available via the Springer Homepage http://www.springer. de or http://Link.springer.de/series/hec/ or http://Link.springerny.com/ series/hec/.

During the last 5 to 10 years there was a growing tendency to include subject matters of societal relevance into a broad view of Environmental Chemistry. Topics include LCA (Life Cycle Analysis), Environmental Management, Sustainable Development and others. Whilst these topics are of great importance for the development and acceptance of Environmental Chemistry Publishers and Editors have decided to keep the Handbook essentially a source of information on "hard sciences".

With books in press and in preparation we have now well over 40 volumes available. Authors, volume-editors and editor-in-chief are rewarded by the broad acceptance of the "Handbook" in the scientific community.

Bayreuth, July 2001 *Otto Hutzinger*

Contents

Foreword

Over centuries mankind has pursued technical progress for the benefit of improved prosperity without simultaneously taking appropriate steps to ensure the environmental friendliness of the involved processes. However, in the middle of the 20th century environmental episodes drew attention to the negative impacts on the environment caused by this progress.

As a matter of fact, concern about the influence of human activities on the environment is neither a new phenomenon nor a new attribute of modern people but has accompanied human society throughout its existence. What is new, however, is the increasing intensity of man's efforts to protect his environment as reflected in a multitude of national and international environmental laws enacted all around the globe.

Life as a whole, and human existence in particular, are characterized by constant movement and changes. This means that living beings need to be mobile to survive. By developing suitable technical means man has enormously increased his mobility – expressed in terms of speed and distance – when compared with other living beings on our planet. The automobile is one of the inventions that has made a decisive contribution to this mobility and it has become an inseparable part of modern human society. In the second half of the 20th century, the automobile developed from a luxury article and prestige object for a few into a basic commodity for millions of people. It is through this widespread use that negative impacts on the environment have become clearly visible. Therefore, since the late 1960s and early 1970s, automotive development has been accompanied by an ever increasing number of strict legal standards, e.g., about the reduction of exhaust gas pollutants, noise emissions, hazardous substances and waste, as well as about improved recyclability of materials and other aspects.

Achievements in improving the ecological characteristics of the automobile are highly impressive: A modern car emits only fractions of the amounts of noise and exhaust gas pollutants produced by its predecessors 30 years ago. Today, 100 modern passenger cars in total emit less of the legally limited exhaust gas constituents than one single car of 1970. The same trend can be found with all the other ecologically relevant automotive features so that the absolute impact of the automobile on our environment is considerably lower today than it was in the past.

The development of the automobile is increasingly linked to deliberations about sustainable development. While this term in the recent past was only related to the aspect of ecological consequences for the environment, it comprises today at least two further essential pillars, namely economic consequences and social responsibility.

When discussing sustainability in the context of automotive development, it must be borne in mind that essential technical elements of the automobile – such as safety, power output, torque, fuel consumption, durability, maintenance intervals, and comfort should not be compromised.

The modern automobile has achieved outstanding performance and superiority compared to its predecessors in all theses elements and will continue to proceed along this evolutionary development path.

This book focuses on ecological aspects related to the development and use of automobiles, leaving many environment-related initiatives towards improvements of the automotive production process out of consideration. It shall, however, be mentioned in this context that also the production of modern cars is not possible without the observance of a wide range of stringent environmental laws. Thus, in order to be allowed to enter the market, a car must not only perform environmental-friendly during its operation but must have been produced to ecological standards as well. Company audits carried out routinely according to EMAS (Eco Management Auditing Scheme) and ISO 14001 show that automotive manufacturers are constantly improving the ecological compatibility of their production processes.

The contributions to this book were written by experts, most of whom have been actively involved in the development of modern automobiles and their combustion engines for more than 30 years. They have participated in all phases of the ecological development of the automobile – from the basic attempts to respond to the first exhaust gas emission control requirements in the USA (1966) and Europe (1970) to the cost-intensive efforts towards meeting the comprehensive and highly demanding emission legislations currently existing and further anticipated worldwide.

As the 20th century ends and the 21st century begins, these experts have summarized their experience and know-how in this book which bears witness to the successful implementation of ecological considerations into automotive development work.

In my capacity as coordinator of the preparatory work for this book I would like to thank my colleagues – Prof. Dr. sc. techn. Hans Peter Lenz and his collaborator, Mr. Stefan Prüller (Dipl.-Ing.) of Technical University of Vienna, Dr. Klaus Borgmann and Mr. Otto Hiemesch (Dipl.-Ing.) of BMW AG and Dr. Wolfgang Berg, Consultant and long-standing collaborator of DaimlerChrysler AG – for their cooperation and valuable contributions.

I would like to express particular gratitude to Dr. Ing. h.c. F. Porsche AG for permission to carry out this project.

Weissach, June 2003 *D. Gruden*

The Handbook of Environmental Chemistry Vol. 3, Part T (2003): 1–13
DOI 10.1007/b11990

Introduction

Dušan Gruden

Dr. Ing. h.c. F. Porsche Aktiengesellschaft, Porschestrasse, 71287 Weissach, Germany,
E-mail: dusan.gruden.@porsche.de

Terrestrial life as it is known today has been made possible by the very special environmental conditions on this planet of ours. Man has failed yet to provide evidence that there are forms of life similar to ours on other planets of the universe. According to our current state of knowledge, the planet earth with its multitude of most different living creatures is unique in the infinity of the cosmos. This unique nature must be preserved – this is one of the greatest challenges to mankind.

Among the creatures on earth it is man who has the greatest inherent capability of bringing about rapid changes to the environment. Man is currently consuming the terrestrial energy and raw material resources in an irreversible manner. In the future we must develop economic systems which reflect our environmental awareness by giving consideration to the environmental potential, the existing resources and the capabilities of nature. The engineer's main focus will have to be on the saving of energy, the efficient dealing with energy being of decisive importance for the success of evolutionary development.

One of the indispensable conditions and basic prerequisites of life is mobility. Today more than ever before, mobility is the key factor of modern society.

Society's need for mobility has resulted in the development of various traffic systems. So, mobility and traffic, though being different from each other, are very closely linked together. In Europe today, road traffic is one of the most important systems for the transportation of passengers and goods. With more than 80% of passenger traffic and 75% of goods transportation being handled by passenger cars and trucks "automobility" has turned into a veritable challenge for human society. Besides its undisputed advantages, road traffic also entails certain environmental problems which are much discussed and frequently argued about. The most frequently cited encroachments are the consumption of land, the cutting-up and separation of landscapes, traffic safety problems, noise and pollutant emissions as well as the using up of energy and raw-material resources.

One of the ecological priorities of modern society is to minimize the environmental burden caused by automotive traffic. What must be done is to create a fair balance between the requirements of economy, industry, the market as such and environment protection.

Keywords. Mobility, Traffic, Transport, Energy, Raw materials, Environmental protection

1 We Have Only One Earth

Terrestrial life as it is known today has been made possible by the very special environmental conditions on this planet of ours, that is the 21% oxygen content of the atmosphere, the fact that 3/4 of the surface of the earth is covered by water, a max./min temperature difference of approximately 100°C (ranging from −50 to +50°C) and an average temperature level of 15°C (Fig. 1).

We believe our civilization to be a very sophisticated one with almost unlimited technical capabilities. Nevertheless, man has failed to provide evidence that there are forms of life similar to ours on other planets of the universe or that there is life out there at all. In other words, according to our current state of knowledge, the planet earth with its multitude of most different living creatures is unique in the infinity of the cosmos. The uniqueness of this nature of ours must

Fig. 1. Planet Earth

be preserved – this is one of the greatest challenges for mankind and a serious obligation to which all of us should commit ourselves.

Among the creatures on earth it is man who has the greatest inherent capability of bringing about rapid changes to the environment. Man began influencing nature thousands of years ago, when he learnt how to make stone axes and how to handle fire. It was then that the delicate ecological balance started to be disturbed and it has been slowly deteriorating ever since.

Since the industrial revolution which started with the invention of the steam engine nature has changed dramatically. Human influence is most strongly felt in densely populated conurbations where many plant and animal species have disappeared for ever.

Quite obviously, the biggest environmental problem is the increasing number of human beings who are crowding our planet. Presently, the earth is home to 6 billion inhabitants. Every second, three babies are born adding another 250,000 to the world's population every day. If this birth rate continues at its present pace, the terrestrial population will double in the next 50 years.

Human beings need befitting living conditions providing them with sufficient space, food and prosperity. For the production of such food and the creation of prosperity energy is needed (Fig. 2).

The worldwide energy consumption per inhabitant varies by the factor of 1000 between 564 GJ max. and 0.5 GJ min. with the average consumption being 56 GJ.

For any type of energy used by man there is an ecological price to pay. None of the energy sources can be exploited without affecting the environment. The only difference is in the extent and intensity of these negative environmental effects.

The enormous number of human beings living on this planet and their ever increasing and unevenly distributed need for energy is the main problem with

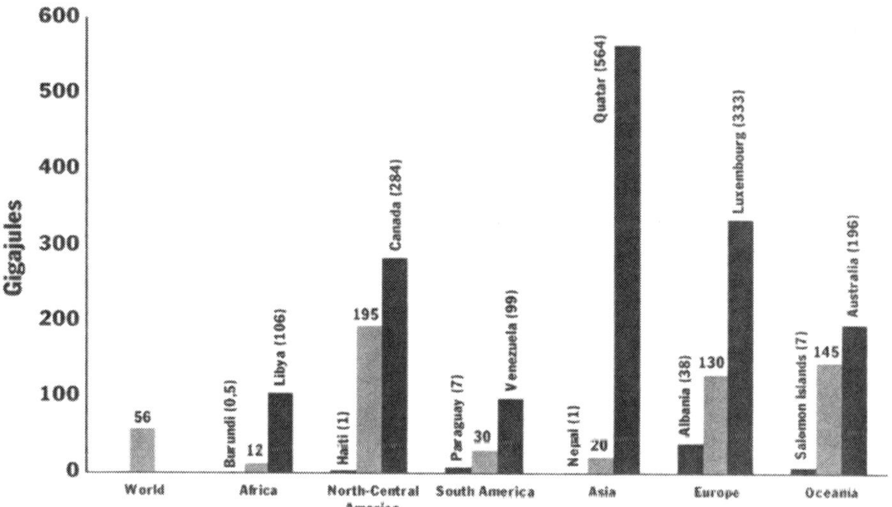

Fig. 2. Energy consumption per inhabitant and year

which environmental protection initiatives are confronted today. If we fail to develop efficient control mechanisms to slow down the population growth while preserving the great variety of peoples, races, languages, customs, cultures and religions, humanity will soon be facing one of its greatest ecological disasters if not the most vital problem of survival in the history of mankind.

Many of the other environmental issues which are publicly discussed with great intensity and emotion are dwarfed by the problem produced by the growth of the world population and the increasing need for energy resulting therefrom.

Man is currently using up the terrestrial energy and raw material resources in an irreversible manner. It is not hard to imagine humanity dying a "thermodynamic death" which may one day be brought about by a rapidly growing entropy.

Is there a solution to this dilemma? Will we be capable of finding ways to preserve life on earth while allowing humanity to continue to evolve and prosper?

Finding answers to those vital questions is an enormous moral obligation, especially for engineers and technicians. And it is one of the greatest challenges of the present.

Our technologies are based on physical laws – and physical laws are laws of nature. What we have to learn is not to use these laws against nature but to harmonize them with the demands of nature. Such aggressive slogans as "man shall fight nature" or "man has conquered nature" which have been considered to be progressive for many centuries, must be taken for what they really are: guidelines leading into disaster. The command we must follow to survive is far more peaceful: Man shall cooperate with nature.

Figure 3 outlines the processes which man must internalize and adapt. In the past, man's activities were focused on the satisfaction of his own needs and the gaining of profit. To this end, various activities and economic branches were cre-

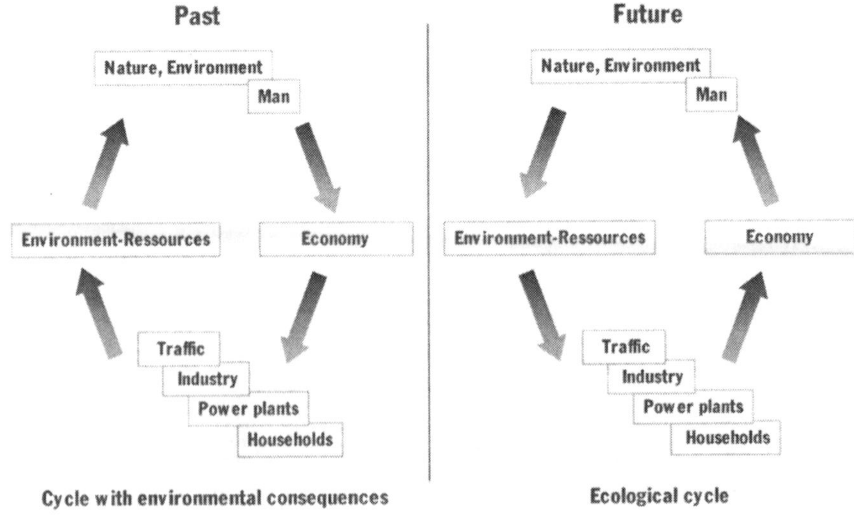

Fig. 3

ated which have contributed to the consumption of terrestrial resources and placed heavy burdens on environment.

In the future, the partners in this global game will be the same. What will have to be changed, though, is the direction in which human activities are moving.

We must develop economic systems which reflect our environmental awareness by giving consideration to the environmental potential, the existing resources and the capabilities of nature.

At present, we are involved in a learning process which will show us how to invert the current negative trend. Given the enormous mass of our global socio-economic system, this change of direction must be done slowly and in an evolutionary way in order to prevent tremendous mass forces from being triggered which might irreparably destroy this system through their shear magnitude. The engineers' main focus will have to be on the saving of energy, the economic dealing with energy being of decisive importance for the success of this evolutionary development.

2
Mobility – Life is a Journey, Not a Destination

There is no life without mobility. Mobility is the indispensable condition and basic prerequisite of life while immobility is considered to be synonymous with death. Among the most deep-rooted qualities of man is his striving for freedom, one of the most important facets of which is the freedom of movement: Citius, altius, cellerius – farther, higher, quicker – those are the key words which prompted man to start improving his mobility.

In the early days, man was a walker. Very soon he discovered that animals could be used to speed up transportation. The invention of the wheel made work easier and also allowed a new means of transportation to be developed – the cart. It did not take man long to find out, though, that driving carts can also be great fun – a quality which has lost nothing of its appeal over the centuries.

The desire for mobility is no invention of modern age but has always been closely linked with the need of mankind to transport people and goods and to exchange information.

Man's wish for mobility and the technical solutions found to satisfy these needs have resulted in dramatic social, economic and environmental changes, mainly in the last century.

Today more than ever before, mobility is the key factor of modern society. Mobility is vital for us to survive. To be mobile is imperative. Those who are mobile can seize opportunities which exist beyond their immediate vicinity. Without mobility there is hardly any possibility to live our family lives, to go to school and to work or to organize our leisure-time activities, to go on vacation, do sports, have fun. Mobility also makes sure that the whole range of products we need is made available at the time and in the place they are needed.

Mobility is an obligation of the entire modern society because all of its many sectors heavily depend on the existence of efficient transportation systems.

3 Traffic – One of the Facets of Mobility

Society's need for mobility has resulted in the development of various traffic systems. So, mobility and traffic, although being different from each other, are very closely linked together. The current traffic scene is the reflection of our complex and diversified modern society in which more and more people and goods must be taken to a multitude of destinations. By allowing people to meet, to act and to discover the world, traffic has become an indispensable element of the economic cycle and is the most obvious feature of our prosperity. The individual transportation systems differ in terms of efficiency, comfort, safety and costs, but none of them is better suited to satisfy man's desire for mobility than the automobile. The 375 billion citizens of the European Union cover approximately 2 billion kilometers per year, which means that each person drives about 6 km per day in his or her car.

In Europe today, road traffic is one of the most important systems for the transportation of passengers and goods. With more than 80% of passenger traffic and 75% of goods transportation being handled by passenger cars and trucks "automobility" has turned into a veritable challenge for human society (Figs. 4 and 5) [4].

Since 1970, passenger car traffic and air traffic have recorded similarly high yearly growth rates of about 3.1%. The contribution of trucks to European goods traffic has continuously increased over the last few decades (Fig. 5) thanks to the inherent growth potential of trucks which are better suited to adapt to current goods traffic needs than any other means of transportation.

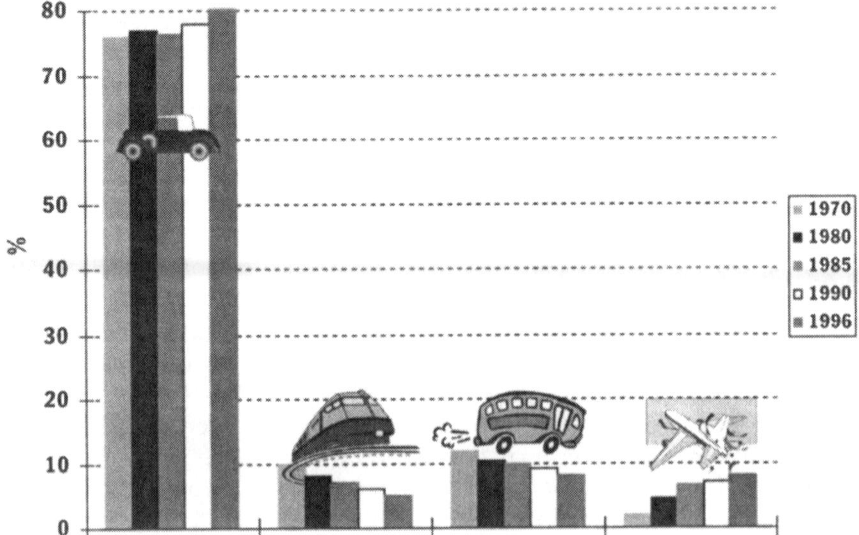

Fig. 4. Development of passenger car traffic in the EU

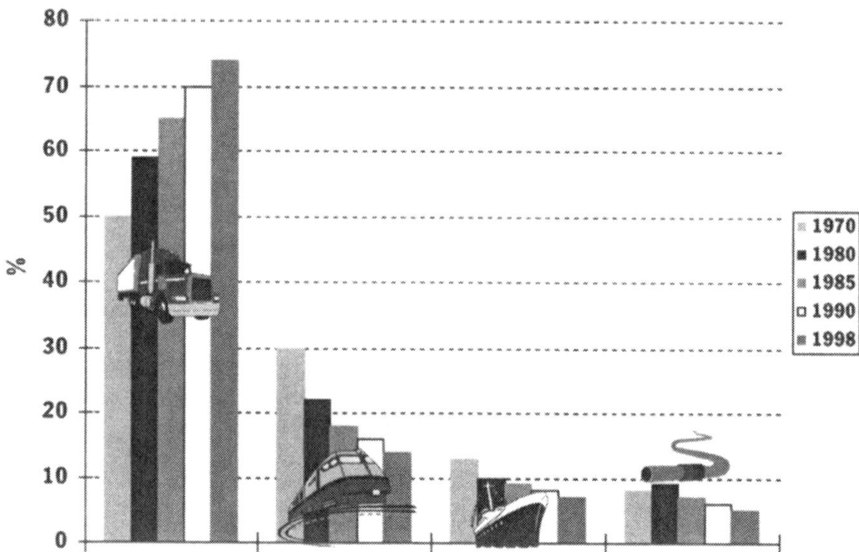

Fig. 5. Development of goods transport in the EU

So, road traffic undoubtedly is the most preferred system for the transport of passengers and goods and as such has greatly contributed to the successful development and strengthening of the individual markets.

The colossal increase of worldwide automotive production from initially 5 million to almost 50 million cars per year today started shortly after World War II. Since 1950, the car output has grown by one million units every year.

The relationship between modern society and automotive traffic is characterized by the fact that the automobile improves the quality of everyday life. Figure 6 compares the number of cars per inhabitant with the gross national products of various countries in the world. Road traffic is the result – or prerequisite – of a society's prosperity.

The automobile has become an instrument which helps us to save time, do our shopping more easily, go to distant places in a comfortable way, organize our excursions and holidays, pick up our children at school and, last but not least, get to our place of work (Fig. 7).

For decades, the time citizens spend on transportation has been essentially the same: 1 hour per day. The average distances covered during that period of time, however, have been constantly growing, mainly as a result of the increasing leisure-time and holiday trips.

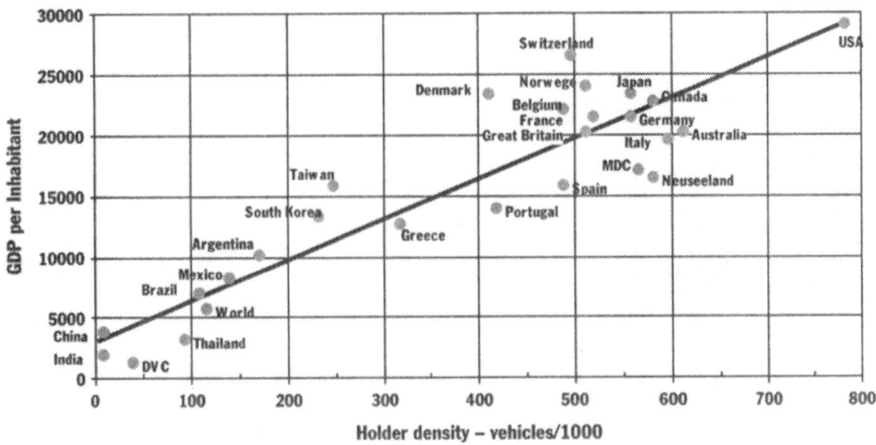

Fig. 6. Vehicle holder density vs. GDI per inhabitant

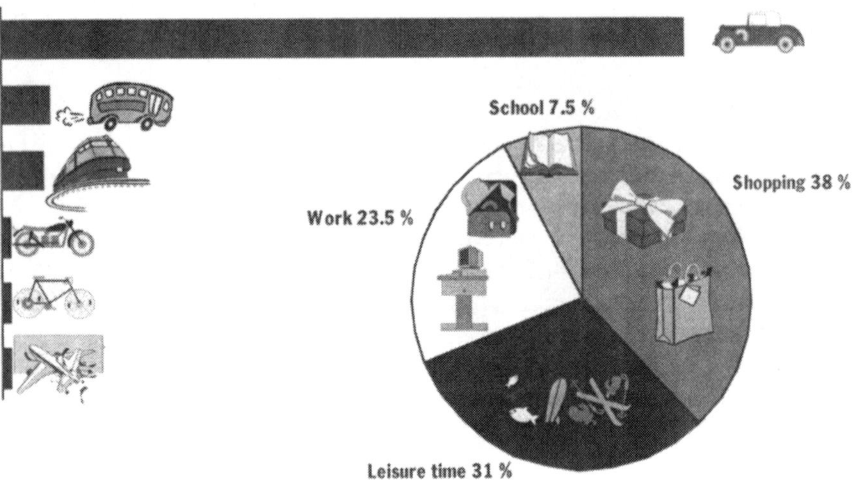

Fig. 7. Traffic means and reasons for journeys of British citizens

4
Environmental Impacts of Road Traffic

Besides its undisputed advantages, road traffic also entails certain problems which are much discussed and frequently argued about. The preferred controversial subjects include the negative effects of automotive traffic on environment, the growing number of car registrations and the resulting pollution through exhaust gases and noise.

Environment is affected by the construction of traffic ways on the one hand and by operation-related effects on the other. The most frequently cited encroachments are the consumption of land, the cutting-up and separation of landscapes, traffic safety problems, noise and pollutant emissions as well as the using up of energy and raw-material resources.

Distinction is made between the local, regional and global repercussions of automotive traffic. Local environmental drawbacks – such as high traffic density, traffic-related noise, HC, CO and NO_x emissions, photochemicals and other substances – make themselves mainly felt in densely populated areas.

Regional environmental pollution occurs in the form of "acid rain", "ozone smog", "forest decay", sprawling dumping sites and the over-fertilization of landscapes through nitrogenous fertilizers.

Global environmental pollution is reflected by the frequently discussed greenhouse effect and the resulting climatic changes, the ozone hole and the using up of natural resources.

One fifth to one fourth of the overall amount of primary energy goes into worldwide mobility. One of the ecological priorities of modern society is to minimize the environmental burden caused by automotive traffic. To condemn mobility wholesale as being little more than a negative side effect of prosperity and progress and to ask for general restrictions and bans would mean to call into question the evolution of society as a whole. What must be done instead is to create a fair balance between the requirements of economy, industry, the market as such and environmental protection.

For many decades, automotive industry – being conscious of its major role in the improvement of the environmental compatibility of its products – has intensively worked on minimizing their ecological impacts.

The multitude of directives concerning the pollutant emissions from engines for passenger cars, light commercial vehicles, transportation and busses introduced since the mid 1960s has greatly contributed to reducing the gas and particle emissions and lowering the noise levels.

Thanks to the intensive further improvement of automotive engines and their exhaust after-treatment systems and the continuous optimization of traffic flows the exhaust gas emissions of modern cars could be lowered to almost zero (Fig. 8).

Since 1978, the numerous innovative technologies introduced into new cars in Europe have helped to cut average fuel consumption and the resulting CO_2 emissions by more than 30% (Fig. 9).

The fuel economy of commercial vehicles has been considerably raised as well: a modern 40-ton truck consumes no more than 34 liters/100 km. The striving for lower environmental aftermaths is also reflected by the reduced noise emissions, improved traffic safety and higher used-car recycling rates.

The environmental performances of today's vehicles are far better than those of their predecessors 20 or 10 years ago. Even so, in public opinion they are still far from being satisfactory.

According to a public opinion poll in France, almost 80% of all citizens are concerned or very concerned about the adverse ecological effects of automotive traffic (Fig. 10). The situation is similar in other EU countries.

60% of the people interviewed believe that the air pollution caused by automobiles is far from being acceptable yet and 44% complain about noise emissions still being excessively high.

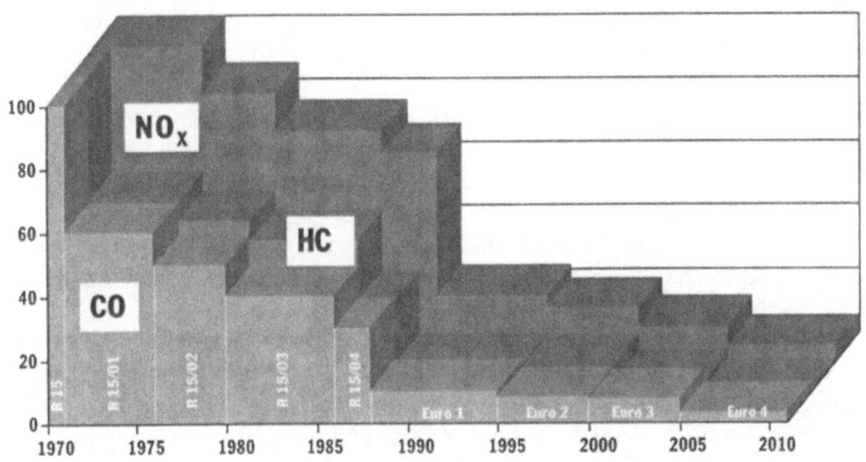

Fig. 8. Development of vehicle emission legislations (displacement >2.0 L) Europe

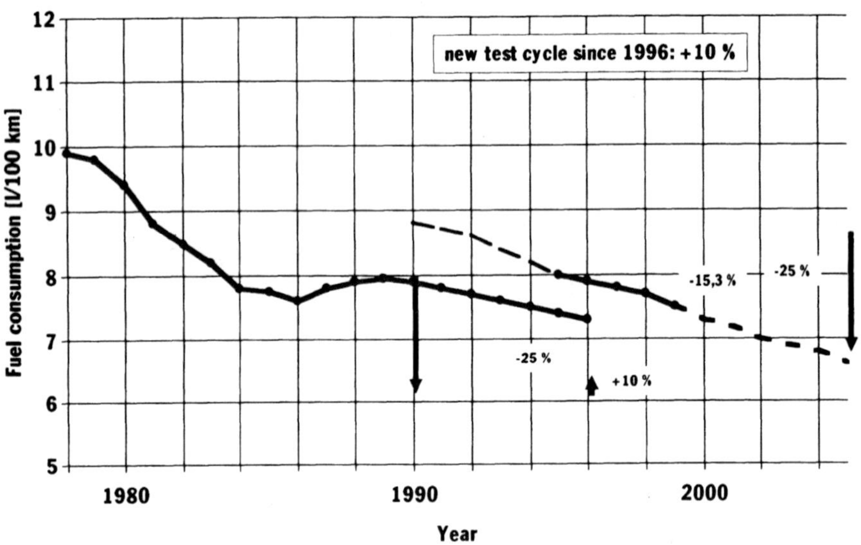

Fig. 9. Development of passenger cars fuel consumption in Germany (VDA-average)

Nevertheless, more than 82% of the population are convinced that the advantages offered by automobiles are greater than their drawbacks.

The repercussions of the continuously growing private and commercial road and air traffic must be harmonized with the environmental demands at local, regional and global levels. Since road traffic is expected to grow by 30 to 40% in the coming 10 to 20 years efforts will have to be made to further minimize the resulting environmental loads.

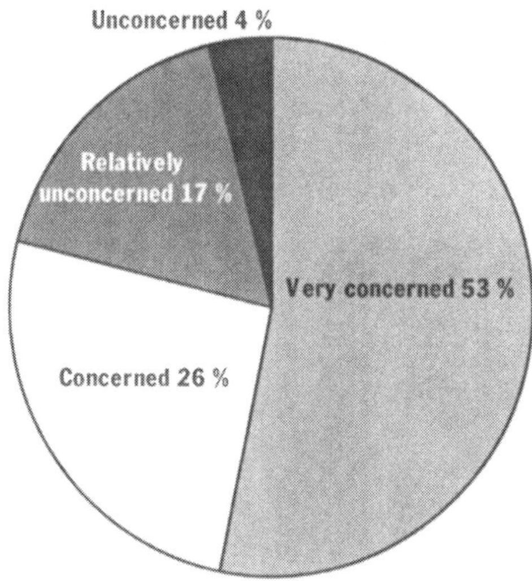

Fig. 10. Concerns about environmental impacts of the automobile

The so-called Kyoto memorandum, prepared and signed by 177 states in 1997, is an effort to lower the greenhouse gas levels and contains a series of relative new targets. The measures to be taken to realize the road traffic goals are the following:

- Further reduction of fuel consumption;
- Use of alternative fuels with an emphasis on regenerative variants;
- Use of different transportation systems (public transportation, shifting of goods traffic from road to rail); and
- Implementation of creative traffic management systems.

5 Intelligent Traffic Management

The ecological burden and energy consumption resulting from the conveyance of passengers, goods and information can be reduced through a more efficient traffic organization and an intelligent traffic control. In the coming 15 years, traffic conditions in the EU will have to be organized in such a way that the transport of goods on public roads can be increased by 50% over the current level. In city traffic, additional supporting measures will be required despite the growing traffic density and technical improvements to the vehicles. Such measures include:

- more efficient traffic control and information systems,
- improved traffic and town planning.

The most important task will be to optimize the cross-linking of the various traffic systems.

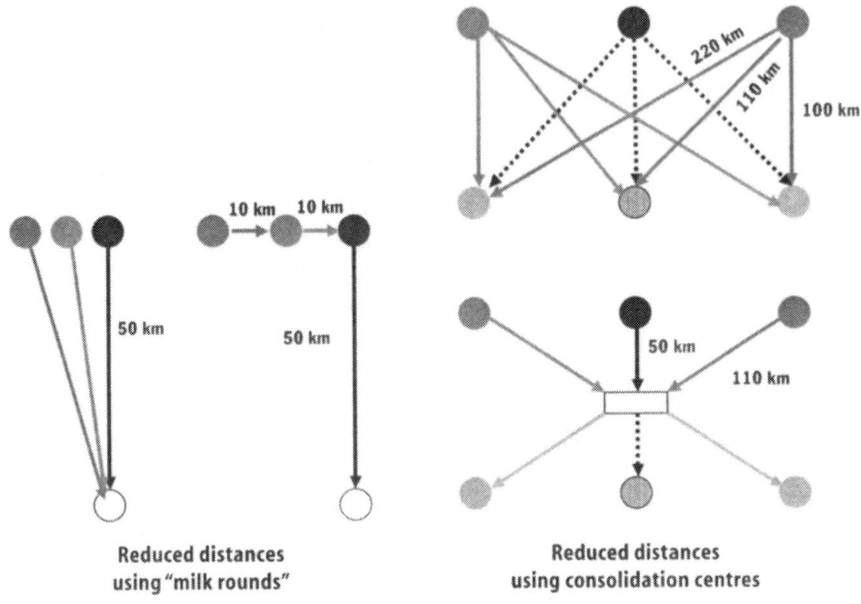

Fig. 11. Reduction of transport ways through route planing [4]

Similar to computer technology where mainframes were replaced by more performing parallel computers and flexible networks, the challenge of this century will be to efficiently interconnect the various traffic systems and traffic flows to make transportation both more efficient and more ecologically compatible. Sustainable mobility can only be guaranteed if the individual traffic systems are considered to be part of a global network. To optimize this network, all of its various elements must be taken into account. It should be possible to handle the same transportation volume with less mileage and less fuel consumption – a target which is beneficial from both an ecological and financial point of view. A most striking example of the efficiency of this approach are the so-called "milk rounds" designed to avoid empty rides and minimize the overall ride distances through the creation of consolidation centers and appropriate traffic control systems (Fig. 11).

One of the instruments allowing traffic and mileage to be uncoupled from each other is the telematics system which combines telecommunication and informatics and thus helps to speed up the cross-linking of the different traffic systems. This modern information technology for traffic control lends new dimensions to road traffic engineering. Telematics allow the driver to be provided with important information about the traffic situation. Private and public traffic systems can be interconnected via intelligent devices. Latest-state information technology can be used to coordinate the goods transportation fleets, optimize the transportation routes and avoid empty rides.

By improving the traffic flows and reducing fuel consumption intelligent transportation systems will make an essential contribution to the lowering of the environmental loads. Exhaust gas emissions are expected to drop as follows:

Carbon monoxide	(CO) by 30%
Hydrocarbons	(HC) by 30–35%
Nitrogen oxides	(NO_x) by 15–30%
Particulate matter	(PM) by 10%
Carbon dioxide	(CO_2) by 15–20%

Simultaneously, the number of traffic accidents is to be cut by 15 to 30%.

The target of intelligent mobility – i.e., solving the transportation problems while lowering the accident rate to almost zero and minimizing the environmental burden – is an enormous challenge for all the people involved.

6 References

1. Steinkohl F, Knoepffler N, et al (1999) Auto-Mobilität als gesellschaftliche Herausforderung. Herbert UTZ, München
2. Seifert U (2000) Mobilität intelligent koordinieren. VDI-Nachrichten, 18.02.2000
3. Europe on the move. ACEA-European Automobile Association. Bruxelles, May 2000
4. Energy, Transport and the Environment. Volvo Car Corporation, 1996
5. Lenz H-P (2000) Visionen zum Automobil aus Sicht eines Forschers. Clusterland Oberösterreich-Innovation durch Kooperation, Linz
6. Mobility, Professional Congress. VDI-World Engineers' Conventions, Hannover, 2000
7. Automotive Technology International, 2000
8. Nutz-Fahrzeuge, für alle auf Achsen. VDA Frankfurt 2000
9. The Automobile and Society. Taylor Nelson Sofres, September 2000
10. Environment, Climate, Health. VDI-World Engineers' Conference, Hannover 2000
11. Environmentally sustainable Transport – EST-OECD-Guidelines, Wien 2000
12. Midlands tests for anti-pollution traffic controls. Automotive Engineer, December 1999
13. Braess HH, Seifert, U (2000) Vieweg Handbuch Kraftfahrzeugtechnik. Braunschweig/Wiesbaden
14. Auto, Annual Report (2001) VDA-Verband der Automobilindustrie e.V. 2001
15. Gruden D, et al (1993) Die ökologische Dimension des Automobils. Expert, Renningen

The Handbook of Environmental Chemistry Vol. 3, Part T (2003): 15–106
DOI 10.1007/b11992

Power Units for Transportation

Dušan Gruden[1] · Klaus Borgmann[2] · Oswald Hiemesch[2]

[1] Dr. Ing. h.c. F. Porsche Aktiengesellschaft, Porschestrasse, 71287 Weissach, Germany
E-mail: dusan.gruden.@porsche.de

[2] Bayerische Motoren Werke Aktiengesellschaft, Hufelandstrasse, 80788 München, Germany

For more than 125 years, gasoline and Diesel engines have prevailed as the exclusive drive unit in road transportation. None of the other power units invented to date has been able to make use of the energy content of mineral oil with the piston engine's same good efficiency.

Combustion is the fundamental process by which the chemical energy of fuels is converted into thermal energy and further into mechanical work. If hydrocarbon-containing fuels were completely burnt, the resulting products would be carbon dioxide and water vapor only. Since it is impossible to obtain a 100% complete combustion the exhaust gases always include a great variety of combustion products, the most important are: carbon monoxide, unburnt hydrocarbons, nitrogen oxides and particulate matter.

During its 125 years of existence, the Otto (gasoline) engine – as it was called after its inventor – has been developed into a mature combustion engine which is characterized by an excellent efficiency and low pollutant emissions. The properties of the gasoline engine strongly depend on the composition of the air-fuel mixtures and ignition parameters. The influence of the so-called engine design parameters on combustion and exhaust emission is no less important.

The emission of many of the exhaust-gas constituents can be influenced and minimized at their place of origin, that is in the combustion chamber by correctly selecting and adapting the relevant engine design and operating parameters. If optimization of engine-internal parameters for further reducing of the exhaust gas emissions are not enough anymore, so-called engine-external measures must be additionally taken. It was found that so-called three-way catalyst reduces the three aforementioned pollutants by clearly more than 90%, provided that a precisely stoichiometric A/F-ratio is used.

Thanks to the strict maintenance of a precise stoichiometric air/fuel mixture the three-way catalyst allows very low HC, CO and NO_x pollutant emissions to be achieved. However, in this operating range, fuel consumption is 8 to 15% higher (with a resulting higher CO_2 emission) than during lean-burn operation.

One of the technically most useful solutions to reduce the fuel consumption and CO_2 emission of gasoline engines is to make them tolerate lean air/fuel mixtures. The future of the lean-burn gasoline engines will almost exclusively depend on the successful development of NO_x-exhaust-gas after-treatment technologies for lean air/fuel mixtures.

Diesel engines are internal combustion units with the highest thermal efficiency. Mixture formation is achieved through high pressure fuel injection. The fuel leads to self-ignition in the highly compressed air of the engine cylinder. The power and torque characteristics of modern Diesel engines are comparable with those of spark ignition (Otto) power units of equal capacity, the fuel consumption however is approx. 20% lower.

The Diesel power unit has achieved a high status in transport. The world wide share of Diesel engines in passenger vehicles is now approx. 20%, whereas in freight transport on the roads and by water the share is approaching 100%, diesel being the only cost effective alternative.

Increasingly, new methods for injection combustion, exhaust gas recirculation and after treatment (NO_x-Cat, Diesel particle filter) are being pursued to meet the ever stricter emission legislations, aimed at limiting the effects on the environment.

Ever since its invention, the 4-stroke reciprocating piston engine has been considered as a rather complex thermal unit which should better be replaced by far less complicated designs.

When summing up all the properties required to smoothly operate cars over wide speed and load ranges and a long lifetime, all alternative concepts have never succeeded in edging the Otto and Diesel engines out of their top positions. Further optimized versions of gasoline and Diesel engines will continue to prevail in the automotive domain in the coming 15 to 20 years. Due to their theoretically high efficiency and low pollutant emissions, fuel cells are among the most promising alternative energy sources of the future.

Keywords. Combustion process, Otto engine, Gasoline engine, Diesel engine, Fuel/air mixture, Power output, Fuel consumption, Exhaust gas emission, Carbon monoxide, Unburnt hydrocarbons, Nitrogen oxides, Particulates, Operating parameter, Design parameter, Ignition, Injection, Compression ratio, Combustion chamber, Valve timing, Exhaust gas after-treatment, Catalyst, Particulate filter, Turbo charging, 2-stroke engine, Alternative engine, Fuel cell, Hybrid drive

1
Combustion Fundamentals and Combustion Products

Dušan Gruden

1.1
General Issues

For thousands of years, horse- or ox-drawn carriages were the main means of locomotion to ensure what is called today passenger or public road transportation. The invention of the steam engine in the late 18th century was soon followed by the appearance of the railway train – a means of transportation which offered one essential advantage over the preceding ones:

Steam engine-powered trains were much faster than all the previous means of locomotion. This attribute was so attractive that it triggered a people movement from slow individual vehicles to this speedy and more comfortable means of mass transportation.

When, at the end of the 19th century, the piston internal combustion engine was invented which was so much smaller and more compact than the big unwieldy steam engine the obvious consequence was to fit it into a horse carriage. In 1886, the first motorized carriage was built in Stuttgart (Fig. 1) which went down in history as one of the first combustion-engine-equipped vehicles. It was

Fig. 1. First passenger car with internal combustion engine built in Stuttgart (1886)

then that the automobile was born. Since, from the very start, motorized automobiles were able to travel at the same speed as – if not faster than – trains they gave rise to another people movement, this time away from mass transportation and back to individual transportation.

Despite intensive research and numerous efforts aimed at developing alternative propulsion systems, the piston engine has prevailed as the exclusive powerplant unit in road transportation. For more than 125 years, gasoline and Diesel engines have been the best answers engineers could find to the cheapest and most convenient terrestrial energy source.

None of the other power units invented to date has been able to make use of the energy content of mineral oil with the piston engine's same good efficiency. Gasoline and Diesel oil being regular by-products of oil refining, Otto and Diesel engines have never been mutually exclusive alternative concepts but have always ideally complemented each other in the efficient employment of mineral oil.

Combustion is the fundamental process by which the chemical energy of fuels is converted into thermal energy and further into mechanical work needed for locomotion.

Combustion in a heat engine consists in the rapid chemical oxidation of HC-containing fuels. This reaction is accompanied by the release of major amounts of heat and luminous radiation. The released heat energy is then transformed into mechanical work by the reciprocating-piston mechanism.

Even though combustion is the basic functional principle of a heat engine, it has not been possible, to date, to define a satisfactory combustion theory which describes the phenomena of combustion in every detail. What we have not got yet is a mathematical method allowing us to precisely calculate all phases of the combustion process taking place in the cylinder of an engine. This lack is due to the fact that combustion is a complicated chemical process characterized by rapidly changing temperatures and pressures and varying concentrations of the reactive substances. The chemical conversions taking place in a combustion engine have little to do with simple chemical reactions. The burning of hydrocarbons triggers chain reactions which are both consecutive and competing with each other. The fuels burnt in the cylinder of a combustion engine are not homogeneous simple hydrocarbons but rather consist of mixtures of hydrocarbons of different structures and highly varying percentages. At the present time, we are far from knowing the whole range of elementary processes going on during combustion.

The velocity of the chemical reactions strongly depends on the chemical and physical properties of the reactive substances. The relationship between the reaction velocity (K) and temperature is given by Arrhenius' law:

$$K = C \cdot e^{-E/RT} \qquad \text{(a)}$$

where:

C constant,
E activation energy,
R gas constant, and
T temperature.

To simplify matters, combustion can be represented as follows:

$$\text{Fuel (CxHy)} + \text{oxygen (air)} \rightarrow \text{chain reaction (combustion)} \rightarrow CO_2 + H_2O + CO + HC + NO_x + \ldots \quad \text{(b)}$$

Fuel combustion consists of chain reactions. According to the chain-reaction theory, the initial substances pass through a number of intermediate states before reaching the end-product condition. A chain reaction mainly depends on so-called active centers (free atoms, radicals, peroxides) which do not enter into contact with the initial compounds or intermediate products.

If hydrocarbon-containing fuels were completely burnt, the resulting products would be carbon dioxide (CO_2) and water steam (H_2O). Combustion products also contain excess oxygen (O_2) and nitrogen (N_2). Since it is impossible to obtain a 100% complete combustion the exhaust gases always include a great variety of other products, too.

1.2 Carbon Monoxide (CO)

Carbon monoxide results from incomplete combustion of the carbons contained in fuel hydrocarbons. Theoretically – in the presence of sufficient oxygen (over-stoichiometric, "lean" mixtures) – the carbon monoxide should be completely burnt to non-poisonous CO_2 and not be present in the combustion products any longer.

However, as CO measurements have shown, the carbon monoxide concentration in the exhaust gas is about 1 vol.-% with stoichiometric mixtures (λ=1,0) with small amounts of CO being detectable also if lean mixtures (λ>1,0) are used. The percentage of carbon monoxide contained in the exhaust gas strongly depends on the reaction temperature: At high temperatures, permanent counter reactions (CO_2 dissociation) take place. Sudden cooling of the combustion gases in the expansion phase "freezes" the balance created at high temperatures thus causing carbon monoxide to be present in the exhaust gas under all operating conditions and A/F ratios.

1.3 Unburnt Hydrocarbons (HC)

Most of the unburnt hydrocarbons an automobile releases into the atmosphere come from the combustion process. The place in the cylinder and the moment at which unburnt hydrocarbons are generated has not yet been precisely determined. They occur even if there is sufficient oxygen for complete combustion, if flame propagation in the combustion chamber is perfect, if there is little residual gas and if there is an efficient distinct charge turbulence.

Most scientists believe that the unburnt hydrocarbons result from incomplete flame propagation, causing the flame to be quenched at the cool walls of the combustion chamber (wall quenching). But the theory of flame quenching explains only part of the generation process of unburnt hydrocarbons. A major portion is generated through incomplete fuel combustion caused by residual gases which

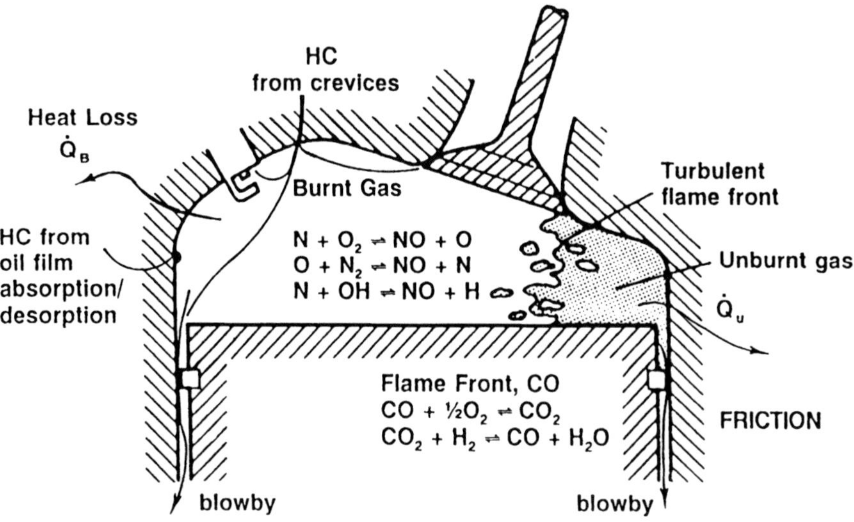

Fig. 2. Sources of unburnt hydrocarbons in combustion chamber

strongly dilute the charge or by low cycle temperatures etc. Unburnt hydrocarbons are also created in those cylinder areas where the mixture cannot be reached by the flame, such as the space between the piston top land and the cylinder wall or the piston ring grooves (Fig. 2) [6].

In the expansion and exhaust phases, the unburnt hydrocarbons mix with the products resulting from complete combustion thus continuing their oxidation the intensity of which depends on temperature, the hydrocarbon and oxygen concentrations and the time available.

The overall amount of unburnt hydrocarbons in the exhaust gas consists of a multitude of individual hydrocarbons. The exhaust gases of gasoline and Diesel engines contain several hundred hydrocarbon compounds with 1 to 9 (and more) C atoms. Unburnt hydrocarbons include paraffins, olefins, aromatic compounds, acetylene and their isomers, partly oxidized hydrocarbons (aldehydes, ketones, alcohols) as well as organic nitrogen and sulfur compounds. Some of these come unchanged from the fuel whereas others are combustion products. Each individual hydrocarbon compound needs a specific temperature to be generated. Any change of the operating conditions will automatically change the respective compound's share in the overall amount of hydrocarbons.

1.4 Nitrogen Oxides (NO_x)

The atmospheric air used for combustion essentially consists of nitrogen and oxygen molecules. Under normal conditions, it is chemically well balanced and very stable.

Under temperatures of several hundred degrees, the two-atom nitrogen and oxygen molecules dissociate into their respective atoms and partly combine to form nitrogen monoxide (NO). The degree of dissociation depends on the temperature and pressure levels and is accompanied by strong energy consumption.

Provided that there is a sufficiently high amount of oxygen, the high cylinder temperatures in a combustion chamber further the partial oxidation of nitrogen from the air forming nitrogen monoxide. The NO concentration in the combustion engine mainly depends on the maximum combustion temperatures, the composition of the air/fuel mixture (A/F ratio) and the reaction time available.

It is generally assumed that the combustion process produces NO only and that other nitrogen oxides such as NO_2, N_2O, N_2O_3, N_2O_4 and N_2O_5 are generated through continued NO oxidation in the expansion and exhaust phases and in the atmosphere. Nitrogen monoxide which has been generated and is then cooled down to ambient temperature will quickly oxidize in the atmospheric air to form NO_2. Further atmospheric oxidation of NO_2 into N_2O_4, for example, is considerably slowed down at ambient temperature. Low temperatures and high dilution with air allow nitrogen oxides to continue to exist in the atmosphere for a long time.

1.5 Particulate Matter (PM)

Besides the gaseous CO, HC and NO_x emissions, Diesel engines also emit particulate matter (PM). Particulates have been defined as solid matter which is detected by diluting the engine exhaust gases with air, passing them through a filter at a temperature of less than 52°C and weighing the resulting residue.

Thus as soot described particulates contained in the exhaust gas is the most obvious form of air pollution caused by combustion engines. The amount of soot measured in the exhaust gas from Diesel engines is a criterion of the quality of both the combustion process and the mixture control.

Soot is an inevitable constituent of exhaust gases resulting from the combustion of organic fuels. Its amount and properties, however, depend on how the combustion process goes.

In the past, distinction was made between three types of Diesel engine smoke emission: white, blue and black smoke.

White smoke is generated if the combustion temperatures are low or if the ignition delay is too long. This kind of smoke occurs after the engine has been started and when the cylinder temperatures are high enough to evaporate but not to self-ignite the fuel.

Blue smoke usually occurs when small amounts of lubricating oil penetrate and are burnt in the combustion chamber.

Black smoke emitted under higher engine loads almost exclusively consists of carbon and other solid combustion products. The smoke is black if less than 1% of the carbon contained in the fuel is emitted in the form of soot.

When analyzing the soot phenomenon, consideration must be given above all to the type of flame used for combustion. In the premixed flame of an gasoline engine, for example, the fuel vapors and the oxygen of the air are closely mixed and in direct contact with each other, so that no soot is generated if the amount of oxygen is sufficiently high ($\lambda \geq 1.0$).

The extremely heterogeneous combustion of a Diesel engine is characterized by the simultaneous existence of a mixture of gases, vapors and liquid fuel in the combustion chamber whose concentrations vary continuously. These heterogeneous conditions (diffusion flame) result in incomplete chemical reactions allowing solid particles as well as unburnt or only partly burnt hydrocarbons to occur in the exhaust gas.

Soot is generated at the flame front under high pressures and temperatures through various chemical and physical processes. It has not been possible to date, to scientifically determine the mechanisms of soot formation with sufficient precision. There are many hypotheses as to the particulates-forming reactions during Diesel-engine combustion none of which is able to provide a complete description of the processes involved. Quite frequently, polymerization is thought to be the primary source of soot formation in a diffusion flame. Other soot-generating reactions are dehydration, condensation and graphitization. An exemplary soot formation model is shown in Fig. 3 [10].

Particulates mainly consist of soot (black smoke). Soot is elementary carbon resulting from incomplete Diesel combustion. The organic compounds (hydrocarbons) settled down on the soot particles – also known under the designation of SOF (Soluble Organic Fraction) – consist of unburnt, partially cracked or polymerized hydrocarbons coming from the fuel and the lubricating oil. In addition, there are sulfates caused by the burning of the sulfur contained in the fuel. Particles also include residues of lubricants and fuel additives as well as settled-down water. Figure 4 shows the typical particle mixture of a Diesel engine at full load.

The results of the particulates analysis suggest that all carbon-containing fuels are susceptible to forming particles. With aromatic compounds, this tendency is greater than with olefins and paraffins. A low hydrocarbon saturation level increases the particle formation trend. This means that the C/H ratio of the fuel is an essential parameter when it comes to evaluating the soot-formation propensity of fuels.

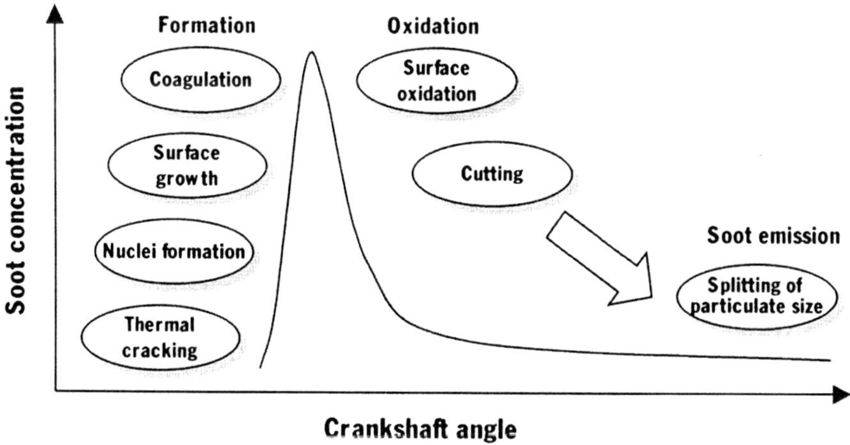

Fig. 3. FVV Project "Soot oxidation model" – soot formation and oxidation in Diesel engines

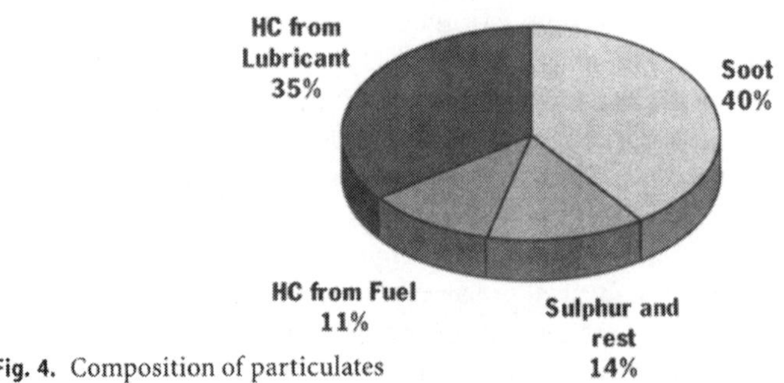

Fig. 4. Composition of particulates

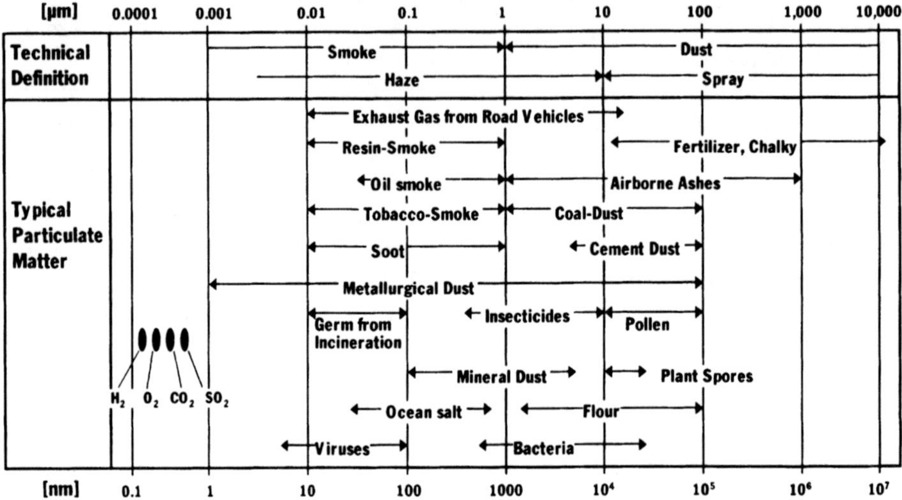

Fig. 5. Size ranges of different types of particulate matter

The first particles have an almost spherical shape with diameters ranging between 0.002 and 0.01 µm. These particles agglomerate very quickly to form chains.

A typical soot particle has a size of about 0.1 to 0.2 µm. With this scatter, Diesel soot is in the same range as numerous other particulates so that it is extremely difficult from a measuring point of view to clearly separate Diesel soot particles and particulates from other sources in the atmosphere (Fig. 5).

1.6 References

1. Woinov AN (1965) Verbrennungsprozesse in schnellaufenden Kolbenmotoren (russ.). Moskau
2. Fristrom RM, Westenberg AA (1965) Flame Structure. McGraw-Hill, New York
3. Bradley JN (1965) Flame and Combustion Phenomena. Methuen & Co Ltd., London

4. Gaydon AG, Wolfhard HG (1970) Flames. Their structure radiation and temperature. Chapman and Hall, London
5. Taylor CF (1985) The Internal Combustion Engine in Theory and Practice. The MIT Press, Cambridge
6. Stone R (1992) Introduction to Combustion Engines. The Macmillan Press, London
7. Warnatz J, Maas U, et al (2001) Verbrennung. Physikalisch-Chemische Grundlagen. Springer, Berlin, Heidelberg, New York
8. Warnatz J (1995) Probleme bei der Simulation von motorischen Verbrennungsprozessen. Symposium Kraftfahrwesen und Kraftfahrzeuge, Stuttgart
9. Polycyclic aromatic hydrocarbons in automotive exhaust emissions and fuels. CONCAWE Report No. 98/55, 1998
10. Pischinger S (1998) Rußbildung und Oxidation im Dieselmotor. FVV-Vorhaben "Rußoxidationsmodell". FVV Frankfurt
11. Moser FX, Flotho A, et al (1995) Entwicklungsarbeiten an Dieselmotoren für den Nutzfahrzeug- und Industrieeinsatz zur Erfüllung der zukünftigen Emissionsanforderungen. Symposium Kraftfahrwesen und Verbrennungsmotoren, Stuttgart

2 The Otto (Gasoline) Engine

Dušan Gruden

2.1 General Issues

When Nikolaus Augustus Otto had his patent registered in 1875, he doubtlessly was unaware of the repercussions his invention was going to have on humanity. During its 125 years of existence, the Otto (gasoline) engine – as it was called after its inventor – has been developed into a mature combustion engine which is characterized by an excellent efficiency. Along with their Diesel counterparts, Otto engines number among the heat engines having the highest combustion efficiency. This has allowed these two reciprocating-piston-engine variants to edge out of the market all other alternative power plant units which have been intensively examined so far as potential substitutes. And everything is pointing to the fact that these two power plant concepts will continue to prevail also in the foreseeable future and far into the 21st century. Of the 750 odd million passenger cars registered world-wide more than 90% are powered by gasoline engines – an indication of the enormous importance this type of propulsion system has had for mankind.

The configuration of an Otto engine depends on the fuel type (gasoline) for which it has been laid out. According to the current state of knowledge, gasolines can only be efficiently burnt in a homogenous gasoline/air mixture. That is why, in the Otto engine, the fuel is injected into the intake manifold (or cylinder) in the suction phase already (Fig. 1).

The intake and compression strokes (360° C.A.), which account for 50% of the working cycles, provide sufficient time to evaporate the fuel and intensively mix the air and fuel vapors.

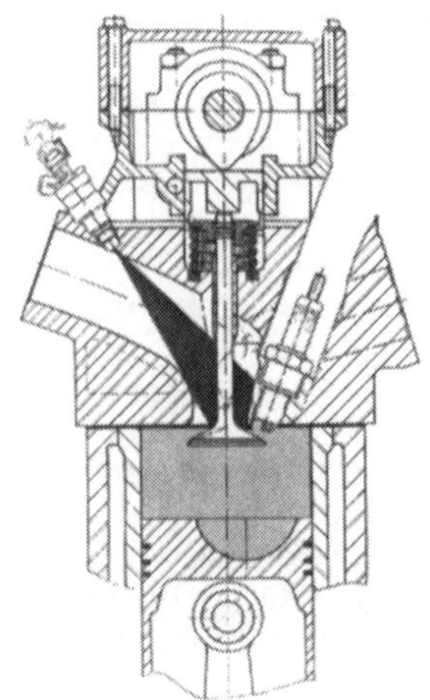

Fig. 1. Mixture formation in Otto engine

As a matter of fact, homogeneous air/fuel mixtures need an external ignition source triggered by a spark-plug in order to be able to burn in a controlled regular manner. Following the ignition of the A/F mixture, the flame spreads throughout the combustion chamber at a velocity of 30 to 50 m/s.

Gasoline engine combustion is represented by the Otto cycle (Fig. 2), consisting of adiabatic compression (T_1–T_2), isochoric heat supply (T_2–T_3), adiabatic expansion (T_3–T_4) and isochoric heat removal and/or gas exchange (T_4–T_1).

The homogenous A/F mixtures in an Otto engine can be burnt efficiently only in a relatively narrow A/F-mixture range about the stoichiometric ratio (λ=approx. 1.0, A/F$\approx$14.5) and require a quantitative engine load control (throttling). With decreasing load, both the amount of fuel and the amount of air must be reduced in order to maintain the A/F ratio at a constant level. This means that the pressure and temperature levels in the combustion chamber at the moment of ignition keep dropping while the engine load diminishes (Fig. 3).

In a Diesel engine, the amount of air sucked in and compressed is practically always the same regardless of the engine load. The pressure and temperature

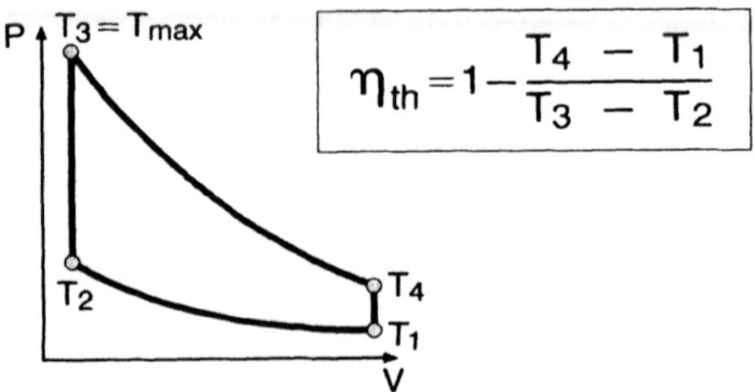

Fig. 2. Thermodynamic cycle (Otto engine)

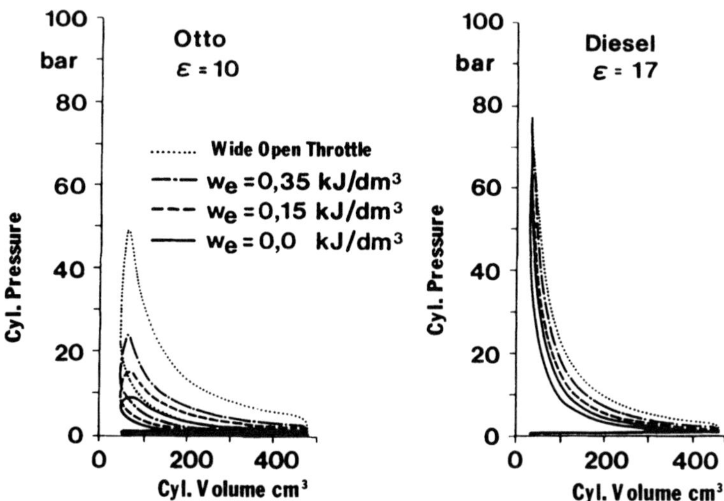

Fig. 3. P-V diagrams at different loads

levels reached at the end of the compression stroke are high and completely independent of the load which is controlled qualitatively (unthrottled) by reducing the amount of fuel injected.

Besides the throttling losses, the unsatisfactory efficiency of the gasoline engine at part load is mainly due to the low pressures and temperature levels during the combustion process. In a Diesel engine, the combustion process takes place always at constantly high energy level. The differences between the partload behaviors of the Otto and Diesel engines are caused, among others, by the differences between their inherent energy potentials at which the combustion processes take place.

2.2 Power Output and Fuel Consumption

The properties of the gasoline engine strongly depend also on the composition of the A/F mixture or the A/F ratio λ. Figure 4 illustrates the dependence of the specific work w_e (mean effective pressure) and the specific fuel consumption b_e on the A/F ratio (λ).

In the event of an air deficiency, homogeneous A/F mixtures can always be safely ignited and burnt in what is called the "rich" mixture range (λ=0.8–0.9). It is in this range that Otto engines reach their highest mean pressures or power outputs. That was also the reason why the early generations of gasoline engines were exclusively operated on rich A/F ratios over the entire operating range from starting through idling to full-load.

These operating conditions made no major demands on engine control. The required amounts of fuel and air were metered in the carburetor; the ignition timing was adjusted via the engine speed by means of a flyweight-controlled regulator in the ignition distributor and via engine load by means of a intake-man-

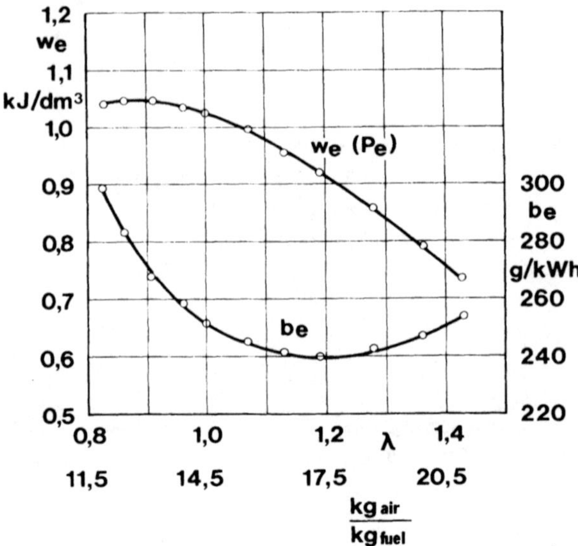

Fig. 4. Influence of air/fuel-ratio on specific work (power) and fuel consumption

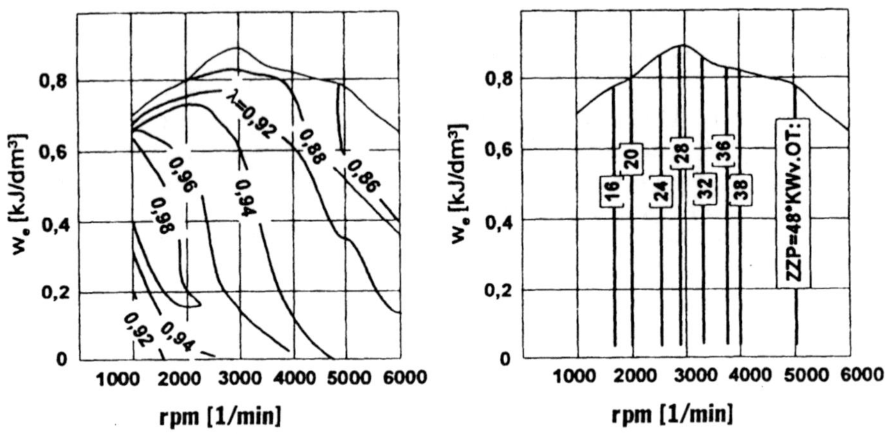

Fig. 5. Air/fuel ratio and ignition timing maps of former gasoline engines

ifold-pressure-controlled vacuum advance unit. Exemplary λ- and ignition-timing maps of a former carburetor engine are shown in Fig. 5.

The low CO, HC and NO_x exhaust emission limits prescribed by environmental legislation as well as the engine manufacturers' constant efforts to reduce fuel consumption resulted in the development of highly complex electronic A/F-mixture and ignition control and regulation systems for modern gasoline engines (Fig. 6).

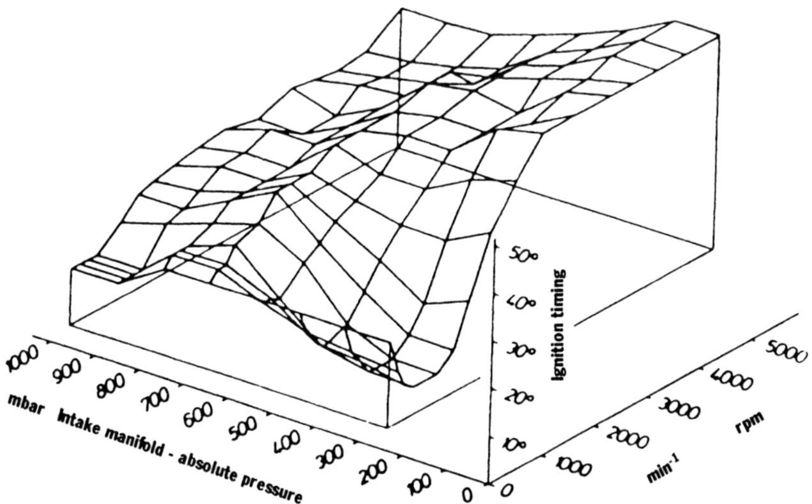

Fig. 6. Ignition timing map of modern gasoline engines

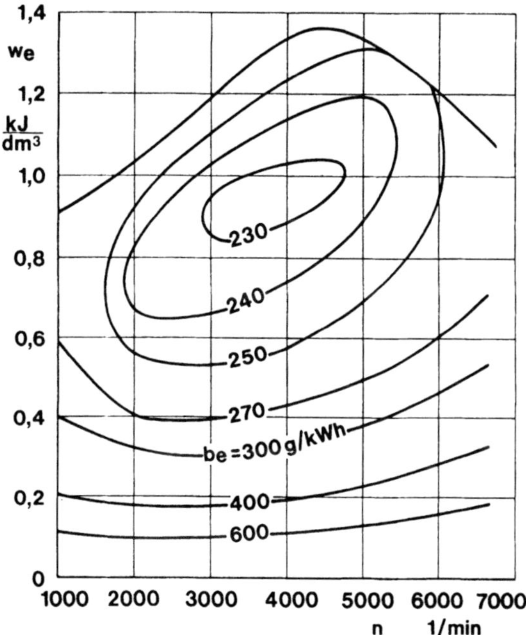

Fig. 7. Map of specific fuel consumption

To reach their maximum power output and torque levels, modern Otto engines use slightly enriched A/F ratios under full-load conditions only, yielding as naturally aspirated Otto engines specific power outputs of as high as

$$P_e = 50 - 65 \text{ kW/l} \quad \text{(a)}$$

and specific torques of

$$Md = 90 - 105 \text{ Nm/l}. \quad \text{(b)}$$

Turbocharged gasoline engines reached specific power outputs of

$$P_e = 65 - 85 \text{ kW/l} \quad \text{(c)}$$

and specific torques of

$$Md = 125 - 170 \text{ Nm/l}. \quad \text{d)}$$

In the part-load range, advanced Otto engines are operated on stoichiometric A/F ratios (A/F=14.5, λ=1.0), in order to create optimum operating conditions for the 3-way catalyst (see Chapter 2.5).

The lowest fuel consumption levels realized with modern Otto engines are about b_{emin}=230–240 g/kWh (Fig. 7).

2.3 Exhaust Gas Emission

The explosive increase of the vehicle population in the industrial countries after World War II resulted in a new problem in the big population centers – with awareness starting in Los Angeles, USA: air pollution through exhaust emissions from combustion engines. First, it was the carbon monoxide (CO) and unburnt hydrocarbons (HC) which were rated as being noxious. Shortly thereafter, nitrogen oxides (NO_x) were added to this group of pollutants. Since that time, the survival of the gasoline engine has depended and will continue to depend on its ability to comply with all the existing and planned regulations meant to reduce the burden on environment.

2.4 Engine-Internal Measures for Pollutant Reduction

For both Otto and Diesel engines, so-called engine-internal measures are the first choice when it comes to reducing pollutant emission. The emission of many of the exhaust-gas constituents can be influenced and minimized at their place of origin, that is in the engine cylinder or in the combustion chamber, by correctly selecting and adapting the relevant engine design and operating parameters.

2.4.1 *Operating Parameters*

2.4.1.1 *Air-Fuel Mixture*

Various investigations of the variables influencing the exhaust emissions of an Otto engine have shown that the amount of individual exhaust-gas constituents mainly depends on the composition of the air-fuel mixture (air/fuel ratio, A/F ratio or λ) (Fig. 8).

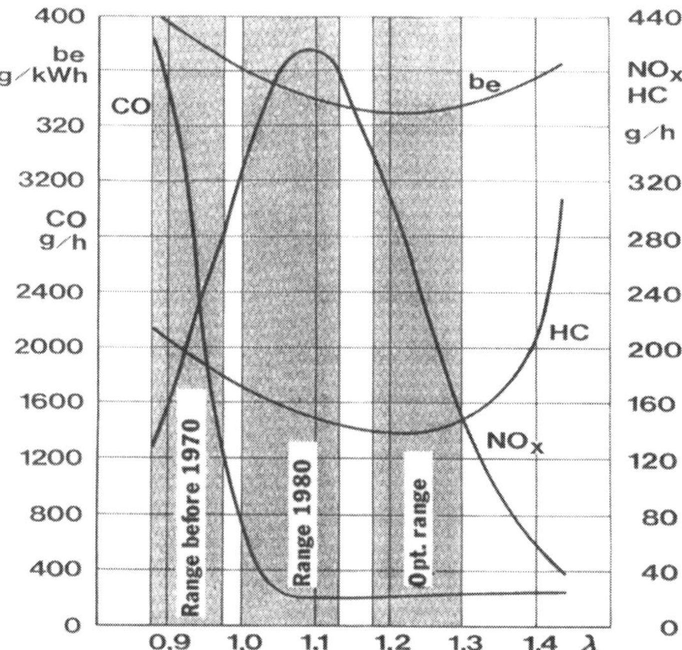

Fig. 8. Influence of air/fuel ratio on fuel consumption and exhaust gas emission

The A/F ratio influences the composition of the exhaust gas far more strongly than any of the other combustion parameters, because it determines with relatively great precision whether the Otto engine is operated on a rich ($\lambda<1.0$) stoichiometric ($\lambda=1.0$) or a lean ($\lambda>1.0$) mixture.

The high levels of CO and unburnt hydrocarbons resulting from rich air/fuel mixtures are due to the fact that the mixture cannot be completely burnt for lack of oxygen. The only way of noticeably reducing these pollutants at their place of origin in the cylinder would be to increase the A/F ratio (mixture enleanment).

The lack of air prevents excessive amounts of NO_x from being generated even though the maximum combustion temperatures are high.

Contrary to the results of corresponding equilibrium calculations, using stoichiometric mixtures ($\lambda=1.0$) does not completely eliminate the CO contained in the exhaust gas, the residue being about 0.5 to 1.0 vol.%. Due to the reaction kinetics of the CO combustion, the exhaust gas contains a certain amount of CO even when $\lambda>1.0$.

The lowest HC levels are obtained with lean mixtures ($\lambda\approx1.1-1.3$) or, in other words, with those A/F ratios at which the highest engine efficiency is reached.

The high combustion temperatures and amounts of air required for the oxidation of CO and HC result in a steeply increasing NO_x concentration. The maximum NO_x level occurs in the same A/F ratio range in which the concentration of unburnt hydrocarbons is lowest.

Further leaning of the mixture deteriorates the combustion conditions: the maximum combustion temperatures and velocities decrease while the combustion time increases. The dropping temperature and deteriorating combustion result in higher HC levels and steeply decreasing NO_x concentrations in the exhaust gas. Excessively lean mixtures frequently lead to sluggish combustion and complete misfires. Misfiring and lack of combustion, however, result in extremely high HC concentrations and increased fuel consumption.

2.4.1.2
Ignition

The operating behavior of an Otto engine – that is its power output, torque, fuel consumption and exhaust-gas composition – essentially depends on the ignition parameters, such as the functional characteristics of the spark plug, its location in the combustion chamber, the electrode gap and the ignition point.

Not every spark is capable of igniting the A/F mixture. For the ignition to be triggered, the spark must have a certain minimum ignition energy which depends on the physico-chemical properties of the mixture next to the spark plug on the one hand and on the state of the electrodes on the other. Quite obviously, an ignition current of I=80–100 mA, a spark duration of t=1.5 to 2.0 ms and an ignition energy of 50 mJ are sufficient to make gasoline engines run also on lean air/fuel mixtures. It is not useful to further increase the ignition energy beyond the above mentioned values.

The position of the spark plug in the combustion chamber influences the octane requirement of the engine, the lean limit of the mixture and the fuel consumption. When optimizing the ignition point (pre-ignition timing) consideration must be given to the power output and torque at WOT (wide open throttle) and to the fuel consumption at part load. Variations of the combustion velocity and

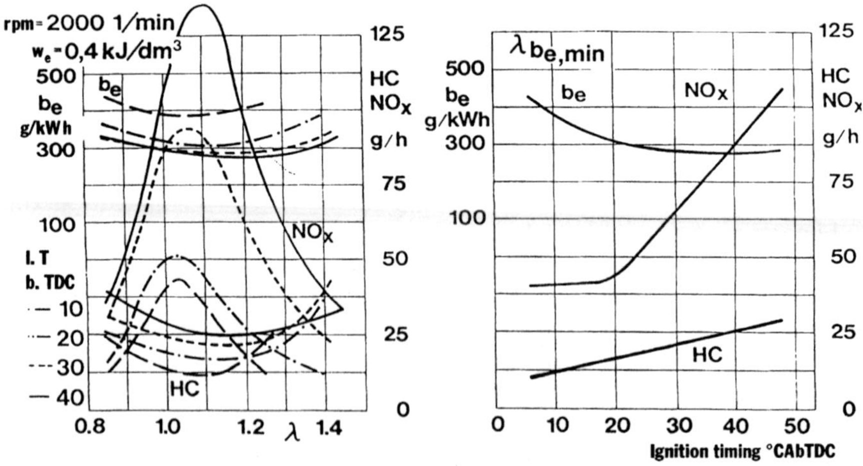

Fig. 9. Influence of ignition timing on fuel consumption and exhaust gas emission, 1-cylinder engine

temperature caused by the ignition timing also have an influence on the exhaust-gas composition (Fig. 9). That is why, for modern gasoline engines, it is essential to maintain the ignition timing stipulated for the instantaneous A/F mixture.

The composition of the exhaust gas is not only influenced by the air/fuel mixture and ignition timing but also by all the other operating parameters, such as the temperature of the charge, engine and coolant, the temperature of the exhaust gas, deposits in the combustion chamber, the amount of residual gas etc.

2.4.2
Design Parameters

The influence of the so-called engine design parameters on combustion and exhaust emission is no less important. These design parameters include the cylinder displacement, S/D ratio, valve timing, layout of the intake and exhaust systems, shape of the combustion chamber, its surface/volume ratio and the compression ratio.

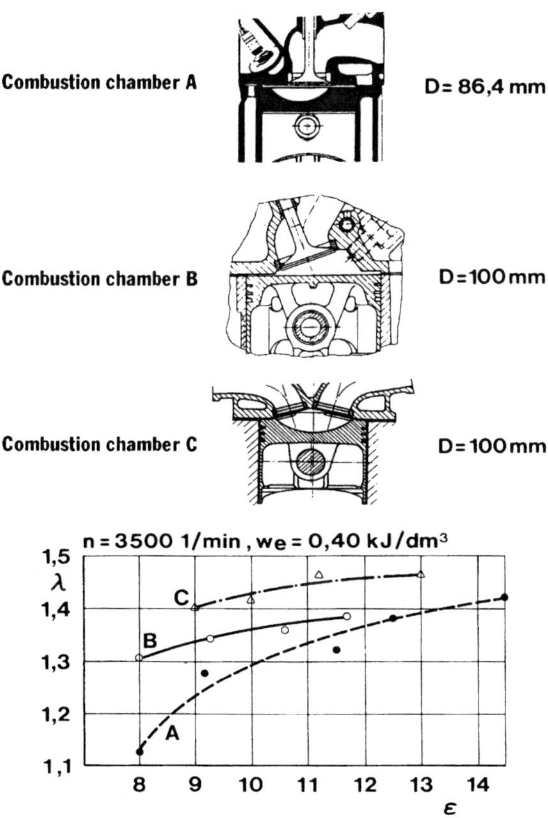

Fig. 10. Air/fuel ratio at misfire limit dependent on combustion-chamber shape and compression ratio

2.4.2.1
Combustion Chamber Shape

The shape of the combustion chamber has a decisive influence on behavior of the gasoline engine mainly in the lean-burn range. The combustion chamber shape determines the movements of the charge (turbulences) upon which the combustion process highly depends.

There are two possibilities to generate charge turbulences in the cylinder: The first solution consists in designing the intake manifold and intake duct in such a way that the charge is made to swirl or tumble during the intake stroke and that this swirl or tumble is maintained throughout the compression stroke. The second solution consists in shaping the combustion chamber in a way so as to realize squish effects which produce intensive turbulences in the combustion chamber at the end of the compression stroke. Since the intensity of the charge movement induced in the intake manifold and intake duct drops clearly during compression stroke, this solution must be combined with the squish effect produced by the combustion chamber shape. This combination allows optimum combustion conditions to be achieved and is particularly suited for lean air/fuel mixtures.

The optimization of the combustion chamber shape is particularly helpful when it comes to shift the lean limit towards higher A/F ratios (so-called "lean-burn" engines) (Fig. 10).

Particularly good results are obtained when using spherical combustion chambers with two intake and exhaust valves each and a central spark plug.

2.4.2.2
Compression Ratio

Increasing the compression ratio is a generally applied method to improve the efficiency of a gasoline engine. To this solution, however, limits are set by the knock resistance of the fuel used. For decades, the compression ratio was chosen taking into account the power output and engine torque only. After the legislations on exhaust emissions had been introduced it was found that the compression ratio can have a considerable influence on HC and NO_x emissions. Today, the compression ratio is chosen with power output, exhaust emissions and fuel consumption in mind.

The compression ratio – which is to be chosen in accordance with the cylinder bore and combustion chamber shape – must be high enough to ensure optimum engine operation mainly with lean air/fuel mixtures. The problem of combustion knock at high compression ratios can be solved by providing for an appropriate layout of the combustion chamber (Fig. 11). To account for all those compromises, the compression ratios of modern Otto engines range between ε=9.5 and 12.

2.4.3
Limitation of Pollutant Reduction by Engine-Internal Measures

The first measures meant to reduce CO and HC emissions, which started in Europe, followed later by the USA, were paralleled by efforts to lower fuel consumption. Be-

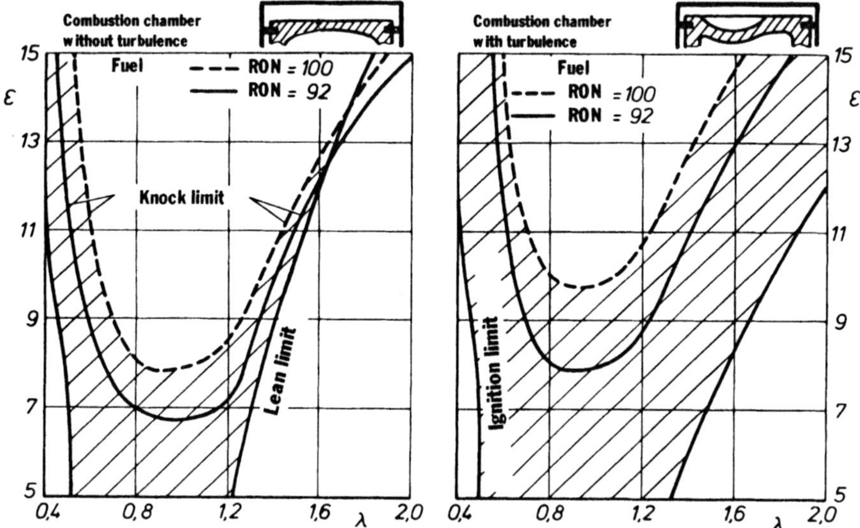

Fig. 11. Influence of combustion-chamber shape and compression ratio

tween R15 – the first emission regulation stipulated in Europe in 1971 – and R1504 which was valid until 1993, the emission limits as well as the fuel consumption of European cars were simultaneously reduced through engine-internal modifications. In doing so, one of the most efficient measures was the leaning of the air/fuel mixtures of gasoline engines from the original A/F ratio of λ=0.8–0.9 of the early 1970s to λ=1.05–1.15 of the early 1990s. Thus, the last pre-catalyst generation of European gasoline engines had been operated on lean air/fuel mixtures.

The following exhaust emission levels could be reached in the ECE test just through engine-internal measures:

CO=6.0–8.0 g/km
HC=1.0–2.0 g/km
NO_x=1.5–2.5 g/km

But these levels were not sufficient anymore to comply with the more and more severe emission limits. The introduction of extremely stringent exhaust gas limits mainly as far as NO_x was concerned put a temporary end to the trend of simultaneously improving both the exhaust emissions and fuel economy. It was not possible any longer to satisfy the legislator's demands by mere engine-internal improvements.

2.5 Engine-External Measures for Pollutant Reduction

If optimizations of engine-internal parameters for further reduction of the exhaust emissions are no longer sufficient, so-called engine-external measures must be additionally taken which, in general, do not have any direct influence on fuel consumption. Indirectly, however, fuel economy is influenced as well, thanks to the engine readjustments required to ensure the functional reliability of the exhaust-gas after-treatment accessories.

2.5.1 *Fuel-Independent Measures*

The first category includes engine-external measures which do not place any special demands on the fuel quality, such as secondary air-injection, exhaust-gas recirculation, portliners and thermal reactors.

2.5.1.1 *Secondary Air-Injection*

This device provides the exhaust-gas system with fresh air to improve the CO and HC oxidation in the exhaust ducts. It is required for and particularly efficient in the presence of the rich A/F ratios (for cold starting, warming up and acceleration) during which the exhaust gas has a very high chemical energy. This system allows the CO and HC emissions to be lowered by 30 to 50% and by 20 to 40%, respectively, during these phases. To be able to meet extremely low emission limits many of the modern catalyst-equipped Otto engine variants must be fitted with secondary-air injection. During the respective operating phase, the secondary air is injected directly into the exhaust port by means of a secondary-air pump. The latter consumes 1 to 3% of the maximum engine power resulting in an increased fuel consumption.

2.5.1.2 *EGR (Exhaust-Gas Recirculation)*

Returning part of the exhaust gas into the cylinder – a process called exhaust-gas recirculation (EGR) – is a service-proven way to reduce the NO_x emission level. The influence of EGR on the combustion process is manifold: it lowers the charge-exchange losses and thus increases the pressure and temperature at the end of the compression stroke. The recirculated gas helps to improve the lean limit by warming up the fresh charge and it influences the flame propagation and thus the HC emission and lean-burn capability of the engine by serving as an inert constituent (residual gas). Therefore, to realize a modern low-NO_x Otto engine, it is essential to provide for a precise control of the EGR system.

2.5.1.3 *Portliners*

The exhaust gas temperature downstream of the exhaust valve should be as high as possible to ensure the secondary reaction of HC and CO mainly if the engine is fitted with a catalytic exhaust-gas after-treatment system. This can be achieved

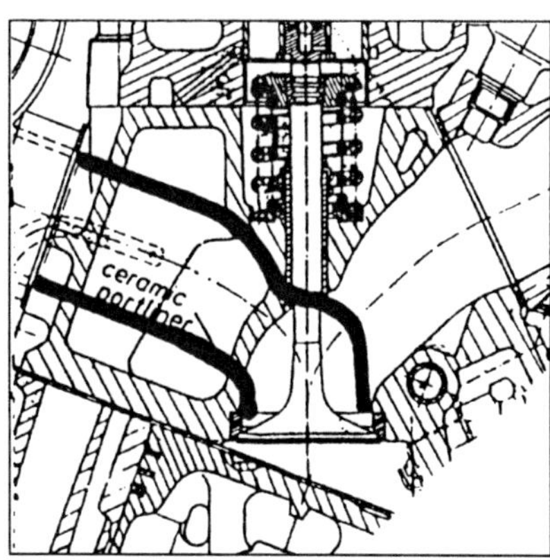

Fig. 12. Portliner

by heat-insulating tubes – so-called portliners – which prevent the heat of the exhaust gas from being transmitted to the cylinder head (Fig. 12). With rich air/fuel mixtures (cold starting, warming up) the portliners offer advantages only if combined with secondary air-injection.

2.5.1.4
Thermal Exhaust-Gas After-Treatment

As far as carbon monoxide and unburnt hydrocarbons are concerned, thermally well insulated exhaust pipes (thermal reactors) were used allowing the combustion process initiated in the combustion chamber to be continued. In the beginning, this approach – which considered the reactor as being a fully integrated constituent of the exhaust system – was thought to be technically correct and useful. However, to achieve efficient conversion rates in the reactors, temperatures of 700 to 800°C are required and it is essential that these temperature levels be reached also at low engine speeds and loads as well as immediately after cold starting. Consequently, the engine had to be tuned for high exhaust temperatures which resulted in excessive fuel consumption increases. So, the thermal reactor developments were stopped soon. In modern low-pollution engines, however, thermally well insulated exhaust pipes are one of the basic elements of catalytic exhaust after-treatment systems of modern gasoline engines.

The engine-external fuel-independent measures for exhaust emission reduction (i.e., secondary air injection, exhaust-gas recirculation, thermal insulation) allow the following emission figures to be reached in the ECE test:

CO=4–6 g/km
HC=0.5–1.5 g/km
NO_x=0.5–1.5 g/km

2.5.2
Fuel-Dependent Measures

The above-mentioned solutions for exhaust emission reduction are not sufficient to comply with the stringent emission legislation.

In the USA and Japan, catalytic exhaust-gas after-treatment systems have been successfully applied since the introduction of more severe emission limits in 1975. For these systems to be operative, unleaded fuel had to be made generally available, because noble-metal catalysts are sensitive to lead, sulfur and phosphorus and undergo rapid aging if exposed to these substances.

2.5.2.1
Oxidation Catalyst

In the catalytic reactors (catalysts) the oxidation of CO and HC is strongly enhanced by the reaction of the catalyst noble materials used such as platinum, palladium and rhodium. Optimum conversion rates are reached already at exhaust gas temperatures of as low as 200 to 250°C. Lean A/F mixtures offer optimum conditions for the reduction of CO and HC whereas with rich mixtures, a secondary-air pump is needed to inject additional fresh air upstream of the catalyst. Oxidation catalysts do not have any major influence on NO_x emissions.

2.5.2.2
Reduction Catalyst

If engine-related measures and EGR do not yield the required low NO_x levels, a so-called reduction catalyst must be used.

To lower the NO_x emission, a low-oxygen atmosphere is required or, in other words, rich air/fuel mixtures must be used. The great amounts of carbon monoxide (CO) contained in rich mixtures make sure that NO is split up into CO_2 and N_2.

$$2\,CO + 2NO \rightarrow 2\,CO_2 + N_2 \qquad \text{(e)}$$

For the oxidation of relatively large volumes of CO and HC during rich A/F mixture operation an additional oxidation catalyst with secondary-air injection is required. Such a catalyst combination, consisting of one reduction and one oxidation catalyst each, is called a dual-bed catalyst. However, this is not a fuel efficient solution as it requires rich air/fuel mixture and permanent secondary-air injection. Therefore, automotive manufactures soon decided to drop this concept.

2.5.2.3
3-Way Catalyst Plus Oxygen Sensor

The legal demand for the drastic reduction of CO, HC and NO_x with simultaneous improvement of fuel economy prompted power plant engineers to search for new technologies. After years of intensive engineering work it was found that so-called three-way catalysts including a precisely defined mixture of platinum (Pt),

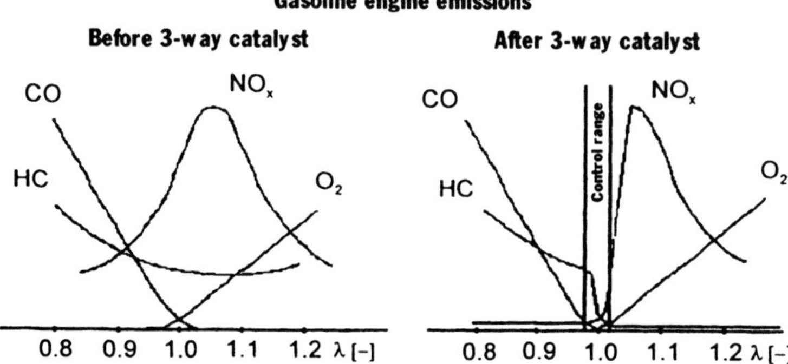

Fig. 13. Exhaust emission of gasoline engines before and after 3-way catalyst

rhodium (Rh) and/or palladium (Pd) reduce the three aforementioned pollutants by clearly more than 90% provided that a precisely stoichiometric A/F ratio (λ=1.0) is used (Fig. 13). By then, a reduction of that magnitude was required to meet the legal standards.

When using a three-way catalyst, it must be ensured that the A/F ratio is fixed at λ=1.0 throughout the major part of the engine map. This is guaranteed by the oxygen sensor or lambda probe developed for the purpose of determining the stoichiometric air/fuel mixture in the exhaust gas on the one hand and by using efficient electronic mixture control systems on the other (Fig. 14).

The pollutant conversion rates of modern three-way catalysts being more than 98%, this technology has firmly established itself because it allows both current and future exhaust-gas standards to be complied with.

Today, all Otto engines come with closed-circuit electronic systems for air/fuel mixture control, one or more oxygen sensors and one or more three-way catalysts.

Modern gasoline engines use stoichiometric mixtures (λ=1.0) over a wide operating range. The mixture is slightly enriched under full-load conditions and during cold-starting only in order to obtain the maximum possible power output from the given engine displacement and make sure that the mixture is correctly ignited and the catalyst protection functions are triggered. Figure 15 shows the A/F ratio map of one of the current gasoline engines with three-way catalyst and with λ-control.

Depending on the required air mass, measured by means of either an air-flow sensor in the intake manifold or via the intake manifold pressure (using the state equation pV=mRT), the amount of fuel needed for each operating condition is precisely dosed and fed into the engine. The ignition timing is programmed in accordance with the respective engine load, speed and temperature and the required exhaust-gas composition while providing for a safe distance from the knock limit.

Thanks to the strict maintenance of a precise stoichiometric air/fuel mixture the three-way catalyst allows very low HC, CO and NO_x pollutant emissions to be achieved which meet the current severe emission limits. However, in this oper-

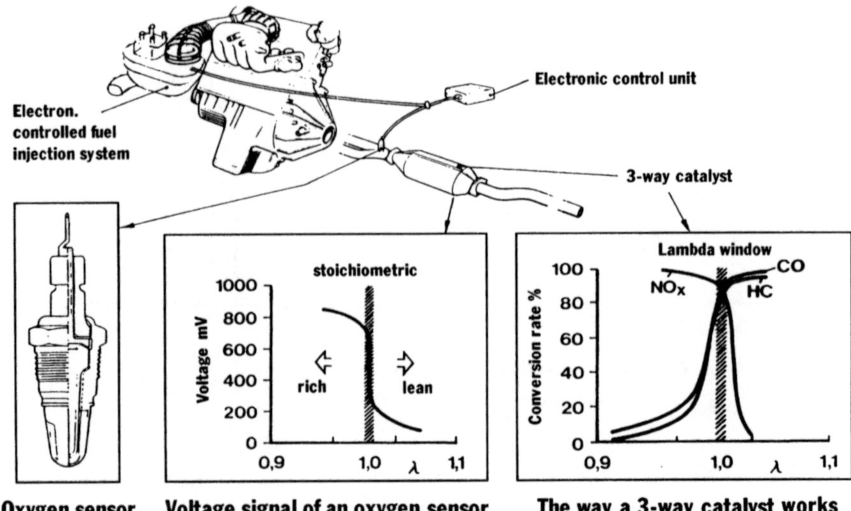

Fig. 14. Gasoline engine with 3-way catalyst and oxygen sensor

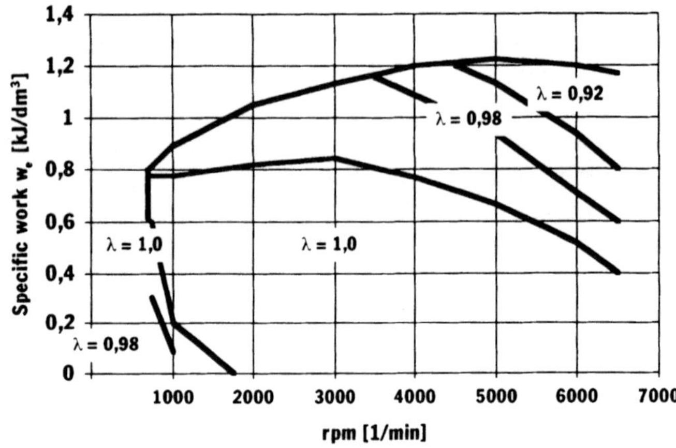

Fig. 15. Air/fuel ratio map of gasoline engine with 3-way catalyst

ating range, fuel consumption is 8 to 15% higher (with a resulting higher CO_2 emission) than during lean-burn operation if no severe NO_x emission limits are to be observed. Gasoline engines reach the best fuel economy figures and lowest CO_2 emissions when operated on lean air/fuel mixtures.

Therefore, current gasoline engine development is focused on achieving optimum engine efficiency or, in other words, operating the engine at optimum A/F ratios at any operating conditions while meeting all the other engine-relevant demands at the same time.

2.6
The Lean-Burn Engine – the Ultimate Target of Otto-Engine Development

One of the technically most useful solutions to reduce the fuel consumption and CO_2 emission of gasoline engines is to make them tolerate lean air/fuel mixtures. When compared with mostly stoichiometric engines, the part-load fuel consumption of lean-burn engines can be lowered by 8 to 15%.

To reach this target, some basic conditions must be fulfilled beforehand:

- Regular undisturbed combustion of externally ignited air/fuel mixtures with a high excess-air coefficient – the so-called lean-burn operation;
- Regular undisturbed combustion of lean air/fuel mixtures also during transient operating conditions such as acceleration, deceleration, starting and warming up;
- Low pollutant emissions with excess air according to current legislation.

To date, the introduction of lean-burn concepts into production has been hampered by the fact that the three-way catalyst is not capable of reducing the NO_x emissions during excess-air operation. Carbon monoxide and unburnt hydrocarbons can be reduced via the service-proven noble-metal oxidation catalyst, whereas no safe, durable and production-ready technology is available yet for the required drastic reduction of the NO_x emissions generated by lean-burn engines.

To date experience has shown that, in order to guarantee the three afore-mentioned prerequisites which are vital for the successful development and market introduction of the so-called lean-burn engine, the engine-internal processes must be precisely controlled. This is mainly true for the air/fuel mixture which must be tuned in such a way that the respective optimum level is maintained for each operating condition.

2.6.1
Problems of Lean-Burn Operation

Gasoline engines show certain combustion irregularities – a phenomenon called "cycle-by-cycle variation" that many generations of engine specialists have had to deal with.

Even though this stochastic problem of cyclic variation is characteristic of gasoline engine combustion, it remains almost unnoticed when rich or stoichiometric mixtures are used. This is due to the resulting stronger torque variations and higher emissions outside of the cylinder are not occurring.

If the mixture is further leaned beyond the stoichiometric A/F ratio, however, the variations of the working cycles intensify, resulting in more and more irregular crankshaft revolutions thus deteriorating the driving comfort of vehicles powered by lean-burn engines.

In Fig. 16, the fuel consumptions as well as the HC and NO_x emissions of a standard Otto engine and a lean-burn engine have been plotted versus λ at the part-load $n=2000$ rpm and $p_{me}=2$ bar. When the engine approaches the lean-burn limit, the working-cycle variations and HC emissions start increasing. The

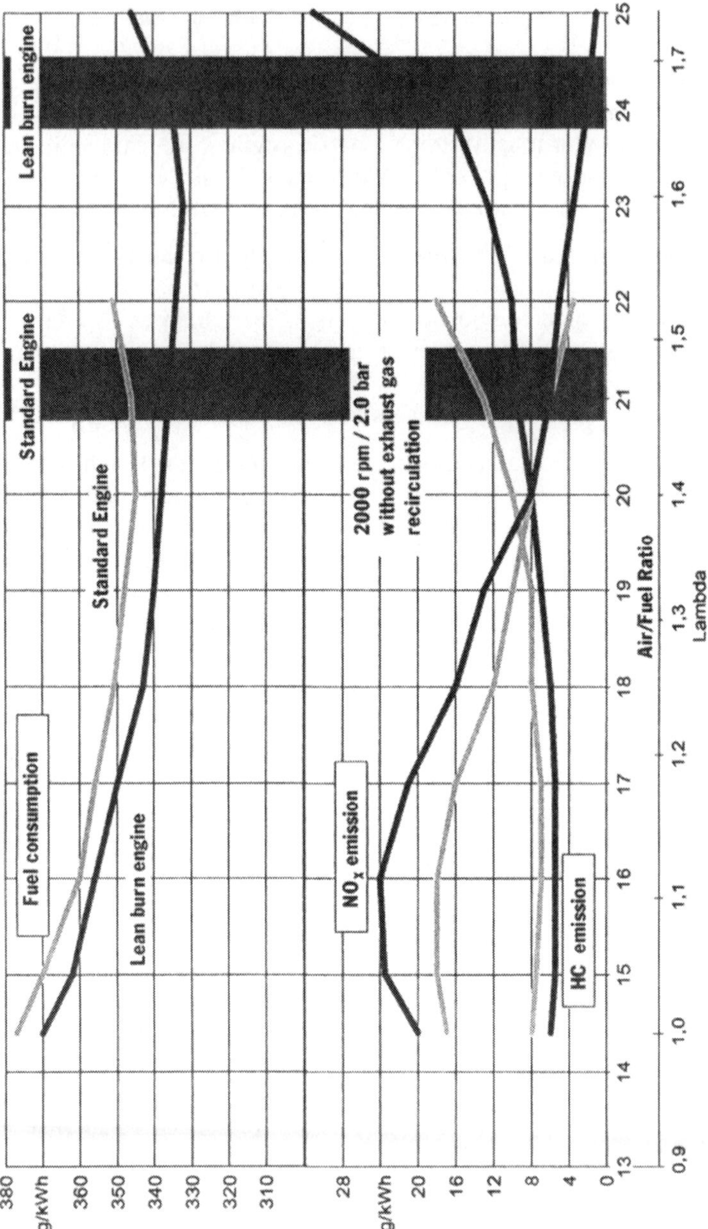

Fig. 16. Air/fuel ratio for gasoline engine

NO_x emissions keep dropping as the process temperature decreases. Fuel consumption drops steeply up to λ=1.3 and continues to decrease towards a minimum limit value if the mixture is further enleaned. The resulting fuel consumption at that load point is about 10% lower than with λ=1.0. If the mixture of a standard production engine is further enleaned beyond λ=1.45, fuel consumption increases again due to the distinct slow-down of the combustion and resulting misfires at the lean-burn limit. A "lean-burn" Otto engine should therefore be operated as closely as possible to the lean-burn limit in order to minimize NO_x emissions but not too close to it either in order not to be penalized by an excessively high fuel consumption increase.

2.6.2
State of the Art

Over the decades, engineers have tried again and again to design Otto engines for lean-burn operation. Different approaches using homogenous and heterogeneous fuel/air mixtures have been proposed and tried out. In the 1970s, work was concentrated on the so-called stratified-charge engine with the only engine of that type installed in a production car being the Honda CVCC unit (Controlled Vortex Combustion Chamber).

However, the classical Otto engine using stoichiometric air/fuel mixtures (λ=1.0) has turned out to be a better alternative. It is mainly due to the severe emission limits that none of the lean-burn engines developed to date has been able to assert itself. New efforts for realizing a production-ready lean-burn engine are being made by Toyota, Porsche, Honda, Ford etc.

There are two distinct trends in modern lean-burn-engine development: using either homogeneous air/fuel mixtures injected into the intake manifold or heterogeneous air/fuel mixtures by direct injection into combustion chamber offering the possibility of charge stratification.

In gasoline engines, operated on homogeneous air/fuel mixtures, the fuel is injected into the manifold upstream of the intake valve. Gasoline engines fed with heterogeneous air/fuel mixtures have direct injection into the cylinder and are called Gasoline Direct Injection or GDI engines. GDI engines can use both homogeneous and heterogeneous mixtures.

Both lean-burn variants use lean mixtures in the part-load range only. At full load and high part load, they, too, are operated on stoichiometric or slightly enriched mixtures in order to achieve the maximum possible torque and power output. For reasons of comfort, stoichiometric mixtures are also applied in the near-idling range.

Figure 17 shows the characteristic A/F ratios map of a modern direct-injection lean-burn engine.

2.6.3
Exhaust Gas After-Treatment for Lean-Burn Engines

The legislator's demands in terms of CO, HC and NO_x emissions at the present time can only be met with the help of three-way catalysts.

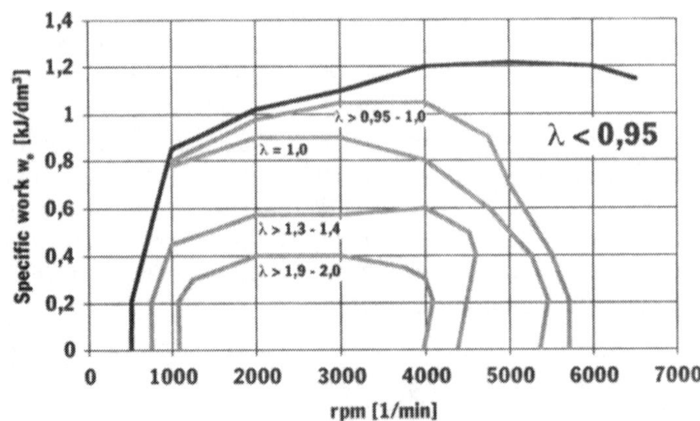

Fig. 17. Air/fuel-ratio map of lean-burn gasoline engine

One of the main problems of lean-burn gasoline engines is the NO_x emission which is greatest at λ=1.05–1.2. Thanks to the high A/F ratio and low combustion temperatures, however, the NO_x raw emissions of lean-burn engines are lower than those of gasoline engines operated on stoichiometric air/fuel mixtures. Due to the strongly oxidizing atmosphere (excess air, 5–9% oxygen content) it is not possible to further lower the NO_x levels with conventional exhaust gas after-treatment systems alone.

When combined with an engine operated on high A/F ratios, three-way catalysts will at best function as oxidation catalysts converting HC and CO into water vapor and carbon dioxide provided that there is sufficient oxygen.

In the presence of lean A/F mixtures, three-way catalysts are not capable of reaching the desired NO_x conversion rates because there is not enough CO available to do so on the one hand. On the other, NO_x reduction is made impossible by the prevailing low temperature levels and the residual oxygen content of the exhaust gas.

2.6.3.1 DeNO$_x$ Catalyst

At the moment, various systems for post-engine NO_x reduction are being investigated and developed which also function in the presence of high A/F ratios. Great hopes are placed in the DeNO$_x$ catalysts which use so-called zeolites, for example, to reduce NO_x with the help of hydrocarbons. There are two potential variants available:

- a noble-metal variant which allows nitrogen reduction at relatively low exhaust-gas temperatures of 180 to 200°C. However, this temperature window is very narrow and there is the problem of simultaneous nitrous oxide (N_2O) generation;
- a noble-metal-free ion-exchange-zeolite variant which seems to be more promising. It has a wide temperature window of about 300 to 600°C which is

better suited to cope with the exhaust gas temperatures of a lean-burn Otto engine. State-of-the-art systems of that kind allow NO_x reductions of about 40% to be achieved even with high A/F ratios of λ=1.5. Nevertheless, the remaining NO_x emissions are up to four times higher than those of an engine running with stoichiometric air/fuel mixture and with an oxygen sensor-controlled three-way catalyst.

The resulting demands on engine management are high: It must provide for optimum operating conditions of the exhaust-gas after-treatment system if the $DeNO_x$ catalyst is to function smoothly. One approach might be to use a management system which adjust the engine in such a way that an optimum HC/NO_x ratio is guaranteed while maintaining the exhaust gas temperature at a level suited for exhaust gas after-treatment.

The development of $DeNO_x$ catalysts is rapidly progressing but there still are some problems which have to be overcome: durability is not satisfactory yet, the conversion rates are insufficient and the temperature window for efficient conversion still is too narrow.

2.6.3.2
NO_x Storage Catalysts

With regard to their comparatively high NO_x conversion rates and the three-way-catalyst-type properties at A/F ratios of λ=1.0 the NO_x storage catalysts are the most promising alternative for the efficient after-treatment of the exhaust gas of combustion engines operated with excess air.

With this solution, the nitrogen oxide produced during lean-burn combustion is not reduced but mainly adsorbed in an NO_x storage catalyst where it is stored for some time.

NO_x accumulating catalysts include an additional storage medium consisting of a basic coating (oxide layer made of alkaline or alkaline earth metals) besides the usual noble-metal three-way layer. The resulting properties under stoichiometric conditions are almost the same as those of the three-way catalysts.

In lean-burn mode, the storage medium fixes the nitrogen oxides in the form of nitrates. Since the system has a limited storage capacity only, a so-called regeneration must be performed from time to time. This is achieved by enriching the mixture to $\lambda \leq 1$. In doing so, the nitrates are split up into their constituents and the released nitrogen oxides are converted by the three-way catalyst (noble-metal layer) during stoichiometric or enriched operation. The main reducing agents used are CO, hydrocarbons and hydrogen.

$$CO + NO \leftrightarrow N_2 + CO_2 \qquad \text{(f)}$$

$$HC + NO \leftrightarrow N_2 + H_2O \qquad \text{(g)}$$

The precise mechanisms of NO_x adsorption during lean-burn operation and nitrate disintegration at $\lambda \leq 1$ have not yet been clearly described. A potential reaction has been illustrated in Fig. 18 using the barium oxide layer (BaO) as a basis.

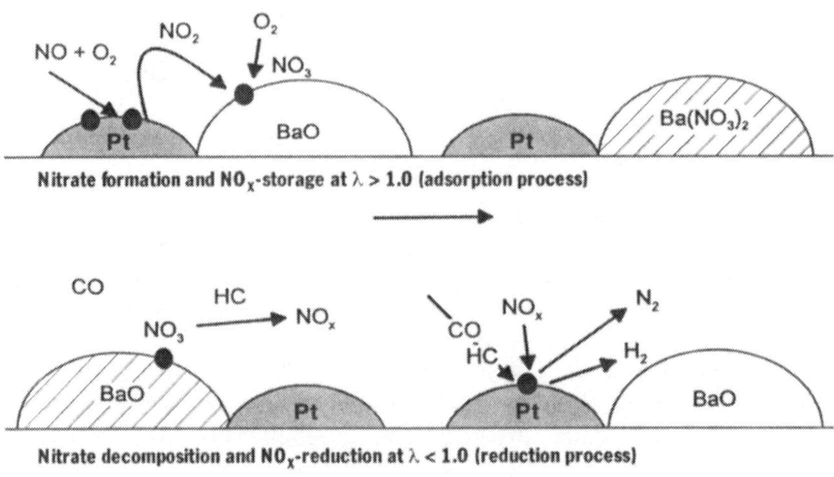

Fig. 18. Adsorption and nitrate formation of NO_x at storage medium barium oxide (BAO) at $\lambda>1.0$ and nitrate decomposition at rich mixture

The adsorber is regenerated with the help of a downstream three-way catalyst during stoichiometric or enriched engine operation.

For this strategy to function there must be a controlled switching between stoichiometric/rich and lean-burn operation phases.

The target is to achieve an optimum balance between lean and stoichiometric/rich phases with maximum fuel economy on the one hand and optimum NO_x reduction on the other and with lean-burn operation prevailing. Under optimum conditions and when using a new unaged catalyst combination (adsorber+three-way catalyst), nitrogen oxide reduction of up to 90% can be achieved according to the current state of the art.

One of the main problem is the sulfur contained in conventional fuels which hampers the smooth functioning of the NO_x storage layer and the catalyst.

Slightest amounts of sulfur – which compete with nitrogen for storage capacity in the lean-burn range – drastically reduce the storage efficiency and the thermal stability of the catalyst.

In addition, the durability and temperature resistance of the NO_x storage catalysts are not sufficient yet for large-scale series introduction.

Since none of the approaches for NO_x reduction – such as $DeNO_x$ catalysts, catalytic converters with additives such as ammonia or urea, selective catalysts, zeolites, adsorbers and so on – have reached production readiness yet, the future of the lean-burn engines will almost exclusively depend on the successful development of NO_x exhaust-gas after-treatment technologies for lean air/fuel mixtures.

2.7 References

1. Eberan-Eberhorst R (1969) Abgasforschung zukunftsweisend für den Fahrzeug- Ottomotor. MTZ-30, 9
2. Daniel WA, Wentworth JT (1967) Exhaust Gas Hydrocarbons- Genesis and Exodus. SAE-Paper 486 B
3. Gruden D (1973) Veränderungen und Grenzwerte der Abgasemission im 4-Takt-Fahrzeug Ottomotor und Wege zur Verringerung des Abgasgeruches. Dissertation der TH-Wien
4. Obländer K, Nagel A (1984) Übersicht über Maßnahmen zur Minderung von Kfz-Emissionen. Staub-Reinhaltung der Luft 44, 9
5. Gruden D, Markovac U, et al (1979) Schadstoffarme Antriebssysteme - Entwicklungsstand, Wirtschaftlichkeit, Kosten. Berichte 2/80 Umweltbundesamt, Berlin
6. Gruden I (2000) Emission und verbrauchsoptimierte Regelung von homogen mager betriebenen Ottomotoren. Dissertation der TU-Wien
7. Stone R (1992) Introduction to internal combustion engines. The Macmillan Press, London
8. Krämer M, Maly T, et al (1995) Emissionsreduzierung beim mager betriebenen Ottomotor. 16. Wiener Motorensymposium
9. Hohenberg G (1997) Analyse der Gemischbildung und Verbrennung am D.I.-Ottomotor. 18. Wiener Motorensymposium
10. Wurster W, Gruden D (1988) Die Verbrennung im Otto- und Dieselmotor mit direkter Einspritzung. 9. Internationales Wiener Motorensymposium
11. Inoue T, Matsushita S, et al (1993) The Development of a High Fuel Economy and High Performance Four Valve Lean Burn Engine. SAE-Paper 930873
12. Iwamoto Y, Noma K (1997) Development of Gasoline Direct Injection engine. SAE-Paper 970541
13. Grebe UD (2000) Zukunft des Ottomotors - Benzindirekteinspritzung oder Laststeuernde. Variable Ventiltriebe. ÖVK-Wien
14. Moser W (1999) Benzin-Direkteinspritzung - ein Beitrag zur Absenkung der CO_2-Emissionen. VDA-Technischer Kongress
15. Noma K, Iwamoto Y, et al (1998) Optimized Gasoline Direct Injection for the European Market. SAE-Paper 980150
16. Baumgarten H, Goerts W, et al (2000) Niedrigstemissionskonzept zur Erfüllung der SULEV-Emissionsstandards für Ottomotoren. MTZ 61:10
17. Glück K-H, Göbel U (2000) Die Abgasreinigung der FSI-Motoren von Volkswagen. MTZ 61:6
18. Freidl GK, Piock W, et al (1997) Direkteinspritzung bei Ottomotoren. Aktuelle Trends und zukünftige Strategien. MTZ 58:12
19. Brandt S, Dahle U, et al (1998) Entwicklungsschritte bei NO_X-Adsorber Katalysatoren für magerbetriebene Ottomotoren. Stuttgarter Symposium FKFS
20. Domes W, Gerwig W, et al (1985) Die neuen 16-Ventil-Motoren für Scirocco und Golf. ATZ 87:6
21. Göbel U, Kreuzer T, et al (1999) Moderne NO_X-Adsorber-Technologien Grundlagen Voraussetzungen, Erfahrungen. VDA-Technischer Kongress
22. Gruden D, et al (1993) Entwicklungstendenzen auf dem Gebiet der Otto-Motoren. Expert Verlag Renningen

3 The Diesel Engine

Klaus Borgmann · Oswald Hiemesch

3.1 General Issues

In a thorough process of research and development, Rudolf Diesel set out to develop a thermal engine with a high degree of efficiency, since he regarded the steam engine serving as the principal drive system in the 19th century, with an efficiency of only about 3 percent, as an enormous waste of energy.

Applying for a patent for his invention in 1892, and publishing his study on the *Theory and Construction of a Rational Thermal Engine* in 1893, Diesel successfully laid the foundation for the thermal engine now referred to as the *Diesel engine* with the highest level of efficiency (currently up to 53 percent) ever achieved. As early as 1897, Diesel's test engine achieved an efficiency of 26.2 per cent which was a sensation at the time. With this increase in efficiency, by almost a factor of 10, the steam engine was very quickly replaced by the Diesel engine which served initially as a stationary drive system.

Thanks to significant progress on fuel injection and turbocharger technologies, the Diesel engine has become used increasingly for transportation purposes in recent decades. In rail-bound and ship transport the Diesel engine has played a dominating role for many years. In utility and commercial road transport (utility vehicles and buses) the Diesel engine is uncontested by any kind of competition. In individual transport (passenger cars) the Diesel engine is looking at an extremely positive future, with registration figures going up steadily, particularly in Europe. With world production of road vehicles continuing to increase (Fig. 1 [1]) the Diesel engine will continue to increase its future market share from the petrol engine.

The Diesel engine, like the petrol engine, uses a controlled burning process to convert the chemical energy in the fuel into mechanical energy. Unlike the petrol engine however, combustion of the fuel injected is initiated by self-ignition in the highly compressed air. Most of the energy lost is thermal energy dissipating through the exhaust, into the coolant and by way of radiation. The process of comparison representing the Diesel engine in theory (see Fig. 2) is the *Seiliger Process* made up of:

- adiabatic compression 1–2,
- isochoric (Q_V) and isobaric (Q_P) heat input,
- adiabatic expansion 3–4, and the
- isochoric heat rejection 4–1.

The p-v diagram shows the useful work L in the area 123′341. In the T-S diagram this work equals thermal energy Q, again in area 123′341.

The fresh charge air in the cylinder has to be compressed to a high level in order to initiate self-ignition of the injected fuel in the Diesel combustion process (see also Fig. 3 in the Chapter 2.1). The Diesel engine has the highest standard of

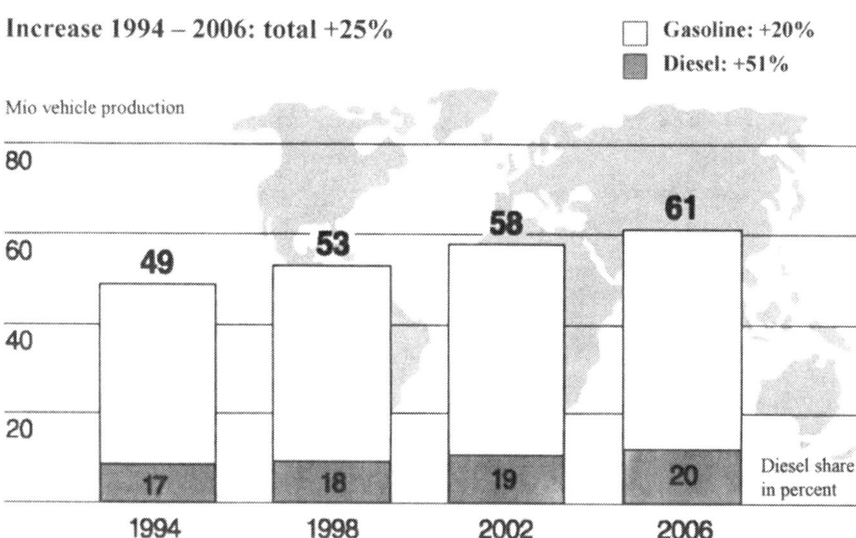

Fig. 1. Forecast of worldwide vehicle production

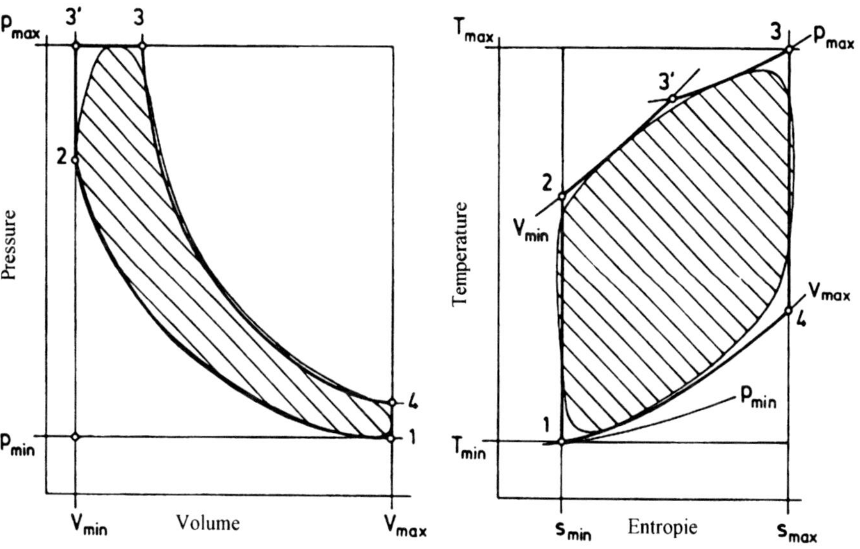

Fig. 2. Seiliger process

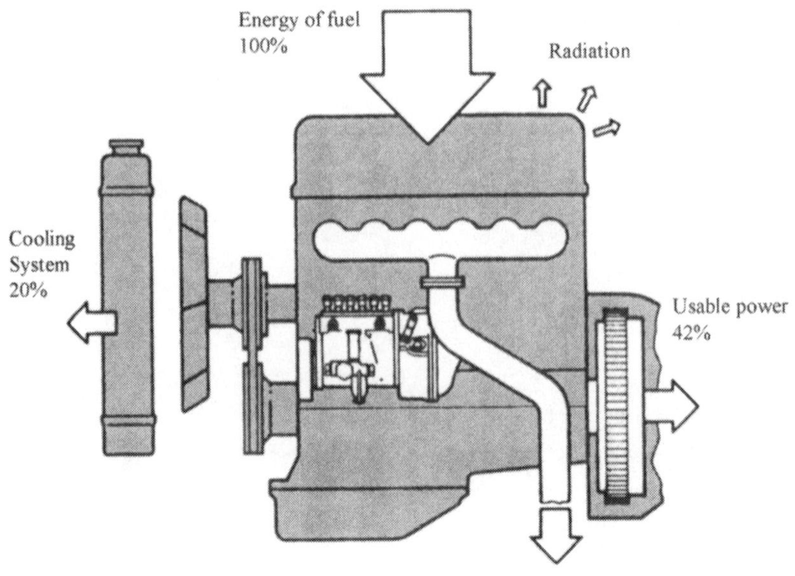

Fig. 3. Thermal balance of a Diesel engine

efficiency of all thermal engines because the loss of energy, as shown in Fig. 3, is kept to a minimum.

Representing a highly complex energy conversion system, the Diesel engine is made up of the following parts and components:

- The fuel/air formation and combustion system consisting of the air and fuel supply units as well as the combustion chamber (or, respectively, the piston combustion chamber in the case of direct fuel injection).
- The engine itself is made up of the crankcase, crankshaft, connecting rods, pistons and cylinder head. In the interest of safe and reliable operation, the lubricating system and cooling are incorporated directly within the engine.
- Seeking to convert energy with maximum efficiency and reduce emissions affecting the environment to a minimum, the exhaust gas path has been developed into an increasingly complex system serving to cut back noise and exhaust emissions to the lowest possible level.

3.1.1
Formation of the Fuel Mixture, Combustion Process

Proper operation of the Diesel engine requires the fulfillment of various criteria:

a) High compression of the air flowing into the charge cylinder in order to reach the self-ignition temperature of the fuel injected. The fuel used for this purpose must be easily ignitable (a factor defined by the cetane number).

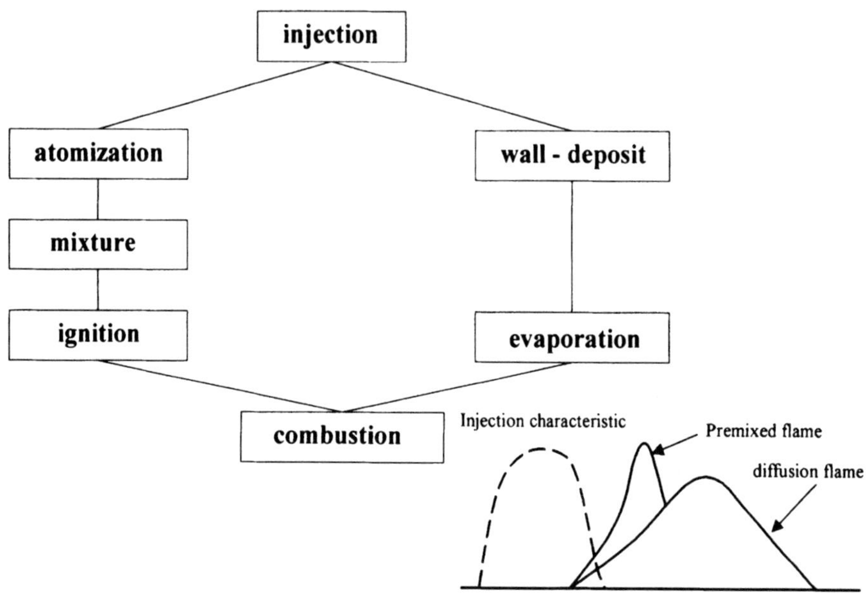

Fig. 4. Main phases of mixture formation and combustion in a Diesel engine

b) A sufficient fuel/air mixture formation process for burning the fuel injected in short time during the operating cycle, i.e., with a high standard of efficiency. With this internal fuel/air mixture formation and the combustion process taking a certain period of time, the timing of the process must be coordinated with the position of the piston in the working cylinder. The pressure curve in the operating cylinder following from the combustion process and the position of the crankshaft ultimately determine the energy transmitted to the crankshaft during an working cycle. Engine output, fuel consumption and exhaust emissions thus depend to a large extent on the formation of the fuel/air mixture and the combustion process. With the fuel being injected in a liquid state directly into the hot, highly compressed air, atomizing into minute droplets, igniting where the conditions for self-ignition are most favorable, and finally burning as a fuel/air mixture, we speak of internal heterogeneous fuel/air mixture formation within the diesel engine. This very complex interaction [2] of fuel injection, mixture formation, ignition and combustion depends on local temperatures, pressure conditions, the concentration of substances reacting with one another, velocity factors, etc. physical processes such as flow conditions, atomization, and evaporation are subject to, and affected by, chemical reactions taking place at the same time (see Fig. 4). Since these processes take a certain time, it is easy to understand that the maximum speed a diesel engine is able to reach is limited and lies at a lower level than the running speed of a spark-ignition engine with external fuel/air mixture formation and ignition. The faster the fuel/air mixture formation process, the higher are the engine speeds achievable.

The main factors crucial to the fuel/air mixture formation process are therefore:

- the kinetic energy of the injection jet depending primarily on injection pressure;
- the controlled movement of air in the operating cylinder, and
- the thermal condition (wall temperature) of the combustion chamber and the compressed air affecting the local temperature and, accordingly, the intensity of evaporation of the fuel droplets as well as the fuel film resting on the walls of the combustion chamber.

Depending on the mode of fuel injection, either directly into the combustion chamber or into a side or ancillary chamber, we distinguish between

- the direct injection Diesel engine and
- the chamber Diesel engine.

Figure 5 presents the various combustion processes as a function of the air and injection energy required in each case. With the development of high-pressure injection systems, the direct-injection engine, benefiting from its higher degree of efficiency, has taken a leading position in all areas and applications.

Running at relatively high speeds (up to 5000 rpm) and with lower noise and exhaust emissions, chamber Diesel processes (see Fig. 6) dominated the passenger car market until the end of the 20th century.

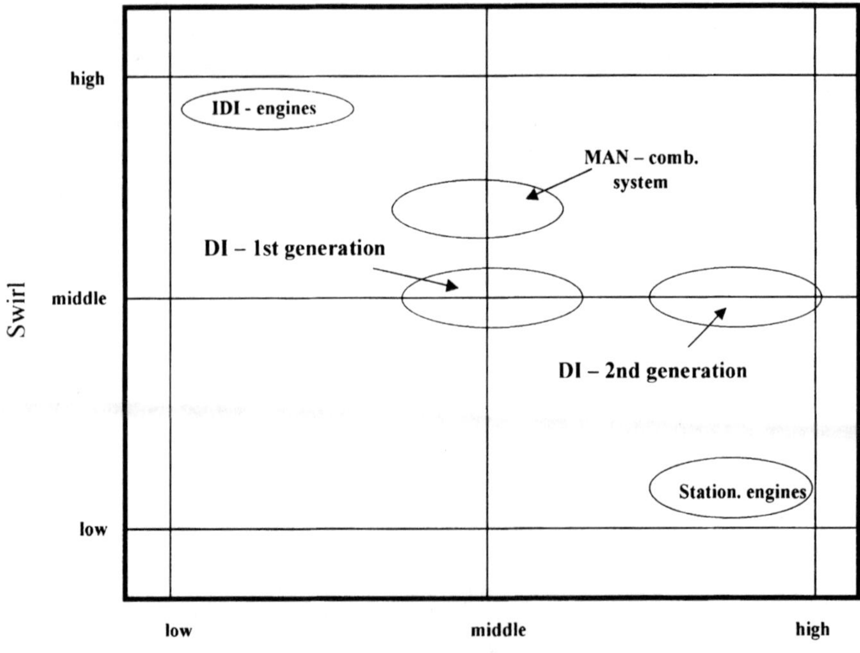

Fig. 5. Combustion process as a function of mixture formation energy

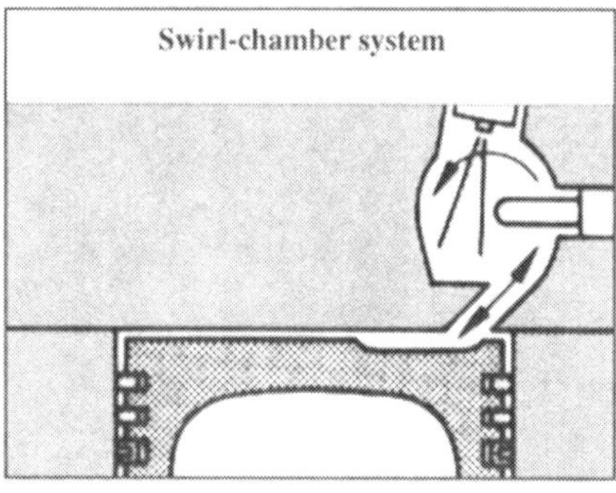

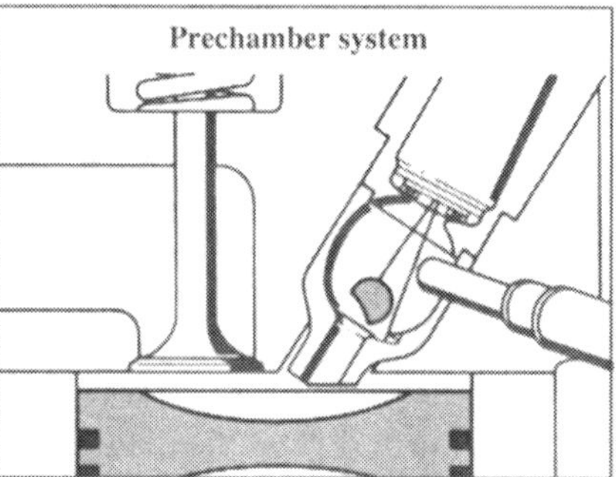

Fig. 6. Diesel chamber combustion process

With drive systems saving fuel and resources in general being widely promoted in the market, the direct-injection Diesel engine has taken on the leading role in all areas.

The direct-injection Diesel engine uses the movement of air in the cylinder to form the fuel/air mixture, depending in each case on the cylinder stroke volume (diameter). This controlled air motion, also referred to as swirl action, serves to atomize the fuel injected even with a reduced number of nozzles. The swirl of air in the cylinder results from the specific design and configuration of the intake ports and is used to provide an appropriate level of variability, primarily on multi-valve engines. On large diesels with a bore diameter of approximately 150 mm (5.90″) or more the combustion process is coordinated through the large number of nozzles and the flatter troughs of the combustion chambers, without any swirl effect (see Fig. 7).

3.1.2
Power Unit

As with every piston engine, the basic design and configuration of a Diesel engine is determined by the

- stroke and bore (giving the cylinder capacity),
- connecting rod ratio, and
- compression ratio.

All other design parameters are varied according to the specific function of the engine and the design chosen.

The main distinctions when compared to a spark ignition engine are the

- higher compression ratio,
- higher ignition pressure load,
- high-pressure injection system with its inherent drive function, and the
- cold starting system.

Specific measures are taken to increase the compression temperature in the interest of smooth and efficient cold starting and warming-up behavior. On smaller

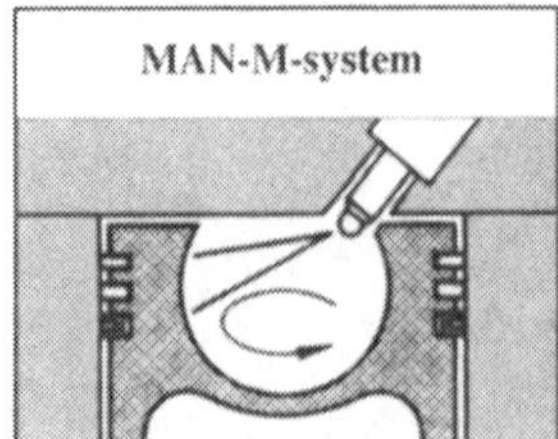

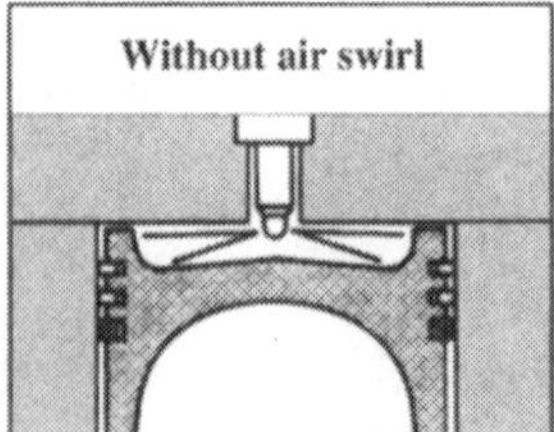

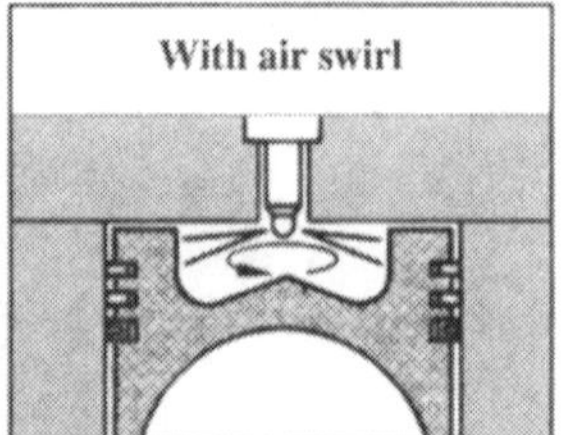

Fig. 7. Diesel direct-injection process

engines in passenger cars, glow plugs extending directly into the combustion chamber have become the technology of choice. Flame starter units heating up the intake air are used mainly in utility vehicle engines. On large Diesels the starting process is ensured by compressed air moving the pistons until they start to run under their own power.

The trend in engine design and configuration on road vehicles points clearly towards a reduction of weight accompanied by an increase in peak pressure. The development and improvement of the materials used is a major factor in this context, with light alloys being used in passenger car engines for the cylinder crankcase to reduce the weight of the engine. New materials are also used for bearings as well as for components subject to substantial thermal loads, such as the turbocharger, etc.

The lubrication and cooling system must be specially adapted to the specific running conditions of the Diesel engine in the interest of enhanced reliability. A point worth mentioning is that the lubricant must, where necessary, be modified to meet the more demanding running conditions with a Diesel engine, for example, with Diesels running on heavy oil, and the contamination of the lubricant caused by solids (particles).

3.1.3
Charge Cycle and Turbocharger Technology

The cylinder charge must be replaced after each operating cycle by fresh air following the actual process of combustion. On Diesel engines in road vehicles this change in the cylinder charge follows the four-stroke principle. On large Diesels running at low speeds the two-stroke principle, with unidirectional gas exchange, has proved to be the method of choice since this significantly increases engine output, while keeping the design and configuration of the engine relatively simple and straightforward.

Four-stroke diesels are controlled by valves, the valve overlap on a Diesel engine being kept to an absolute minimum by the start-up process and the specific design of the combustion chamber within the pistons themselves. On direct-injection Diesels, with a swept volume of up to approximately 1.5 liters/cylinder, the intake port(s) (see Fig. 8) is/are used to generate the desired swirl effect. Here it is preferable to vary the swirl effect by switching on or switching off the intake port as a function of the engine control map.

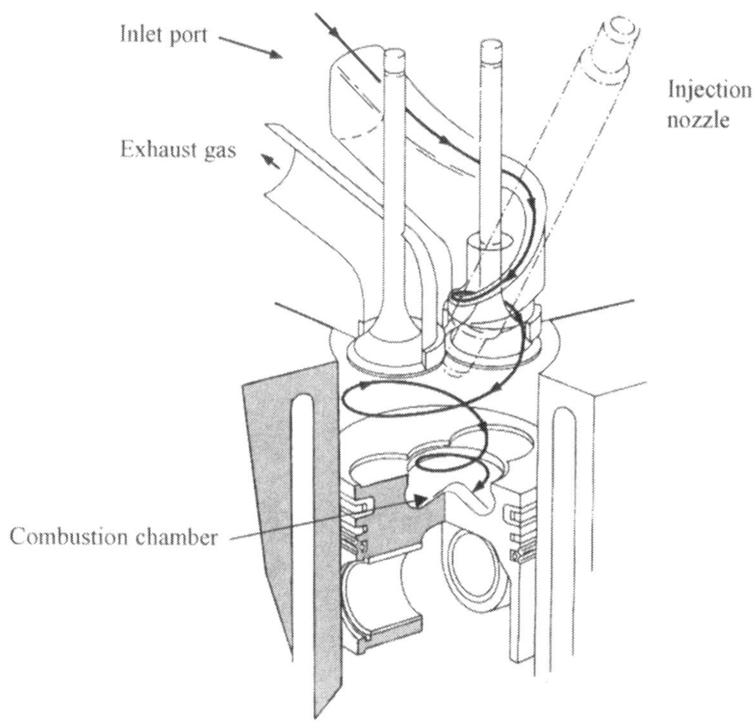

Fig. 8. Generation of the swirl effect by the specific design of the intake port

On turbocharged Diesel engines [3] the charge cycle is influenced by the specific turbocharger technology used. Turbocharging significantly increases the supply of fresh air to the cylinders, ensuring an appropriate increase in output over a normally aspirated engine with virtually no increase in engine size. This explains why nearly all Diesel engines today are turbocharged, with the added benefit that this technology not only means more power, but advantages in noise management and a reduction in fuel consumption.

The standard technology is exhaust gas turbocharging together with an intercooler (Fig. 9). To optimize the turbocharging effect, particular attention is given to the configuration of the inlet manifold/ports (cross-sections, volume, etc.) as well as the exhaust gas ports.

Particularly in Diesel engines driving road vehicles, the response of the turbocharger system is crucial to the engine's transient running behavior. To improve this response, most manufacturers use exhaust gas turbochargers with variable geometry (VNT) (see Fig. 10). This variability ensures not only a more rapid ramp-up of the turbine when accelerating under transient conditions, but also a higher standard of turbocharger efficiency under full load. This allows better tuning of the engine with high torque at low speeds and greater fuel economy at high speeds. Variable turbocharger also provides advantages in the management and reduction of exhaust gases thanks to more flexible exhaust gas recirculation.

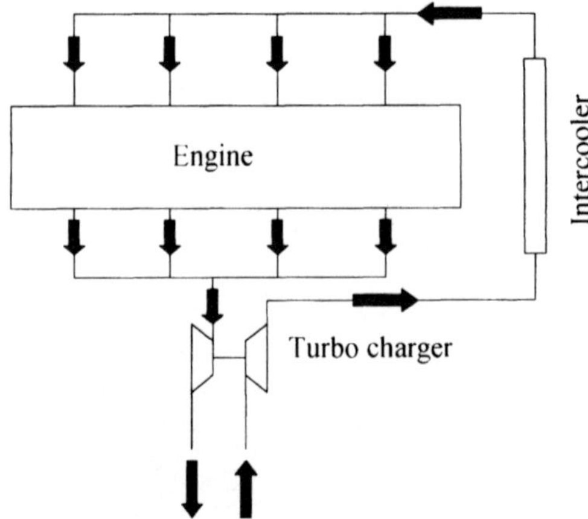

Fig. 9. Exhaust gas turbocharging with intercooling

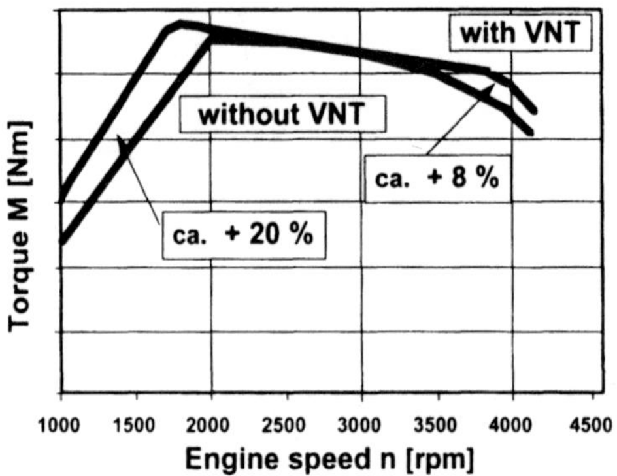

Fig. 10. Full load curves with variable turbocharger geometry

The following specific turbocharger technologies may provide advantages in certain niche applications:

- Two-stage turbocharging to increase engine output and reduce fuel consumption even further.
- Turbo-compound charge technology serving to make additional use of exhaust gas energy in a low-pressure turbine ensuring a higher standard of efficiency (used mainly on engines running under stationary conditions).
- The comprex principle with fresh air being compressed by pressure waves in the exhaust gas generated by a cell wheel. This method has, however, not succeeded in achieving, let alone exceeding, the benefits of exhaust gas turbocharging.

3.1.4
Fuel Injection Systems

The injection system on a Diesel engine serves a number of purposes:

- To dose the amount of fuel injected as a function of the current operating point and conditions (quality control).
- To inject fuel at exactly the right time during the operating cycle.
- To form an appropriate fuel/air mixture by way of atomization and local distribution of fuel in the combustion chamber.

The three most important subsystems within a fuel injection system are as follows:

- The low-pressure system - made up of the tank, fuel delivery pump, filter, pipes, etc.
- The high-pressure system - building up fuel pressure and conveying this high pressure via the injector into the combustion chamber.
- The electronic control unit - determining the right amount and timing of fuel injected as a function of engine running and environmental conditions.

It is particularly the high-pressure system, where a wide range of different fuel injection technologies has developed, as a function of specific requirements and applications [4] (Fig. 11), which is of significance with the Diesel engine.

These injection systems are as follows:

- The inline pump - with a separate pump element for each cylinder, with the injectors connected by a high-pressure pipe.
- The distributor pump - with central pressure generation and a distributor system rotating from one engine cylinder to the next. Following the axial principle, these pumps operate at an injection pressure of up to 1,000 bar, however thanks to their operating principle, radial pumps achieve almost twice this injection pressure.
- The unit injector - without a connection pipe between the pump and the injector itself, meaning that each engine cylinder forms one unit driven preferably by the engine camshaft.
- The insertion pump - usually housed in the cylinder crankcase and driven by the engine camshaft, with a very short pipe to the injector in the cylinder head.
- The common rail (CR) system - with a high-pressure pump supplying fuel to a pressure reservoir connected to the injectors by separate pipes. Electrical pulses (transmitted to electromagnetic, or piezo-controlled injectors valves) are able to initiate the injection process with a very high degree of flexibility and accuracy, achieving a positive effect on the fuel/air mixture and combustion process as well as the final treatment of exhaust emissions.

The CR system also offers the advantage of pressure being built up in the reservoir more or less independently of engine speed, allowing free choice of injection pressure as a function of engine map management. Particularly on vehicle engines (see Fig. 12), this provides the option to build up sufficient injection pressure for high torque also at low engine speeds.

feature \ type	in-line injection pump	distributor injection pump	Unit injector	single plug pump	common rail injection system
symbolic representation	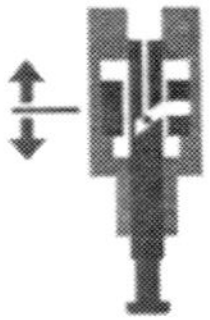			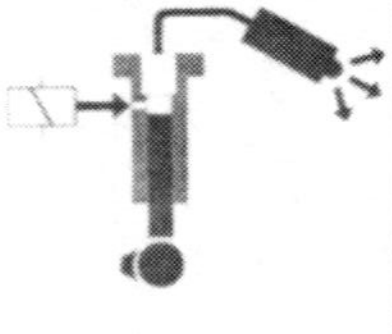	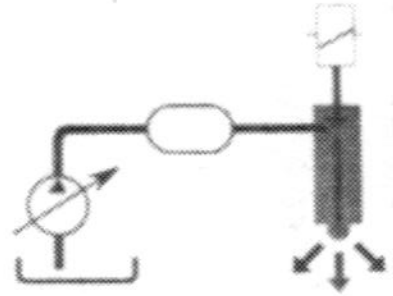
operational area	Truck engines	PC- and light truck engines	PC engines	Trucks and marine engines	PC, trucks and marine engines
injection pressure [bar]	1200	1500	2000	2000	1600

Fig. 11. Injection systems

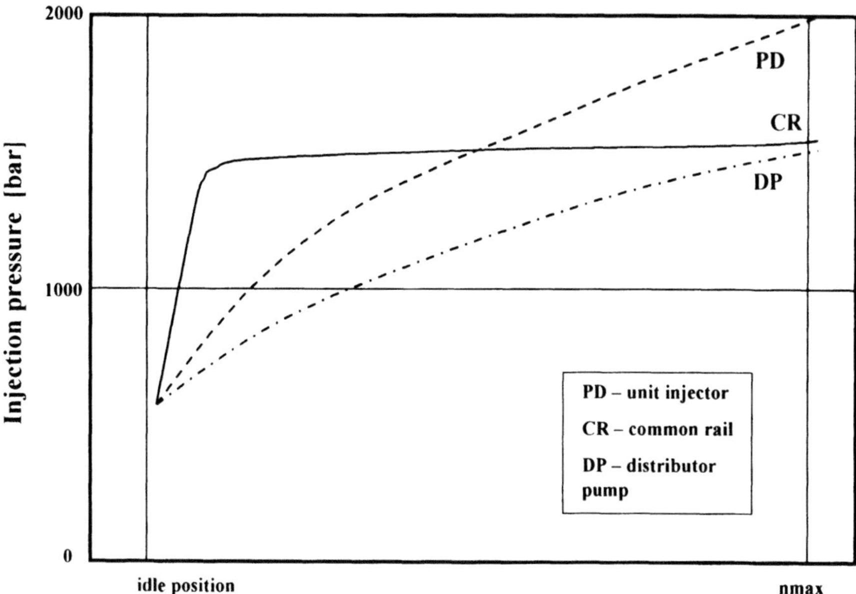

Fig. 12. Pressure characteristics of various fuel injection systems

Starting in 1998, CR technology has helped to significantly increase the performance of Diesel engines in passenger vehicles. Additional functions, pre- and post-injection as well as multiple injection provide a significant potential for individually optimizing the injection curve, reducing noise and exhaust emissions from the start.

The CR system presented in Fig. 13 allows simple configuration and operation of the high-pressure pump on the engine with lower drive torque peaks and free choice of the number of cylinders.

This relatively new technology has a significant potential for further development. Volume control on the high-pressure pumps will help to reduce fuel consumption, piezo-injectors are able to control the injection process more precisely and flexibly.

3.1.5
Injector Support and Injector Nozzle

The injector is the single most important unit crucial to the fuel/air mixture formation. It atomizes and distributes the fuel throughout the combustion chamber. Being fitted directly in the cylinder head, the injector is exposed to substantial pressure and temperature in the combustion chambers. Efficient cooling is therefore required to limit the injector's operating temperature, and ensure that the fuel will not coke in the injector nozzles so crucial to fuel/air mixture formation.

Two types of fuel injectors (Fig. 14) have proved most successful in the market, depending on the specific diesel combustion principle:

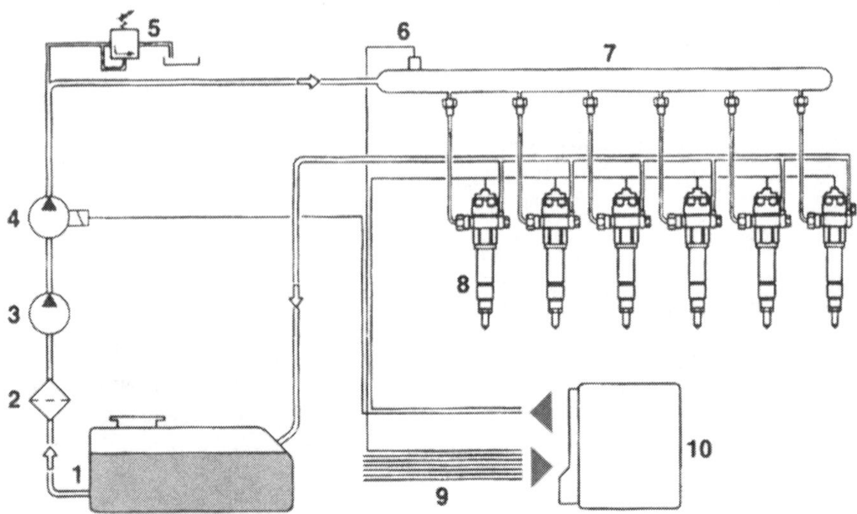

Fig. 13. Common rail system. 1 Fuel tank, 2 Filter, 3 Primer pump, 4 High-pressure pump, 5 Pressure-release valve, 6 Pressure sensor, 7 Fuel rail, 8 Injectors, 9 Sensors, 10 Electronic control unit

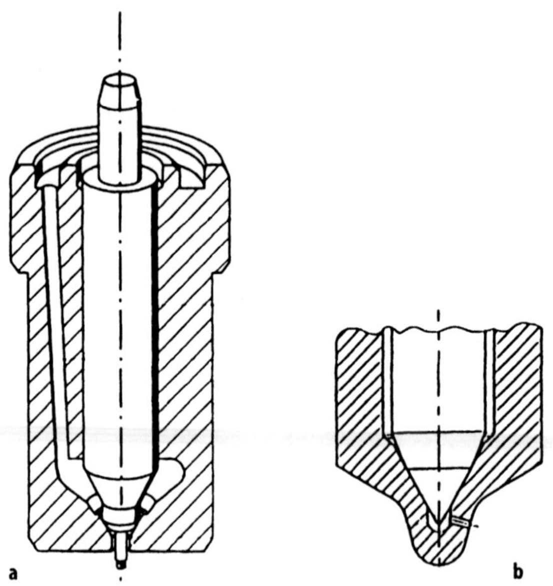

Fig. 14 a, b. Different types of injectors. **a** Throttling pintle nozzle; **b** Multi hole nozzle

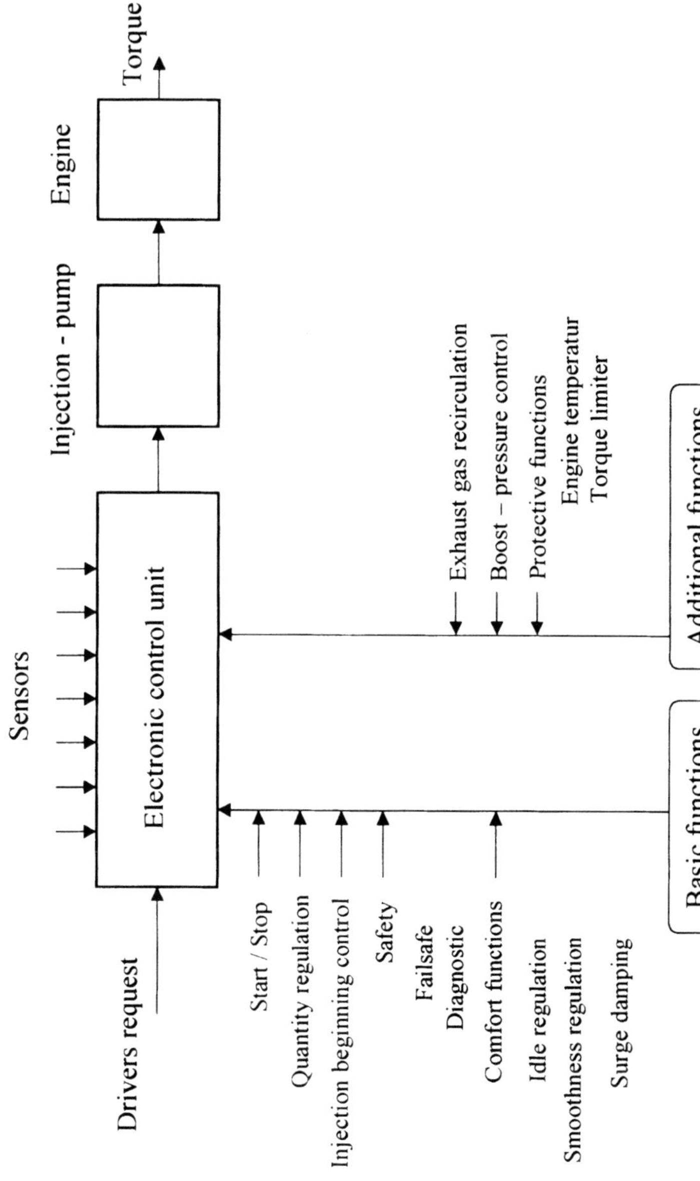

Fig. 15. Electronic engine management system

- The throttle pin nozzle on chamber engines, where the effective cross-section of the injector is determined by the shape of the throttle pin.
- The multi-hole injector on direct injection engines.

Hole geometry is of utmost significance to the atomization of fuel. By improving the production of injector nozzles and using new technologies (switching over, for example, from drilling to spark eroding or a hydraulic erosion process rounding off the nozzles, etc.), thus making the nozzle smaller (down to approximately 0.1 mm) and increasing injection pressure in the process. The injector nozzle has become the single most important factor in improving the process of fuel/air mixture formation and reducing particulates dramatically.

Without electronic engine management (Diesel control) the many functions expected of a Diesel today, as well as the high standard of comfort demanded, would not be possible. As indicated in Fig. 15, the basic functions of conventional mechanical injection systems have been supplemented by a wide range of additional functions [5]. The major benefits of electronic systems are; the high standard of precision in engine management, the option to combine numerous parameters as well as engine and vehicle systems with one another, the freedom in setting the engine to specific operating conditions shown in the engine control map, and the lasting consistency of such electronic systems. In the evolution of modern Diesel engines, the development of these functions and the application of these systems is of great significance.

In general, injection systems show an ongoing trend towards higher injection pressure and more flexible injection processes. This must be seen in the light of the greater demands now made of modern power units in terms of both precision and emission control.

3.2 Current Status of Modern Diesel Engines and Future Trends

In the area of transportation Diesel engines are used mainly for

- driving passenger cars (personal transport),
- driving utility vehicles (trucks and buses), and
- driving ships (marine Diesels).

The following table presents the current standards in terms of cylinder capacity, specific output, engine running speed, optimum efficiency and the specific power-to-weight ratio (Table 1).

Table 1

	Car engines	Utility vehicle engines	Two-stroke marine Diesels
Cylinder capacity (Single) [Liter]	0.5	2.0	up to 1800
Specific output (kW/liter)	50	30	3
Max speed (rpm)	4000	2000	100
Effective degree of efficiency (%)	42	45	53
Specific power-to-weight ratio (kg/kW)	1.4	approx. 3	approx. 30

While these standard figures reflect the current status of the Diesel engine, they cannot cover the entire range of applications, particularly in the case of large Diesels characterized by the wide range of designs and configurations. Ultimately it is always the customer (user) who decides on the appropriate drive system, taking a wide range of criteria into account such as packaging, output and performance, weight, the cost (investment) involved in purchasing the vehicle, its service life, cost of ownership, etc.

Figure 16 presents the most important milestones in the development of the Diesel engine. Assuming a steady increase in worldwide transport in the years to come, the Diesel engine, as shown in Fig. 1, will have an over-proportional share in the market place thanks to its unparalleled fuel economy.

Based on some Diesel engines already in existence, the following presents the current status of development of modern engines currently in use today.

3.2.1
Passenger Car Diesel Engines

In a direct comparison with the spark-ignition engine, the Diesel engine used in the passenger car has developed in recent years from a less powerful and rather loud drive technology into a refined and dynamic drive system, thus becoming a genuine alternative equal in its qualities to the spark-ignition/petrol engine.

All modern Diesel engines are direct-injection power units with highly developed injection systems (in most cases using common rail technology) and variable turbochargers. Introduced in 1999, the vast proportion of direct-injection Diesel engines are sold in the midrange market sector, however they are now becoming more acceptable in the luxury (with the first "three-liter car" being introduced in the year 2000) and even performance segment of the market.

With the share of pre-combustion engines in new car registrations having dropped dramatically, as seen in Fig. 17, reference should be made here to a typical pre-chamber Diesel engine in passenger cars, and to the four-valve series 600 power units built by Mercedes-Benz [6]. Featuring five cylinders displacing a total capacity of 2497 cc, this power unit develops maximum output of 110 kW or 150 bhp at 5000 rpm with the help of turbocharger and intercooler technology. With normal aspiration, this engine series features some interesting technical solutions such as a register resonance charger, intake pressure control, etc.

An outstanding representative of swirl-chamber engine technology is the six-cylinder power unit with turbocharger and intercooler [7] introduced by BMW in 1991. Displacing 2500 cc, this engine develops a maximum output of 105 kW or 143 bhp at 4800 rpm. The modified swirl chamber combustion principle with a V-shaped piston crown in conjunction with the turbocharger, exhaust gas recirculation, oxidation catalyst and Digital Diesel Electronics gives this power unit its particular exhaust emission control concept. With its modern configuration and design, this engine also ensures very good acoustics and vibration management. An additional, temperature-controlled capsule reduces noise emissions when starting the engine cold and in the warm-up phase to an even lower level.

Direct-injection Diesel technology made its breakthrough into the passenger car market through the introduction of innovative injection systems providing

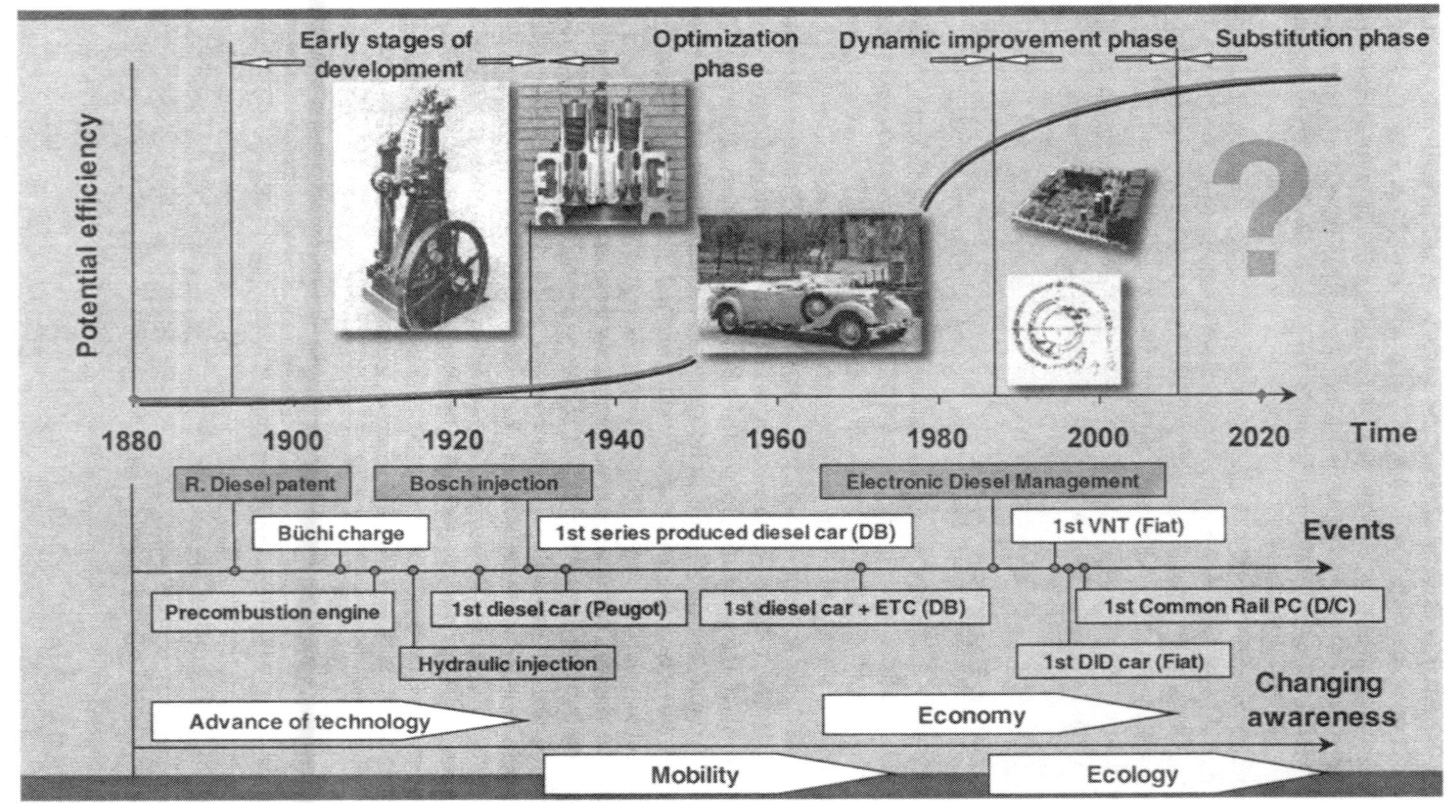

Fig. 16. Milestones in the development of the Diesel engine

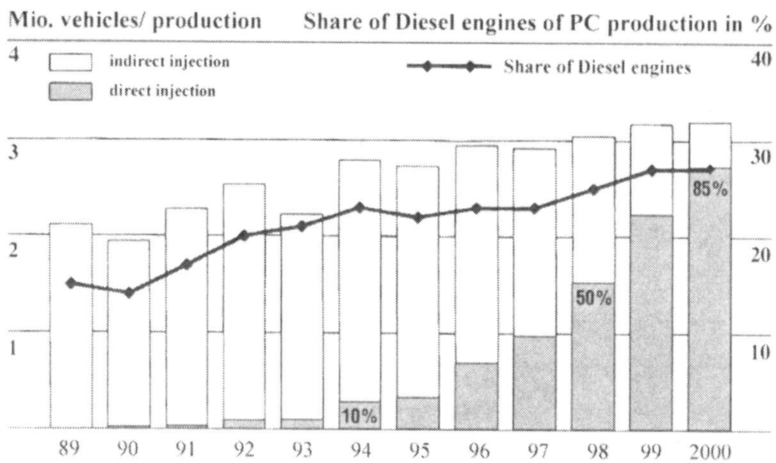

Fig. 17. Share of direct-injection engines in the passenger car Diesel engine segment

an injection pressure of more than 1500 bar and allowing additional functions such as pre-injection. The technologies worth mentioning here are the radial piston pump, the unit injector and, in particular, common rail (see the chapter on injection systems). The pioneers in direct injection Diesel engines for the passenger car were Fiat (1987 in the Fiat Croma) and Audi (1989).

Today particularly European car manufacturers have a wide range of direct-injection Diesel engines extending from the smallest three-cylinder power units (such as the Smart Diesel engine displacing 0.8 liters and developing maximum output of 30 kW/41 bhp) all the way to the most powerful luxury performance cars (with V8 and V10 Diesel engines built by AUDI, BMW and Mercedes-Benz). Without doubt, the reason for building such Diesel-powered passenger cars is their low fuel consumption, Fig. 18 clearly shows how Diesel cars dominate the market of very low consumption models.

As a typical representative of the new generation of direct-injection Diesels, BMW's four-cylinder diesel [8] displacing 1.95 liters and developing a maximum output of 100 kW/136 bhp at 4200 rpm has all the features of such a future-oriented power unit. Figure 19 presents a longitudinal and cross-sectional view of the engine. Standing out in particular is its integrated intake manifold with intake ports coming in from above (swirl port) as well as the filling port entering the cylinder head from the side, also the radial-piston distributor pump (VP44), with increment cams intentionally slowing down the combustion process in the initial phase. Figure 20 illustrates the full load curves and the fuel consumption control map of BMW's four-cylinder direct injection engine which is equipped with a variable geometry turbocharger (VNT) giving the specific model involved, the BMW 320d, particularly dynamic and agile performance.

It should be noted that leading car manufacturers use mainly common rail direct-injection engines, since it is agreed that this system has the greatest potential for ongoing development.

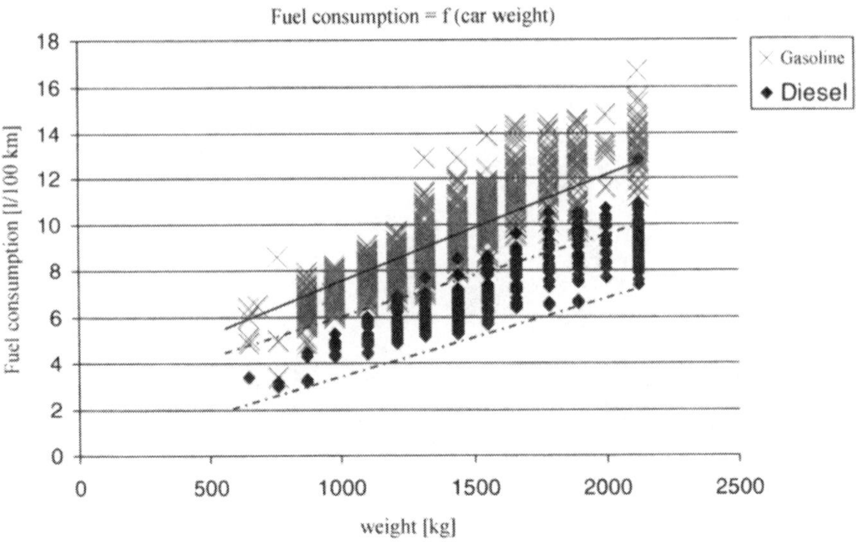

Fig. 18. Passenger car fuel consumption

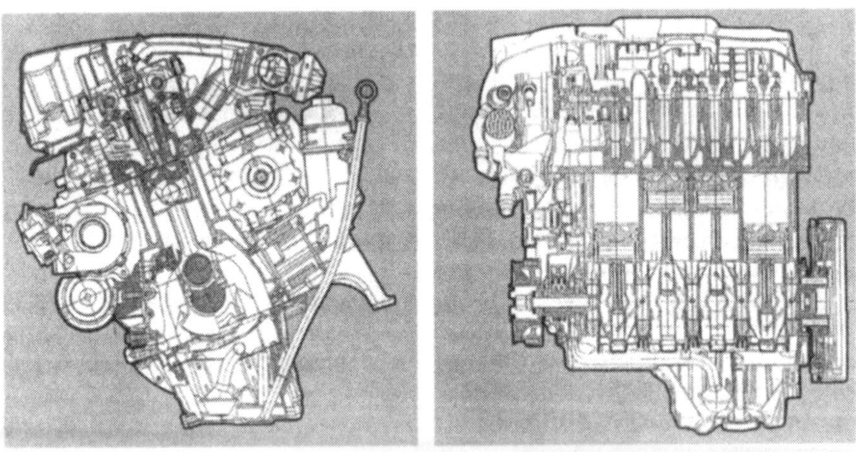

Fig. 19. Longitudinal- and cross-section

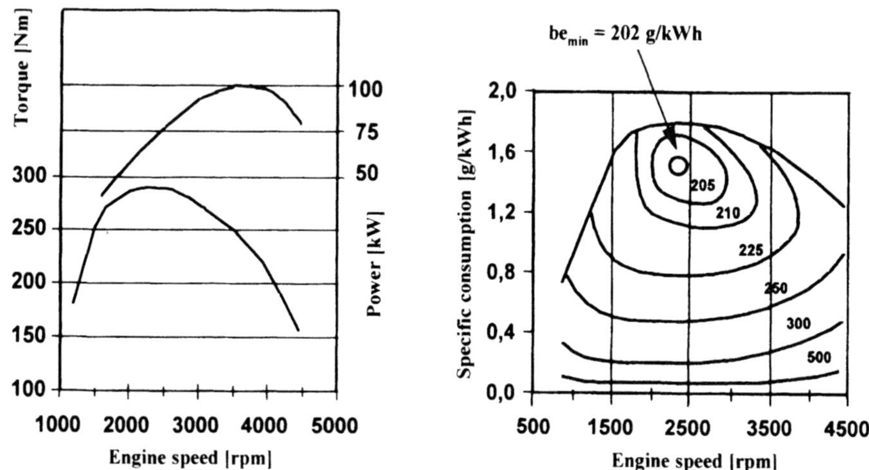

Fig. 20. Full load operation points and fuel consumption map

3.2.2
Utility Vehicle Diesel Engines

Utility vehicles serve to ensure safe and rational transport of goods and passengers. The range of such vehicles extends from small vans all the way to heavy trucks weighing 40 tonnes and more as used in the construction industry. Diesel engines are, without any doubt, the main type of powertrain for utility vehicles, offering unchallenged economy, environmental compatibility, a long service life and low-cost, easy maintenance.

Increasing engine output and reducing fuel consumption were the main objectives in the ongoing evolution of such engines until the 1970s. In the 1980s, the reduction of exhaust emissions became the focal point of development, noise emission standards (see EC regulation 70/157/EWSG) also becoming stricter in the process. Significant progress has been made by modification of the engine and encapsulation of the engine compartment within the vehicle itself. Compared with technologies used in the 1980s, limited components in exhaust emissions, for example, are now down by approximately 80 percent.

Engines used by all major manufacturers of utility vehicles are described in detail in the relevant literature [9, 10].

3.2.3
Marine Diesels

Following a very wide range of different types of engine and engine configurations in the past [11], three types of marine Diesels have been dominating the ship market for more than 30 years:

- High-performance marine Diesels running at high speeds (up to approximately 2000 rpm).

- Four-stroke engines with immersion pistons running at medium speeds (approximately 1000 rpm).
- Two-stroke Diesels in cross-head configuration running at low speeds (approximately 100 rpm).

The demands made of these engines are very good fuel economy, a long running life and a very high standard of reliability. Keeping down-times, due to maintenance, to an absolute minimum, with concepts ranging from preventive maintenance to maintenance only in the event of repair.

In terms of exhaust emissions, new ships have been subject, since 1 January 2000, to the IMO regulations with maximum NO_x emissions of <5 g/kWh. Another requirement is that marine Diesels must be able to run on low-quality fuel such as heavy oil, which demands the utmost from the lubricants used.

Marine drive systems, with four-stroke engines running at high and medium speeds, require a reduction transmission in order to achieve the best propeller speed, while two-stroke engines running at low speeds are able to make do with direct propeller drive. Ultimately, therefore, the user of each vessel must decide which type of engine to use, since this decision depends on a wide range of different criteria.

The highly charged two-stroke engine running at low speeds has the highest effective efficiency of all thermal engines with approximately 53 percent. Using the exhaust heat as well as the heat stored in the coolant, such engines are even able to achieve an overall efficiency rating of more than 80 percent. Detailed descriptions of such engines are to be found in the relevant technical literature.

3.2.4
Future Trends in the Use of Diesel Engines

Future trends in the use of Diesel engines include:

- Increase in engine output in terms of both overall power and power concentration (specific output), combined with light-weight material technology;
- Enhancement of overall economy by the ongoing reduction of fuel consumption;
- Fulfillment of future emission standards (noise and exhaust emissions) including recyclability;
- Increase in service life and reliability (including greater ease and lower cost of maintenance).

The need to increase engine output and reduce fuel consumption while at the same time fulfilling current and future exhaust emission standards places increasing demands on the development engineer. Since specific, individual improvements alone are no longer able to meet these demands, there is a clear need for all-round concepts based in particular on the efficient use of electronics.

Promising technologies of the future are the extension of functions by way of greater variability (engine map-specific control) or the interaction of vehicle systems (drive engine, transmission, car stability systems, energy recycled from the brakes, etc.).

Taking modern high-performance car engines as an example [12], we see this emerging area of modern production technology and new, even greater potentials in future. Top models already offer a specific output of 50 kW or 68 bhp/liter, and a new high-end power unit with bi-turbo technology built in small numbers on the basis of a 3.0 liter production engine (61 kW or 83 bhp/liter and 170 Nm or 125 lb-ft/liter) currently represents the cutting edge in this development. This elevates the passenger car Diesel into a performance range so far exclusive to the high-performance spark-ignition engine, at the same time allowing a process of down-sizing with a great potential for reducing fuel consumption.

The following methods can be used to realize a further power increase:

- Raising the charge pressure increases the air mass in the cylinder. To stay within the maximum permitted engine cylinder pressure limits (currently approx. 160 bar for passenger cars) an adaptation of the compression ratio is necessary (see Fig. 21);
- A further increase of injection pressure (see Fig. 22) improves the ignition process and therefore the thermodynamic efficiency; and
- Reduction of losses in the low pressure process through careful configuration of the turbocharger (see Fig. 23) offers considerable potential for power increases.

A good example for these power increasing solutions can be seen on the impressive race engine [22].

The high-end and racing engine technologies presented above are currently not available for volume production purposes. The concepts presented in following table reveal the significant potential for extra output and torque still provided by appropriate modification of the base engine, with higher ignition pressure, highly flexible injection systems and charge systems with enhanced efficiency (Table 2).

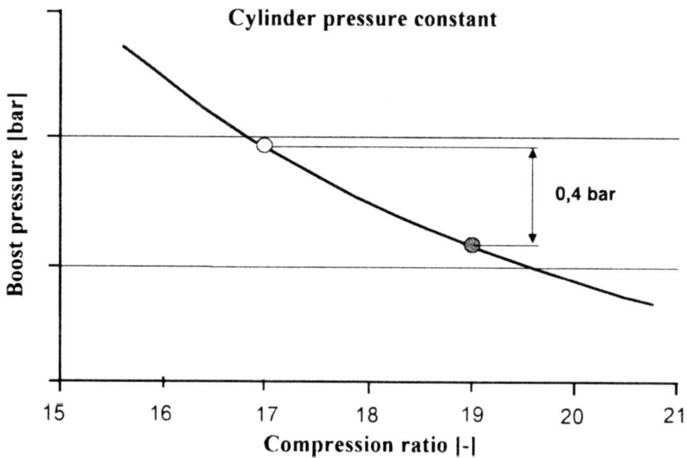

Fig. 21. Potential of higher charge pressure

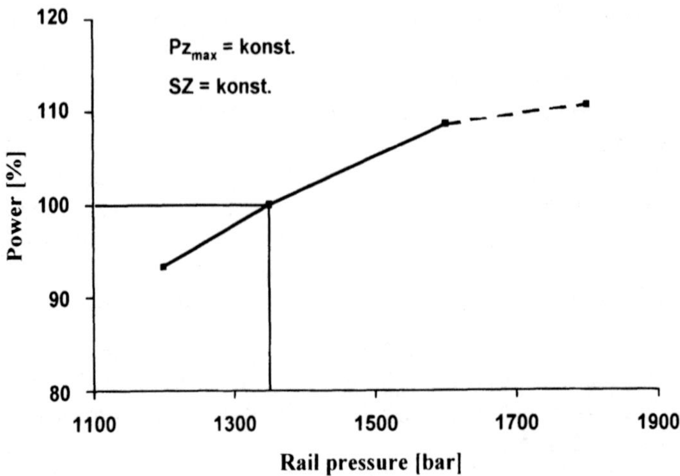

Fig. 22. Increase in engine output by higher injection pressure

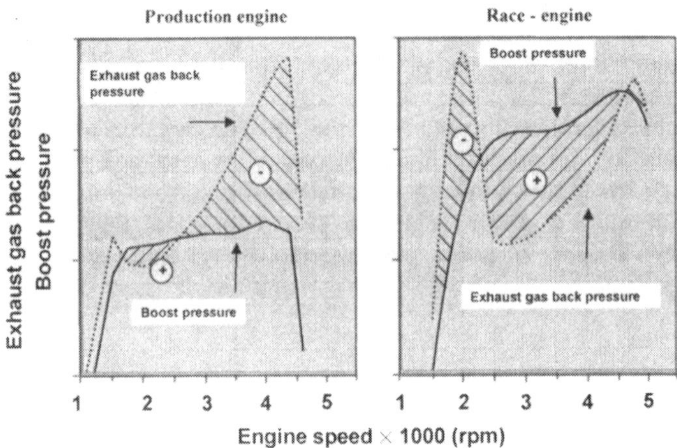

Fig. 23. Comparison of turbocharger configuration – production engine versus racing engine

Table 2

Features	Volume model	High-end	Racing version
Engine capacity (liter)	1.95	2.925	1.95
Specific output (kW/liter)	51.3	61	90
Specific torque (Nm/liter)	144	170	205
Compression ratio	18:1	16:1	15:1
Injection system	VP44	CR	VP44
Injection pressure (bar)	1750	1600	1900
Turbocharger technology	VNT	Bi-VNT	VNT
Max charge pressure (bar)	2.1	2.3	2.7

Transmission concepts with "short" setting-off ratios and "soft" converters when starting off support this trend.

3.3 Fuel Consumption

Current passenger car direct-injection Diesel engines have the highest efficiency of all drive concepts. When running at their optimum point, such engines offer specific fuel consumption of approximately 200 g/kWh, equaling an effective degree of efficiency of approximately 42 percent.

The comparison of CO_2 emissions of spark-ignition and Diesel engines with the same output, as presented in Fig. 24, is based on the direct-injection Diesel engine. As can be seen from this comparison, CO_2 emissions with a Diesel engine are 25 percent lower than with a spark-ignition power unit. Presenting further potentials for reducing fuel consumption, Fig. 25 analyzes the energy required by a 1400 kg vehicle developing 100 kW/136 bhp maximum output in the European driving cycle.

This analysis shows that optimization of the engine alone only provides part of the future potential. The rest must be ensured by appropriate concepts within the overall engine package, such as down-sizing, start/stop systems, etc.

A further option for reducing fuel consumption is to decrease the loss of power. Figure 26 shows how fuel consumption can be reduced noticeably in the mean term by making the transition from pressure-controlled to volume-controlled common rail systems.

The examples chosen do present a potential of approximately 5–10 percent provided by carefully optimizing and enhancing subsystems and individual engine units. It is however, becoming, increasingly important to consider the overall system in order to achieve an even higher standard of fuel economy in future with down-sizing, new transmission concepts or even hybrid drive systems with energy recycling allowing further progress.

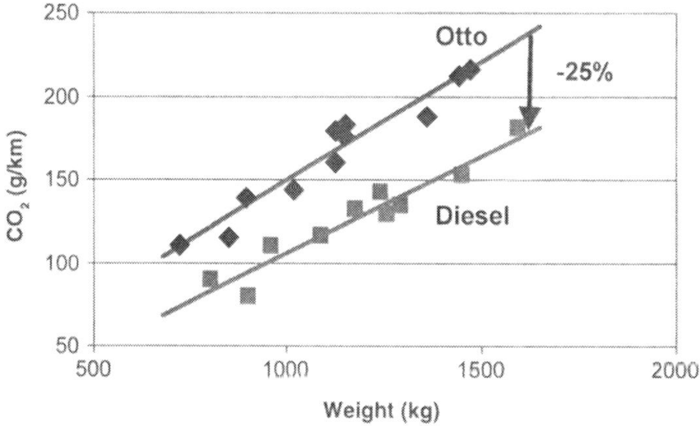

Fig. 24. Comparison of CO_2 emissions – spark ignition versus Diesel engine

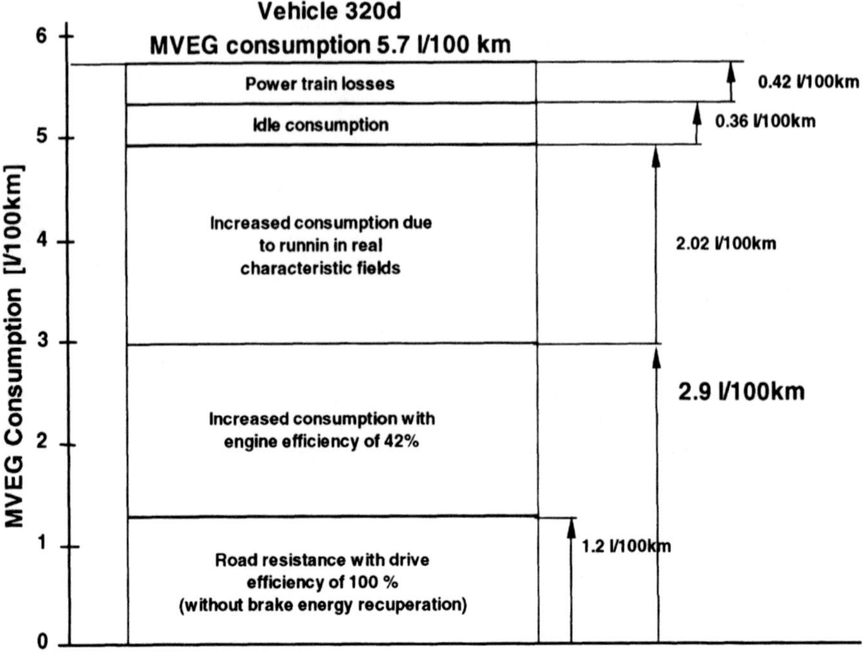

Fig. 25. Energy uptake in the MVEG test

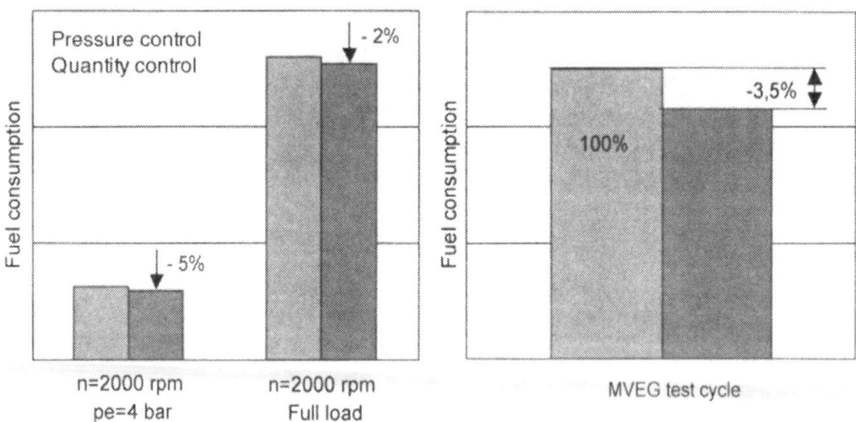

Fig. 26. Reduction of fuel consumption by volume-controlled generation of high pressure

By and large, these points apply to all other types and applications of Diesel engines [13].

3.4 Exhaust Emissions

Combustion engines generate exhaust emissions resulting from the chemical reaction of air and fuel.

As described in detail in the chapter "Measures for the reduction of exhaust gas emission", HC, CO, NO_x and particulates emissions are determined in representative test program (in Europe, for example, in the EC driving cycle for passenger cars).

Legislation already resolved leading all the way to EU4 requires a drastic reduction in road vehicle emissions. The EU3 emission standards already imposed significantly reduce road traffic emissions in general – with both passenger cars and utility vehicles – leading to the less overall volume of emissions after the year 2000 as in the period prior to 1970, with far fewer cars on the road at the time. Air quality forecasts [15] show that after introduction of the EU3 standard for road traffic there is hardly any reason to expect any infringement of the World Health Organization standards, which therefore will hardly be exceeded any more. Introduction of the EU4 standard halves these limits once again, although the improvements are now relatively small compared with EU3, due to the extremely low absolute ratings already achieved.

The principle of heterogeneous fuel/air mixture formation (see the chapter on fuel/air mixture formation) and the lean-burn principle already reduce the level of untreated HC and CO emissions from the Diesel engine to a very low point. The development of emission control technology, therefore, concentrates mainly on NO_x and particulates.

The **generation of nitric oxides** during reaction processes within the combustion chamber is favored by the level of temperature and the local supply of oxygen. Reaction processes change in terms of both time and space during combustion, meaning that all modifications serving to reduce temperature (e.g., intercooling) and the local oxygen content (e.g., exhaust gas recirculation) also reduce NO_x.

The **generation of soot particles** is mainly attributable to a local lack of oxygen. Reacting quickly and efficiently, hydrogen released from the fuel molecules burns more quickly than the carbon atoms forming larger carbon molecules and leaving the engine as soot particles separated in the exhaust emission test by a defined filter and subsequently analyzed for their properties.

A soot particle is made up largely of carbon possibly forming compounds with hydrocarbons (e.g., PAHs), sulfates (resulting from the sulfur content in the fuel), water and metal compounds. The mass of particles is currently determined gravimetrically by weighing the filter samples obtained under partial flow conditions during the exhaust emission test cycle. The soluble organic and inorganic share (including the sulfate content) can then be extracted from the filter samples, the final product remaining on the filter being almost exclusively soot (carbon).

The share of fine dust caused by traffic [16, 21] in Germany is less than 10 per cent of the total volume of dust emissions. Examinations by the German Federal Office of the Environment nevertheless show that Diesel particle emissions have a greater influence on the concentration of dust hovering in the air around the average citizen than, say, emissions generated by power stations. Precisely this is why the German Federal Office of the Environment calls for a reduction in Diesel engine emissions to the same level as with spark-ignition engines. Engine manufacturers are therefore working all-out on the particularly effective Diesel particle filter, seeking to enhance this technology to production level while maintaining the superior fuel economy of a Diesel engine. In the year 2000 a French car maker introduced a Diesel particle filter with a regeneration process supported by a fuel additive on its top model.

The main approaches taken in reducing Diesel engine exhaust emissions can be subdivided into specific, individual areas:

- modifications within the engine itself,
- after-treatment of exhaust gases,
- improvement of fuel quality.

Modifications within the engine itself and the treatment of exhaust gases are considered in greater detail in the following, while the improvement of fuel quality [13, 17], which also has a very significant potential, is described in Chapter “Fuels”.

3.5 Engine-Internal Measures for Reducing Exhaust Emissions

Optimization of the fuel/air mixture and the combustion process represents a significant step in reducing exhaust emissions.

The combustion of a Diesel engine is determined by the following parameters:

a) constructive (fixed) parameters such as the compression ratio (ε), stroke volume (V_H; S; D), the shape of the combustion chamber, the position of the injectors, the injector geometry (hole diameter, number of injector holes), etc., as well as
b) variable parameters (adjustable operating parameters), such as the degree of charging (turbocharger level), air motion (swirl), injection pressure, timing of the injection process, injection rate, etc.

Some of the variable parameters may be chosen freely and are determined on modern Diesel engines by the electronic engine management as a function of the engine’s operating points.

The influence of both fixed and variable parameters on an engine’s output, combustion, noise and emission behavior is very complex, individual parameters often acting against one another (Fig. 27). The process of optimizing an engine and the various parameters therefore increases exponentially with the number of factors involved. New simulation and optimization processes, in some cases using neuronal networks, open up new potentials in this area and provide options so far not available.

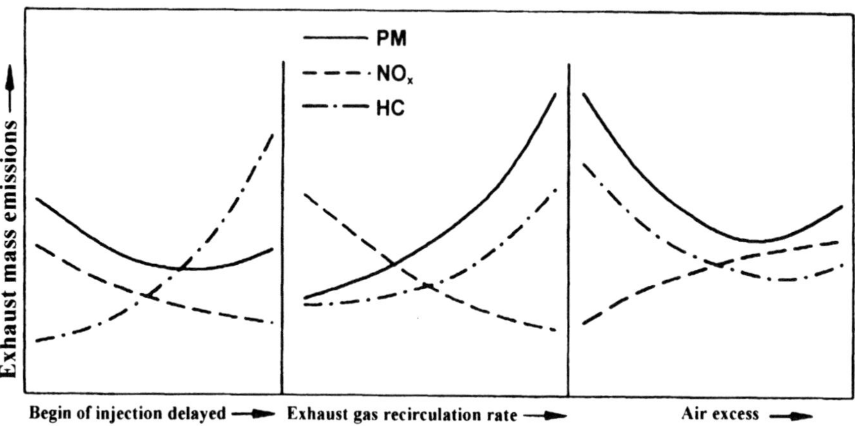

Fig. 27. Different trends with exhaust gas components

3.5.1
Development of the Combustion Process

Development of the combustion process in order to minimize untreated emissions while maintaining the greater fuel economy typical of a Diesel represents the most important task and requirement for the future. Effective solutions focus on the injection system as well as air motion and the design/configuration of the combustion chambers, combining to provide the biggest step towards a further reduction in emissions. Diesel engines with subdivided combustion chambers (pre-chamber and swirl-chamber engines) have a high level of kinetic mixture-formation energy in the air flowing into the ancillary chamber. The fuel jet flowing into this hot and turbulent air is processed very quickly, with combustion starting right away. The limited amount of oxygen in the ancillary chamber then slows down the combustion process (which is why this is called the first stage in combustion). The increase in pressure then forces the mixture into the main combustion chamber where combustion in air with a large share of oxygen continues in the second stage – which is why this is referred to as a two-stage combustion process.

The ratio between the chamber volume and the main combustion chamber, the geometry of the transition sections and the shape of the combustion area within the pistons have a significant effect on the process of fuel/air mixture formation and combustion. With the combustion process in chamber engines being controlled in this manner, the temperatures are not as high as with a direct-injection engine, keeping NO_x emissions to a relatively low limit. Now attempts are being made to achieve the same effects of controlled combustion also in the direct-injection engine.

In the direct-injection engine with a powerful air swirl, the swirl effect acting on the fuel injected by the injectors into the combustion chamber is even greater in order to avoid any local lack of air (see Fig. 8). With cylinder diameter increasing, the intake swirl effect becomes smaller, since a larger number of injec-

Table 3

	Fuel consumption	NO_x	Particles
Injection pressure	+	-	+
Injection starting late	–	+	–
Injection hole geometry (→ smaller)	–	+	+
Injection rate	o	+	+

tion holes also means a flatter combustion chamber. This is a concept experts refer to as air-distributed fuel dosage.

The **injection system** is a key factor on the Diesel engine and is therefore also crucial to the engine's emission behavior. Its functions in providing and dosing the flow of fuel are described in the appropriate chapters of this book. The main parameters crucial to exhaust emissions are injection pressure, injector hole geometry (diameter and length, as well as the hydraulic throughput), the actual timing of fuel injection as well as the injector rate.

The following table presents these factors and their general effect in terms of fuel economy, NO_x emissions and the formation of particles (Table 3):

The table shows how these injection parameters act against one another. High injection pressure (Fig. 28), for example, ensures extremely good atomization of the fuel thanks to the use of smaller injector opening cross-sections and, as a result, low particulates emissions, while at the same time postponing the commencement of fuel injection with a dramatic reduction of NO_x emissions. Lower emission of particulates is also conducive to the use of exhaust gas recirculation processes serving, in turn, to reduce NO_x emissions. Pre-injection, on the other hand, serving to enhance noise control and management, may be counter-productive with a negative

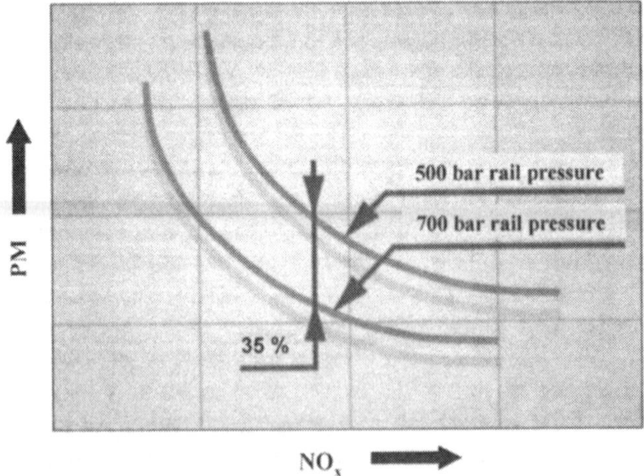

Fig. 28. Influence of injection pressure on particle NO_x emissions

effect on particulates emissions and exhaust gas recirculation. The volume tolerance limits for pre-injection are extremely tight if this technology is to be used without any disadvantages in terms of exhaust emissions. While new injection systems such as common rail technology already provide significant improvements in this respect, great efforts are still required in order to meet the EU4 standards.

The gas exchange cycle has a significant impact on the emission of Diesel engines not to be underestimated in practice. Throughout their entire operating range, turbocharged engines have a much higher air surplus than normal-aspiration power units. In combination with intercooling this provides additional options in choosing the right configuration, with a particularly positive effect on exhaust emissions.

Figure 29 presents the gas exchange system of a V8 passenger car Diesel [3] consisting of the following components:

- intake silencer,
- hot film air mass sensor,
- exhaust gas turbocharger compressor,
- intercooler,
- air collector,
- ports and valves,
- exhaust gas turbocharger turbine with variable geometry,
- pre-catalyst,
- underfloor catalyst.

Figure 29 presents also the exhaust gas recirculation system with its influence on the gas exchange cycle. The main parameters to be optimized in the interest of

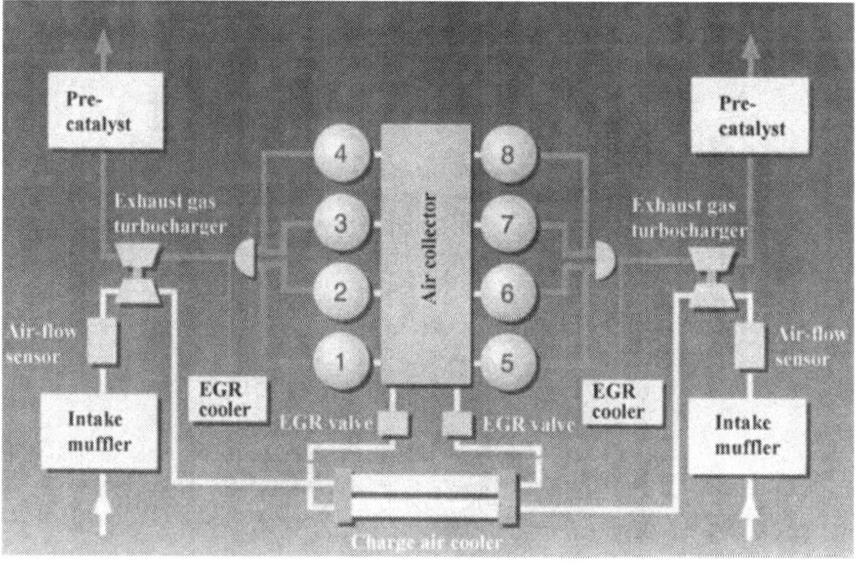

Fig. 29. Charge cycle of a V8 Diesel engine

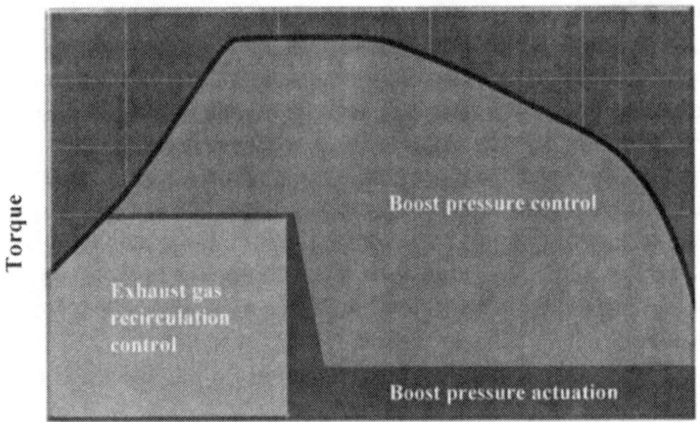

Fig. 30. Exhaust gas recirculation and charge pressure control operating areas

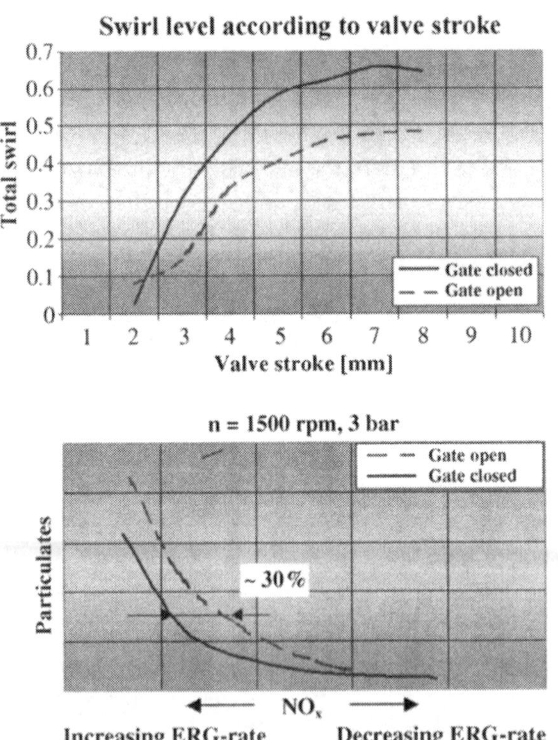

Fig. 31. Effect of variable swirl

minimum emissions are charge pressure and the intercooler, which are able, with the right dimensions and configuration as well as air cooling, to reduce the volume of emissions. As is to be seen on Fig. 30, map-specific charge pressure control provides a very wide range of positive effects.

An additional option on four-cylinder engines is to appropriately control the swirl level, that is the mixture formation energy on the air side, in order to reduce untreated emissions to a minimum. Figure 31 shows that this can reduce NO_x emissions by up to 30 percent under low loads, with the level of particle emissions remaining unchanged. This is done by using a flap closing the supply of air to the cylinder head filling port and, as a result, directing the entire flow of air through the swirl duct into the cylinder. There is still a potential for optimization by the continuous adjustment of this flap using a broader range of engine running conditions.

3.5.2
Exhaust Gas Recirculation

Exhaust gas recirculation (EGR) is the most effective way to reduce NO_x on a Diesel engine. As indicated by Fig. 30, EGR reduces the air surplus with a fresh cylinder charge under part load to such an extent that NO_x emissions decrease significantly. This effect results from the reduction of local oxygen and the corresponding delay in combustion, avoiding local temperature peaks. Excessive exhaust gas recirculation nevertheless increases HC and CO emissions as well as particulates.

Maintaining a specific geometric configuration, the variable turbocharger as presented in Fig. 32 is particularly suitable for influencing the rate of exhaust gas recirculation. EGR cooling, as presented in Fig. 33, also boosts the EGR effect. Particular attention must be given in this context to the dynamic management of exhaust gas recirculation, since the EGR valve must be closed as quickly as possi-

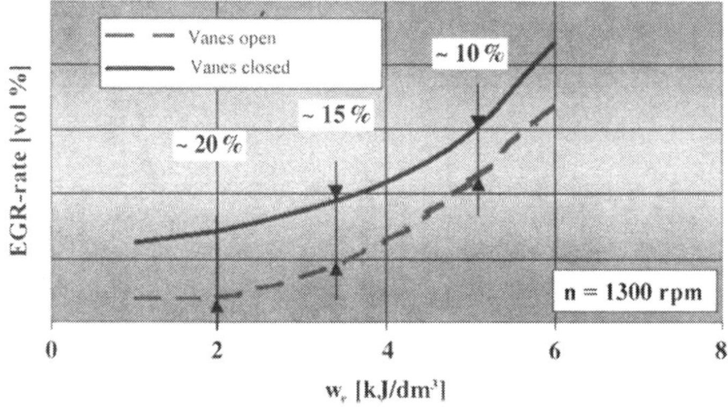

Fig. 32. Influence of turbine blade position on the possible EGR rate

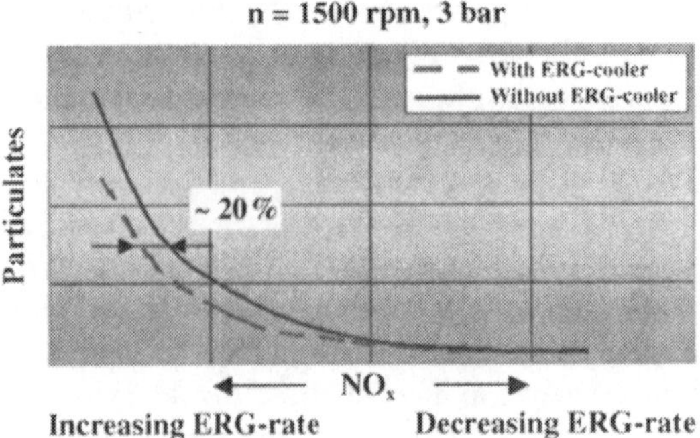

Fig. 33. Effect of EGR cooling

ble when accelerating all-out from low engine loads with a high rate of exhaust gas recirculation in order to avoid any undesired emission of smoke (see Fig. 34).

Electronic engine management (Diesel control) and the fast actuators already available today (e.g., electric EGR valves) are able to precisely coordinate and harmonize the various control areas and systems, but involve much more complicated applications and processes.

It is fair to say in summary that the injection system will remain the key parameter in the further development of combustion processes. The potentials of flexible common rail systems in terms of their injection curve (pre-injection, multiple injection), as well as variability of the injector itself (optimization of cross-sections), are still far from exhausted [20]. Control systems such as lambda control, as specified in Fig. 35, are required to provide more precise air ratio management, reducing tolerances and discrepancies in production units [17] and, as a result, cutting back the level of untreated emissions quite significantly. Innovative combustion concepts such as homogeneous diesel combustion also show encouraging results in this area.

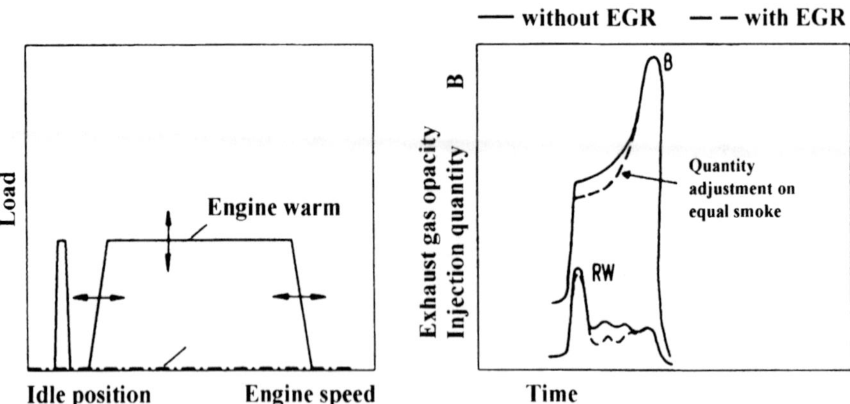

Fig. 34. Dynamic set-up of exhaust gas recirculation

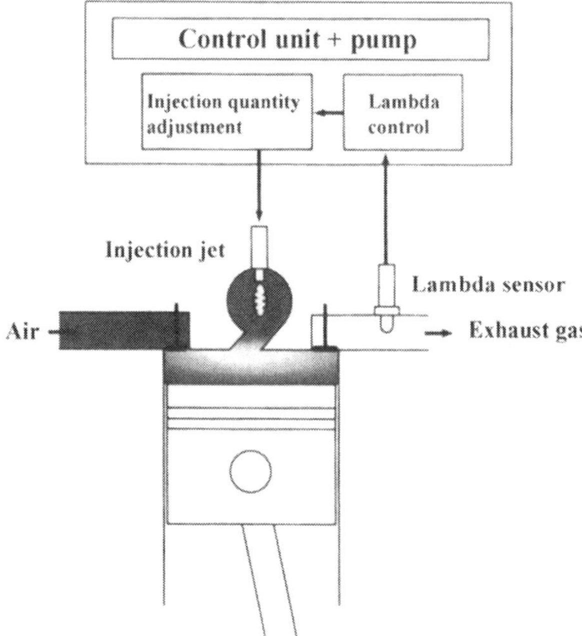

Fig. 35. Control concept with oxygen sensor

3.6 Exhaust Gas After-Treatment

With emission standards becoming stricter all the time, internal measures within the engine itself are no longer sufficient to fulfill exhaust regulations on the basis of untreated emissions alone. It is therefore fair to assume that new systems for the after-treatment of exhaust gases (see Fig. 36) will be required in specific areas. And since the Diesel engine, on account of its underlying principle, always runs in the lean-burn mode, new catalysts are required in order to reduce NO_x emissions.

The Diesel particle filter is very effective in ensuring the ongoing, dramatic reduction of particle emissions. The oxidation catalyst already used by all manufacturers of Diesel engines since approximately 1989 will support the various exhaust emission control concepts being introduced in the market.

3.6.1 *Oxidation Catalyst*

On account of its heterogeneous mixture formation the Diesel engine always runs with an air surplus, meaning that there is always sufficient oxygen in the exhaust gas to convert CO into CO_2 and HC compounds into CO_2 and H_2O in an oxidation catalyst. A pleasant side-effect is that hydrocarbons in soot particles are also burnt in part in this process, reducing the mass of particles accordingly.

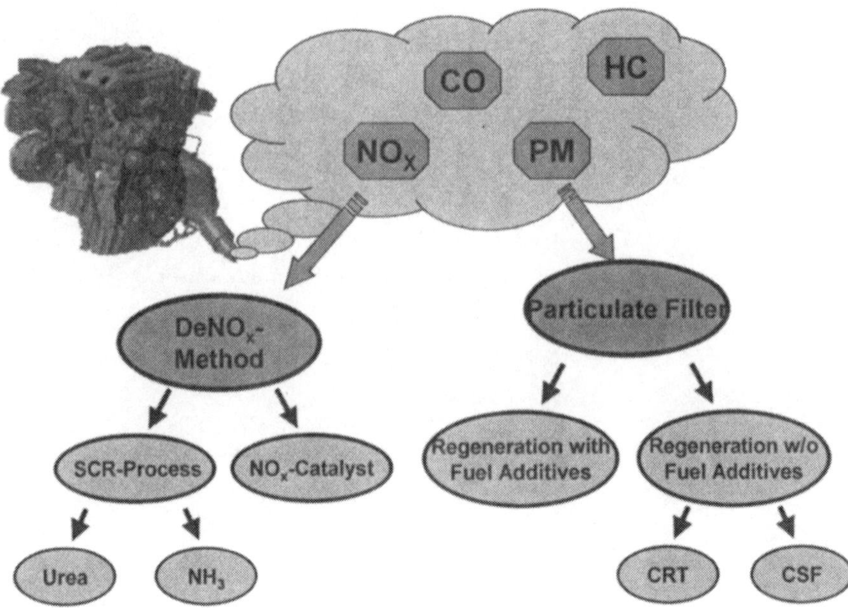

Fig. 36. Possibilities of exhaust gas after-treatment

The oxidation catalyst has become the method of choice in the meantime for EU2 and EU3 passenger car emission management concepts. Through their effect they make it possible to retard the injection process and, as a result, to optimize injection for lower NO_x emissions. Their particular configuration (application of coatings and layers), in turn, is based on the following criteria:

- conversion behavior (for low exhaust emission temperatures),
- response in the warm-up phase,
- stability against aging, sulfur, etc.,
- suppression of sulfate formation,
- temperature stability and long-term behavior.

Figure 37 shows the influence of catalyst layers on HC conversion.

3.6.2 *DeNO$_x$ Catalyst*

Since the three-way catalyst with lambda 1 control featured on the spark-ignition engine is not appropriate for the Diesel, the task facing engineers working on the lean-burn engine is to introduce the DeNO$_x$ catalyst. Like the DeNO$_x$ catalyst on lean-burn spark-ignition engines (see the chapter on spark-ignition engines), this technology also plays a key role in further reducing the emissions generated by a Diesel.

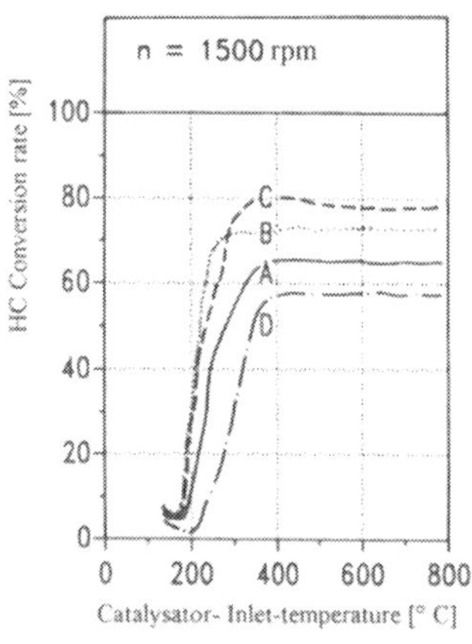

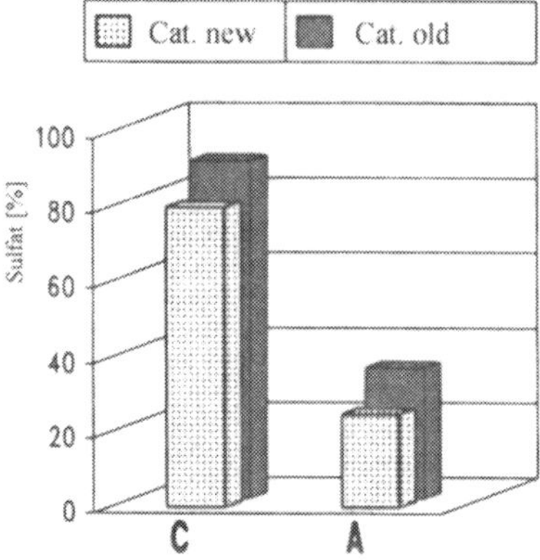

Fig. 37. Influence of catalyst coating

The following processes are being developed in order to reduce NO_x through the after-treatment of exhaust emissions, although it is important to note that so far none of these technologies is suitable for general, unrestricted use in the passenger car:

- NSCR (non-selective catalyst reaction) catalyst,
- SCR (selective catalyst reaction) catalyst,
- NO_x storage catalyst.

The **NSCR catalyst** uses unburnt hydrocarbons as the reduction agent in the following reaction process:

$$4\,NO + 2\,CH_2 + O_2 \rightarrow 2\,N_2 + 2\,H_2O + 2\,CO_2 \tag{a}$$

The catalyst uses zeolite coatings doped with copper or precious metals. The disadvantage is that the desired reaction takes place only within specific temperature windows (approximately 200–250°C) not large enough for practical operation of the vehicle. A further point is that results obtained under stationary test conditions which were in part quite satisfactory were not matched in the slightest under real-life driving conditions. With the additional disadvantage of such catalysts not providing appropriate temperature resistance and long-term behavior, the NSCR catalyst is not an attractive solution for the Diesel engine.

The **SCR process** is applied to remove nitric oxides from exhaust gases coming out of large power stations. The reduction agent used for this purpose is ammonia (NH_3) reducing the nitric oxides in the exhaust gases to N_2. Since ammonia cannot be used for mobile applications, Siemens have developed the so-called $SiNO_x$ process using urea as the reduction agent. As shown in Fig. 38, this system requires a hydrolytic catalyst recovering ammonia from urea for the subsequent reduction of NO_x. A further requirement is a dosage system providing the necessary supply of ammonia without allowing any NH_3 to escape into the environment.

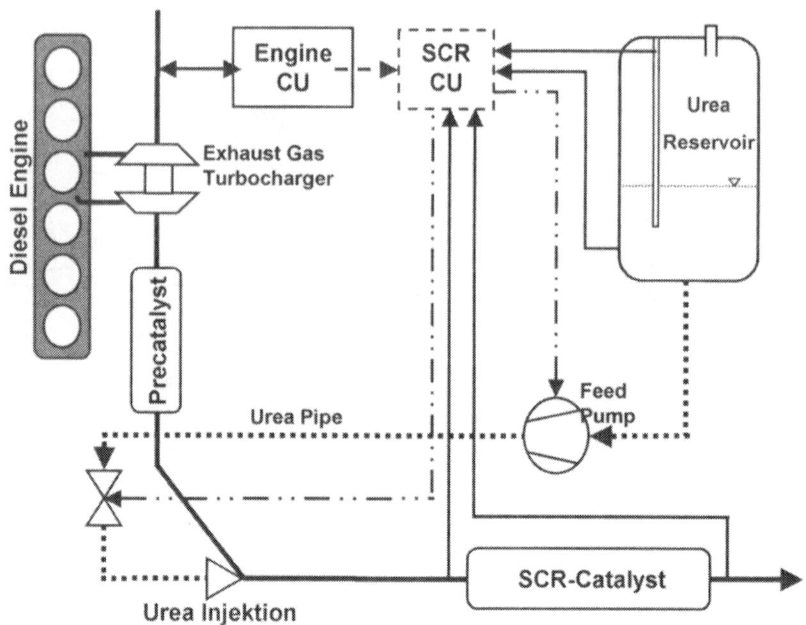

Fig. 38. Selective catalytic NO_x reduction (SCR)

$SiNO_x$ systems are currently used in small numbers in utility vehicles where there is also a significant potential for further reducing fuel consumption since the SCR process allows the engine to be tuned for maximum fuel economy.

The **NO_x storage catalyst** (see Fig. 39) on the Diesel engine follows the same principle as with the lean-burn spark ignition engine. An aggravating factor in the Diesel engine control strategy is that the emission of particles in the short rich-burn phases (see Fig. 40) required for regenerating the storage catalyst also has to be taken into account. This technology would currently appear to be the most appropriate method for reducing NO_x in the passenger car provided the fuel is sulfur-free. Even a tiny amount of sulfur such as 10 ppm residual sulfur resulting from the production process, however, requires a complicated desulfatization strategy in order to maintain the long-term function (see Fig. 41) of the storage catalyst.

3.6.3
Particle Filter

Particle filters apply the principle of retaining particles mechanically within the filter matrix. A typical particle filter system is shown in Fig. 42. To avoid any inappropriate increase in exhaust gas counter-pressure (and, accordingly, fuel consumption) in the filter it is essential to ensure regular combustion of particles, enabling the filter to regenerate in the process.

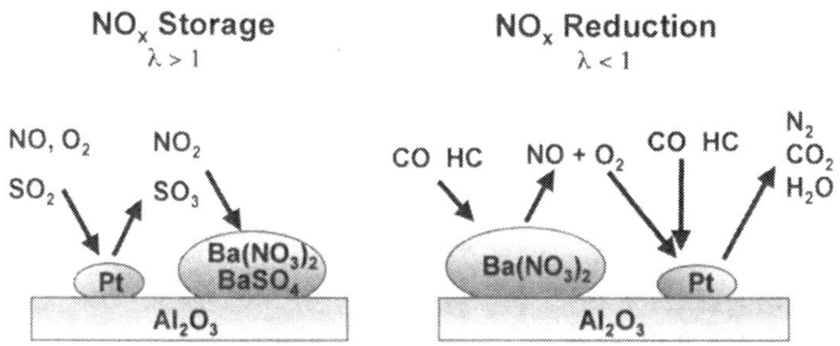

Fig. 39. Function principle of the NO_x storage catalyst

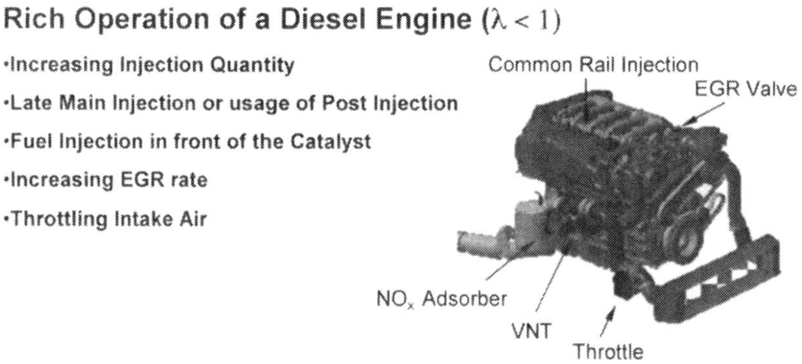

Rich Operation of a Diesel Engine (λ < 1)

- Increasing Injection Quantity
- Late Main Injection or usage of Post Injection
- Fuel Injection in front of the Catalyst
- Increasing EGR rate
- Throttling Intake Air

Common Rail Injection
EGR Valve
NO$_x$ Adsorber
VNT
Throttle

Fig. 40. Sub-stoichiometric diesel operation

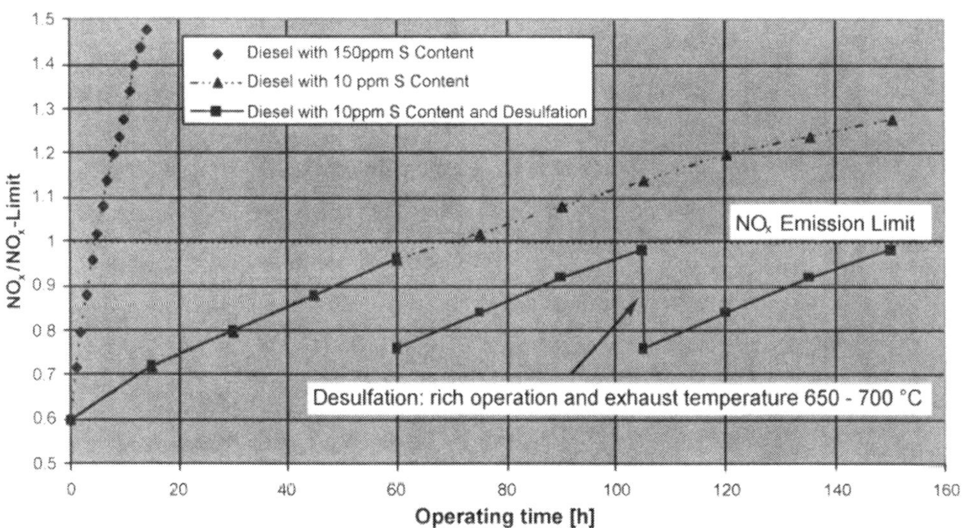

Fig. 41. Influence of sulfur on catalyst NO_x efficiency

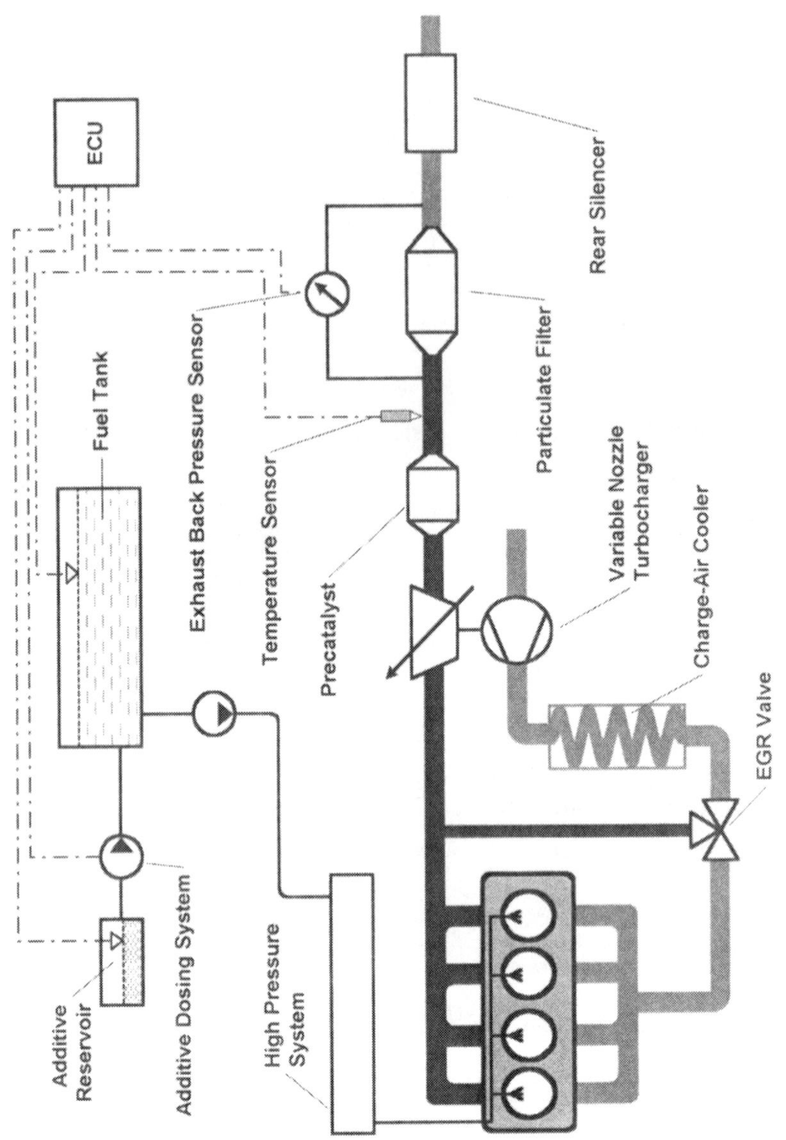

Fig. 42. Diesel particle filter system

The exhaust gas temperature of more than 500°C (930°F) generally required for this process can be reduced through the use of fuel additives, such as ferrocene, cerium, etc. Ancillary burners in the emission train are also being tested in particle filters used in city buses or utility vehicles. Large-scale tests already carried out with utility vehicles show that in principle this technology is suitable for practical purposes.

Regeneration of the particle filter is a particular problem with the passenger car due to load curves generally remaining at a lower level. Only common rail fuel injection, offering the possibility of subsequent fuel injection, is able to increase emission temperatures to the level required for regeneration of the filter throughout a wide range of engine operation. In combination with suitable fuel additives, this is able to prevent any clogging of the filter even when running under very low loads and in cold weather. The electronic engine management must perform these additional functions dosing the supply of additive from the additive tank.

Currently no manufacturers have production experience with systems of this kind proving their safe and effective operation throughout long-running periods. The current concept requires removal of ash deposits every time the car is serviced.

The CRT (continuously regenerating trap) system shown in Fig. 43 represents a special type of particle filter combining an oxidation catalyst with a Diesel particle filter in one common process. In its operating principle, the CRT system is based on continuous oxidation of NO into NO_2 in the oxidation catalyst, this NO_2 then acting as a fuel additive to burn particles separated in the filter at relatively low temperatures of approximately 300°C or 570°F.

Conditions for CRT

➢**Sulphur Free Diesel**

➢**NO_x/Soot – Ratio > 8[g/g]**

➢**Exhaust Gas Temperature ≥ 300 [°C]**

Chemical Reaction:

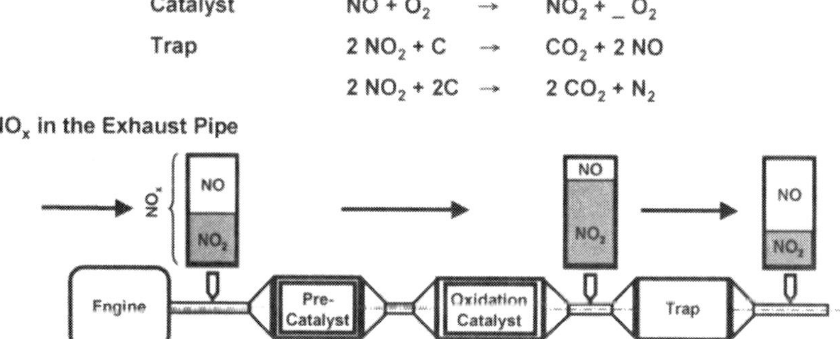

Fig. 43. CRT system

The combination of CRT and SCR technology allows simultaneous treatment of exhaust gases for CO, HC, particles, and NO_x.

3.7 Exhaust Gas Concepts and Outlook

In the course of the last 20 years, development engineers have made great efforts to fulfill emission standards, as a result, a new vehicle today has only about 10 percent of the exhaust emissions generated by a vehicle built in 1990. Depending on the specific area of use and application (passenger cars, utility vehicles, marine engines, etc.), internal measures within the engine and exhaust gas after-treatment technologies are combined to provide suitable emission management concepts.

In the **passenger car segment,** dominated by pre-chamber Diesel engines in the days of EU1 and EU2, the exhaust gas management concept typical today incorporates exhaust gas turbocharging with an intercooler, exhaust gas recirculation, an oxidation catalyst (underfloor) and electronic Diesel management.

Ever since the introduction of EU3 for passenger cars registered as of the year 2000, passenger car engines have been direct-injection power units in virtually all cases. The exhaust gas concept in such advanced Diesel engines incorporates exhaust gas turbocharging with variable turbine geometry, four-valve technology and a variable swirl effect, high-pressure fuel injection with the injector positioned in the middle and common rail technology, electronic Diesel management, and an exhaust gas catalyst near the engine with improved coating.

As of the year 2003, Diesel passenger cars in Europe will have to fulfill the EOBD (European On-Board Diagnosis) standard, Fig. 44 presenting the measures required for this purpose and their current state of development.

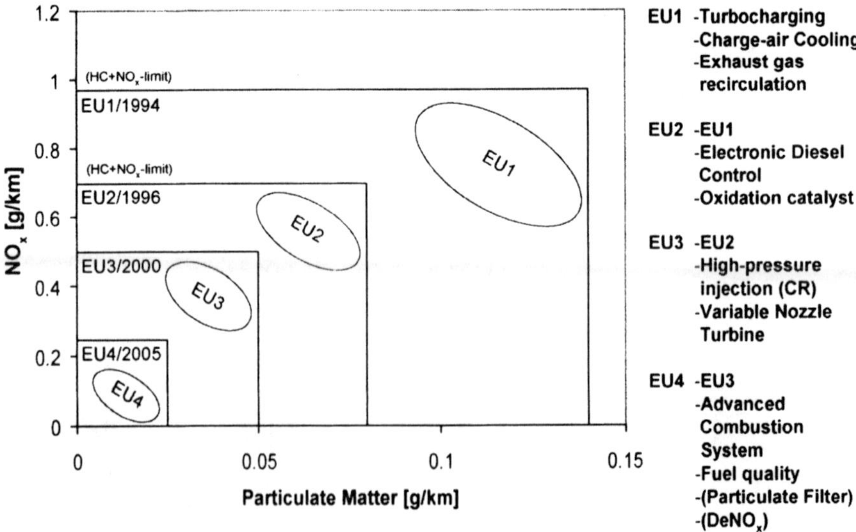

Fig. 44. Emission concepts for passenger car Diesel engines

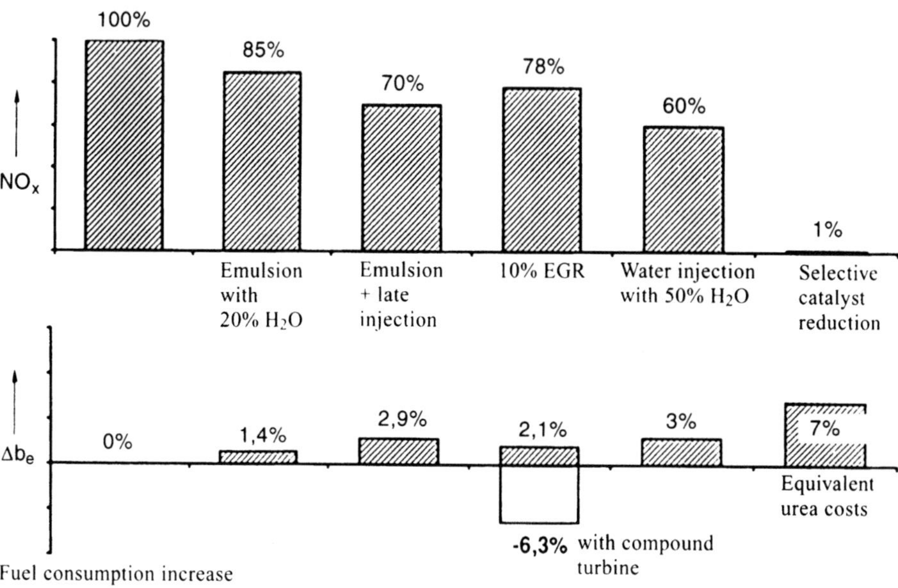

Fig. 45. Measures for reducing emissions on large Diesel engines

Diesel engines **in utility vehicles** have so far relied mainly on improvements of the fuel injection system and modifications in the design of the combustion chambers in order to fulfill the standards required. In this case the trend will continue towards even higher injection pressure and low- or no-swirl combustion processes with multi-jet injectors (with a tight hole cross-section) and flat combustion chambers. Exhaust gas recirculation and the injection rate will also be part of the future exhaust gas concept for utility vehicle engines. After-treatment of exhaust gases [23] by way of combined SCR particle filter systems is required, for example, in order to comply with the strict US 2007 emission standards already proposed today.

Marine Diesels, finally, are subject to the ISO 8178 standard applicable to engines not used in road vehicles (as well as construction and rail-bound vehicles, earth-moving machines, etc.). Figure 45 [11] presents possible measures for the reduction of NO_x. Since fuel consumption has a very great influence on the overall economy of these engines, hardly any ocean-going ships so far have been fitted with NO_x after-treatment systems.

3.8 References

1. Krieger K (1999) Das Einspritzsystem (The Injection System) Special Edition of MTZ – Ten Years of the Audi TDI Engine
2. Hiemesch O, Reibold G (1988) Möglichkeiten zur Begrenzung der Partikelemission bei PKW-Dieselmotoren (Possibilities of Limiting Particle Emissions in Passenger Car Diesel Engines). Paper presented at the Haus der Technik, Essen, Germany, February 1988

3. Steinparzer F, Neuhauser W (2000) Gestaltung des Ladungswechsels zur sicheren Einhaltung. Brüne, H.J der EU3-Emissionsgrenzwerte am Beispiel der neuen BMW Direkteinspritzmotoren (Configuration of the Charge Cycle to Reliably Maintain the EU3 Emission Standards, Taking BMW's new Direct-Injection Engines as an Example). Aufladetechnische Konferenz 2000, Dresden, Germany
4. Paper published by Robert Bosch "Systeme für die Dieseleinspritzung" (Diesel Injection Systems). 1 987 709 359
5. Ertl C, Hiemesch O (1995) Elektronisches Motormanagement beim Diesel-PKW zur Erfüllung zukünftiger Anforderungen (Electronic Engine Management of the Passenger Car Diesel to Fulfill Future Requirements). International Vienna Engine Symposium
6. Conrad U, Feucht MJ Klingmann R, Krause R (1993) Die neuen Vierventil-Dieselmotoren von Mercedes-Benz. Mercedes-Benz' New Four-Valve Diesel Engines. MTZ No 7/8
7. Anisits F, Hiemesch O (1991) Der neue BMW Sechszylinder (BMW's New Six-Cylinder). Kratochwill H, Mundorff F (eds), Special print of MTZ in No 10/11
8. Anisits F, Borgmann K, Kratochwill H, Steinparzer F (1998) Der neue BMW-Vierzylinder Dieselmotor (BMW's New Four-Cylinder Diesel). Special print of MTZ
9. Schittler M, Heinrich R, Kerschbaum W (2000) Mercedes Benz Baureihe 500 – eine neue V-Motorengeneration für schwere Nutzfahrzeuge (The Mercedes- Benz 500 Series – a New Generation of V-Engines for Heavy Utility Vehicles). ATZ 9
10. Harmeier R, Rieck G (2000) Die Trucknologie Generation TG-A der MAN Nutzfahrzeuge AG (The TG-A Trucknology Generation of MAN Nutzfahrzeuge AG). ATZ 9
11. Mollenhauer K (1997) Handbuch Dieselmotoren (Manual of Diesel Engines). Springer, Berlin, Heidelberg, New York
12. Borgmann K, Steinparzer F, Stütz W (2000) Potenziale und Grenzen moderner PKW-Hochleistungsdieselmotoren (Potentials and Limits of Modern High-Performance Passenger Car Diesel Engines). The Aachen Colloquium on Vehicle and Engine Technology
13. Zukunftsperspektiven des Verbrennungsmotors (Future Perspectives of the Combustion Engine). Special Issue 60 Years of MTZ / 1999
14. Berger H, Eichlseder H (1998) Das Abgaskonzept des BMW-Vierzylinders (The Exhaust Gas Concept of BMW's Four-Cylinder Engine). MTZ
15. Metz N, Seiko M (1998) Die Luftqualität in Europa bis zum Jahr 2010 mit und ohne Euro IV-Grenzwerte (Air Quality in Europe up to the Year 2010 with and without the Euro IV Limits). 19th International Vienna Engine Symposium
16. German 1999 Annual Report (page 55). Federal Office of the Environment
17. Anisits F (1991) Der Kraftstoffeinfluss auf die Abgasemission von PKW- et al Wirbelkammermotoren (The Influence of Fuel on the Exhaust Emissions of Passenger Car Turbulence Chamber Engines). MTZ
18. Wieland J, Vent G, Wirbeleit F (2000) Können innermotorische Maßnahmen die aufwendige Abgasnachbehandlung ersetzen? (Can Measures within the Engine Replace Elaborate After-Treatment of Exhaust Gases?). 21st International Vienna Engine Symposium
19. Allanson R (2000) The Use of the Continuously Regenerating Trap (CRT) on SCRT Systems to Meet Future Emission Legislation. 21st International Vienna Engine Symposium
20. Dürnholz M, Wintrick T, Polach W (2001) Potenzial der Einspritzung zur Reduzierung der Schadstoffemission von Dieselmotoren (Potential of Fuel Injection to Reduce Harmful Emissions of Diesel Engines). 22nd International Vienna Engine Symposium
21. Metz N, Resch G, Schönberger K, Steinparzer F (2000) Size Distribution and Characteristics of Soot Particles from Modern Diesel Engines. MTZ
22. Steinparzer F, Brüne HJ (1998) BMW Diesel in Pole Position. Haus der Technik, Meeting No E-30-315-056-8, March
23. Moser FX, Sams T, Cartellieri W (2001) Der Nutzfahrzeug-Dieselmotor unter Druck (The Utility Vehicle Diesel Engine under Pressure). 22nd International Vienna Engine Symposium

4 Alternative Propulsion Systems

Dušan Gruden

4.1 Introduction

Ever since its invention, the 4-stroke reciprocating piston engine has been considered as a rather complex thermal unit which should better be replaced by far less complicated designs.

Today, the internal combustion piston engine is looking back at a history of more than 125 years. During that time, there have been many initiatives to substitute piston engines for simpler and more efficient alternative powerplant units.

The potential alternatives can be subdivided into two major groups: The first one consists of thermal power units, such as gasoline (Otto) and Diesel engines which use combustion to convert the chemical fuel energy into mechanical work. The best known representatives of this group are

- the 2-stroke engine,
- the Wankel engine,
- the gas turbine,
- the Stirling engine, and
- the steam engine.

The second group includes drive units designed to store energy or to directly convert chemical fuel (or mechanical) energy into electric energy, such as

- electric batteries,
- fly-wheel accumulators, and
- fuel cells.

As far as certain individual criteria are concerned, most of these drive systems perform better than the four-stroke combustion piston engine, so that, in the past, opinions about their chances as alternative passenger car powerplants were very optimistic (Fig. 1).

However, when summing up all the properties required to smoothly operate cars over wide speed and load ranges and a long lifetime, these alternative concepts have never succeeded in edging the Otto and Diesel engines out of their top positions.

4.2 Thermal Engine with Discontinuous Combustion

4.2.1 *Two-Stroke Engine*

The thermal unit which yields the highest efficiency of more than 53% is the low-speed two-stroke marine Diesel engine. This great performance combined with the relatively simple design of these cross-flow two-stroke engines has induced

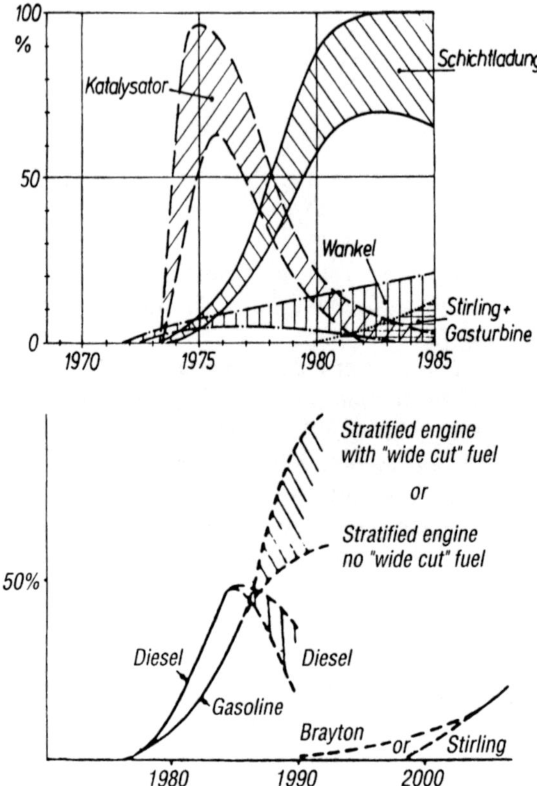

Fig. 1. Previous prognosis about alternative powerplants. Source: Eaton Corp. 1973, Jet Prop. Lab. 1975

engineers at all times to try and develop a two-stroke engine for automotive application.

The last intensive research on the 2-stroke automotive engine concept was carried out by Orbital, an Australian company and ended with the conclusion, in the mid 1990s, that a 2-stroke engine will not be able to meet the demands on modern passenger cars in terms of power output, fuel consumption, NO_x emissions and durability.

4.2.2 *Wankel Engine*

The rotary-piston Wankel engine is the best of the solutions found, when looking for other ways of realizing the Otto cycle. The Wankel concept solves the inherent mechanical problems of the piston engine in an astonishingly simple manner: no reciprocating pistons – and thus no vibration, no valves – and therefore less noise.

However, the real challenge in combustion engine development is not in the mechanical configuration but in the combustion process itself. The fact is that combustion is far easier to influence in a reciprocating-piston engine than in a rotary-piston Wankel engine, whose crescent-shaped combustion chamber with its big surface/volume ratio can be optimized only within rather tight limits. Since the Wankel engine was not capable of solving the urgent problems of exhaust gas pollution, of fuel consumption and durability, it finally disappeared from the gross production market [4].

4.3 Thermal Engines with Continuous Combustion

With the gas turbine, Stirling engine and steam engine, continuous fuel combustion takes place in an external burner and not in the working medium itself.

This configuration strongly facilitates the optimization of the combustion process when compared with a reciprocating piston engine. This is an advantage, which is highly appreciated and comes along with another welcome quality of such thermal engines: their multi-fuel capability.

4.3.1 *Gas Turbine*

It was in 1950 that a gas turbine was used for the first time to power a (Rover) passenger car. The investigation of gas turbine technology reached its climax in the 1970s. Its major drawbacks are the poor transient behavior and high fuel consumption. High efficiencies can only be achieved if the working medium has a very high temperature when it enters the turbine. With modern turbines, maximum inlet temperatures of T_{max}=1100 to 1350 °C are realized. At such temperatures, fuel consumption is 30 to 80% higher than that of a gasoline engine. Under laboratory conditions, the following exhaust emission values are reached:

$$\begin{aligned} HC &= 0.16 \text{ g/km} \\ CO &= 2.10 \text{ g/km} \\ NO_x &= 0.25 \text{ g/km} \end{aligned} \qquad \text{(a)}$$

The manufacturing costs of a gas turbine are twice as high as those of a gasoline engine. According to current knowledge, gas turbines cannot be considered as appropriate substitute units for passenger cars.

4.3.2 *Stirling Engine*

The Stirling engine has been known since 1816. Ever since, it has been intensively investigated for its suitability for various applications. The greatest advantage of this concept is that it is noiseless. It is problematic, however, with regard to the sealing of the working medium – which can be either hydrogen or helium. Engine cooling and engine weight cause problems as well.

More than 60% of the heat supplied must be dissipated via radiators. The fuel consumption of a Stirling engine is higher than that of a gasoline or Diesel unit.

The exhaust emission levels are as follows:

$$\begin{aligned} CO &= 1.20\ g/km \\ HC &= 0.14\ g/km \\ NO_x &= 0.30\ g/km \end{aligned} \qquad (b)$$

The manufacturing costs of a Stirling engine are 5 to 8 times higher than those of an Otto or Diesel unit.

The Stirling engine, as it is today, is not being considered as a suitable drive unit for passenger cars.

4.3.3
Steam Engine

In the early 1960s, eight different steam engine concepts were investigated in the USA for their potential as future passenger car drive units [9]. The best known variant was the so-called Leer engine. Low exhaust emissions and noise levels, however, can only be obtained at the expense of high costs, additional accessories such as a burner, heat exchanger, condenser and various control systems, and a poor overall efficiency. Despite repeated and intensive testing, steam engines are not expected to enter the passenger car domain.

4.3.4
Common Characteristics of Continuous Combustion Engines

With all continuous-combustion engine concepts, the heat is generated in the burner and mostly fed to the working medium through the walls. Thus, it is upon the thermal strength of the wall material that the maximum temperature of the working medium depends. The maximum temperature tolerated by current materials is about 1250 to 1350°C.

When compared with other thermal engines, reciprocation piston engines (gasoline and Diesel) offer an essential advantage: the heat is directly fed into the working medium (air) in the combustion chamber whose walls are cooled from outside with wall temperature being approx. 250°C. This allows much higher maximum cycle temperatures (higher than 2000°C) to be reached than in other thermal engines, where the temperature of the working medium is limited by the thermal properties of the wall material. The thermal efficiency η_{th} of a heat engine, however, mainly depends on the maximum temperature level of working medium that can be reached (Fig. 2).

The higher T_{max}, the better is the thermal efficiency. Thus, the alleged drawback of the 4-stroke piston engine – that is the intermittent combustion process – actually turns out to be its greatest benefit. The internal combustion process with its continuous exhaust and refill permits high maximum combustion temperatures of the working medium to be achieved with comparatively low combustion-chamber wall temperatures. Therefore, the reciprocation piston engine with internal combustion will always have a better thermal efficiency and consume less fuel than heat engines with external continuous combustion.

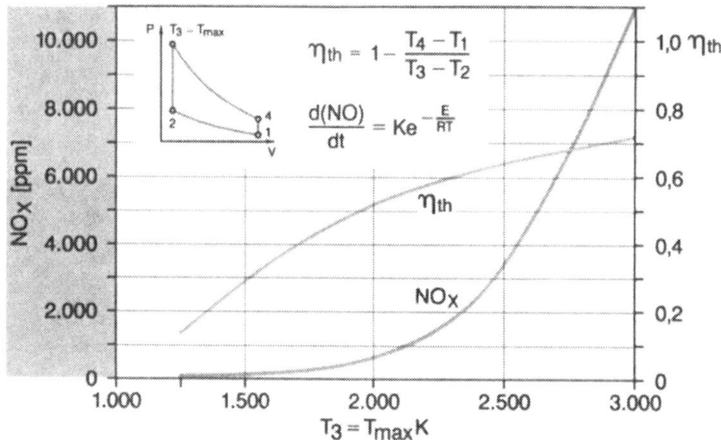

Fig. 2. Peak combustion temperature, thermal efficiency and NO_x emission

4.4 Electric Motor

The first electric motor-powered car was built in London, in 1886. At that time, its chances of success were supposed to be far better than those of the cars fitted with gasoline and Diesel engines which had just been invented.

The first electric cars yielded a top speed of 106 km/h and had an operating range of about 80 km. By the mid-nineties of the 20th century, various efforts had been made to develop a competitive electric car, but despite intensive research and development, there is not much difference yet today between modern electric cars and their earlier ancestors: recent versions reach top speeds of v_{max}=120 km/h and their operating range is between 70 and 100 km.

The main handicap of electric cars is the low energy density of the batteries used (Fig. 3).

From the very beginning, lead-acid (Pb) batteries have been used whose energy density is about 30 Wh/kg. To yield the same amount of energy as a gasoline fuel tank with a capacity of 70 L and a weight of 57 kg, a lead-acid battery of almost 3000 L and 6.5 tons would be required. Other alternative batteries such as sodium-sulfur, nickel-cadmium, lithium-ion and nickel-metal-hybrid units are too heavy and too expensive. Their storage capacities and lives are insufficient and their charging times of 6 to 8 hours are excessively long.

Electric cars cannot be considered as zero-pollution cars either, since exhaust emissions do occur during the generation of the electric energy required to operate them (Fig. 4).

Therefore, the Californian CARB authority would actually have to revise its plans for the compulsory introduction of electrically powered cars under the name of "Zero-Emission Vehicles" (ZEV) in California [23].

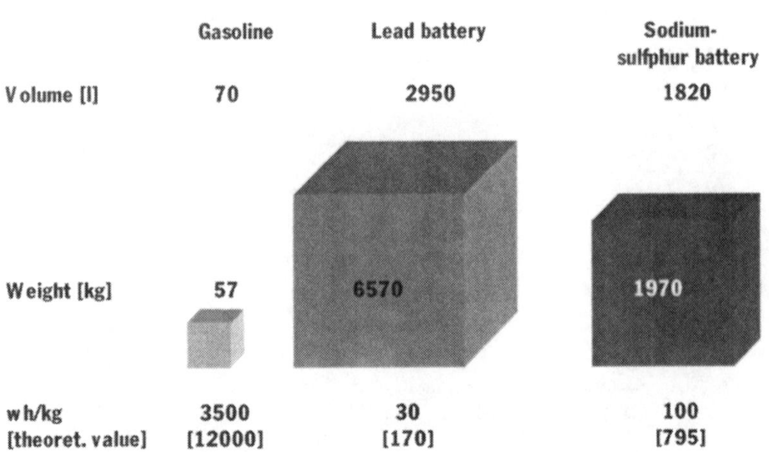

Fig. 3. Fuel versus batteries (energy content corresponding to 70 L of fuel)

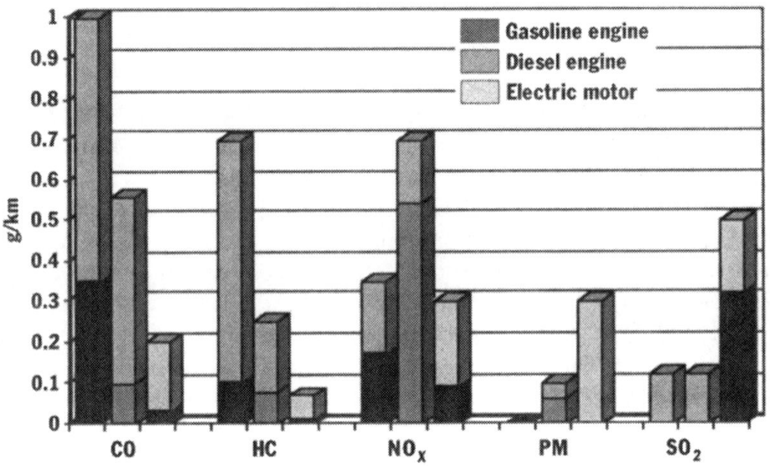

Fig. 4. Exhaust emission levels: gasoline and Diesel engines versus electric motor

4.5 Flywheel Storage System

The basic idea is a fascinating one: not to "destroy" – that is transform into heat – the kinetic energy, resulting from braking maneuvers, but to store it instead in an energy accumulator for transformation into different useful types of energy – or, in other words, to recuperate the brake energy. The most suitable devices for energy recuperation are flywheels, which rotate in a vacuum and transform stored mechanical energy into electric power.

Systems of that kind are useful for vehicles, such as trains or busses, whose routes are predefined, which run on a lane of their own and whose driving conditions and brake distances are known. For all other cars with random and arbitrary accelerations and decelerations there is no confirmed benefit from such concepts.

4.6 Outlook on the Future

During the last century, alternative drive systems were intensively investigated, but to date, none of those systems has been capable of pushing the 4-stroke piston engine out of its privileged position (Fig. 5). Surprisingly enough, it is not some exotic design but the good old reciprocating piston engine in its well known configuration, which will continue to dominate the market also in the foreseeable future.

The reciprocating piston engine has displayed an astonishing potential for improvement and an extraordinary adaptability to all new demands placed upon modern cars. Figure 6 illustrates the forecasts for passenger-car drive-system development in the coming 15 to 20 years. As can be seen, further optimized versions of the gasoline and Diesel engines will continue to prevail in the automotive domain.

From today's point of view, hybrid and fuel-cell powerplants are the only alternative systems expected to gain a foothold in the automotive market.

4.6.1 *Hybrid Drive*

Frequently, the opinion is held that the best way to solve the problem of fuel consumption and of pollution from automotive emissions would be to fit vehicles with combined drive systems, consisting of electric motors, coupled with small highly efficient internal combustion engines (hybrid drives) (Fig. 7).

The fuel economy potential of hybrid vehicles is summarized in the following three points:

- Energy regeneration during deceleration (conversion to electrical power);
- Electric motor propulsion in low-load conditions and as assistance in acceleration, enabling an internal combustion engine to run within the efficient regions of its operation;
- Engine stop during vehicle stops.

The hybrid vehicle system also has a high potential for emission control:

- Reduced gas volume resulting in lower catalyst deterioration;
- Engine stop and propulsion by electric motor to reduce HC emissions especially in lower load ranges;
- Power assist by the electric motor during acceleration, allowing the air/fuel mixture to be controlled closely around Lambda=1 in all engine operating conditions by reducing engine load deviations;
- Engine quick-start by high motor power avoids unstable combustion during starting.

Drive Concept	Outlook 1970	Current status	Outlook 2000
Internal combustion piston engine (SI, diesel)	- expensive combustion unit - unable of meeting legal requirements - will be substituted by the end of the century	↑	- expensive combustion unit - no substitute available in the foreseeable future - great inherent evolution potential - most economic combustion-engine variant
Gas turbine	- well-controlled steady-state combustion - low pollutant emissions	↓	- of no importance for automotive engineering
Stirling engine	- low pollutant emissions - noiseless operation	↓	- of no importance at the time being
Rotary-piston engine	- simple design - low weight - low price	↓	- engineering discontinued worldwide
Electric engine	- no pollution - low-noise operation	↓	- energy-storage problem unsolved - might be suited for niche vehicles
Hybrid drive systems	- expensive - costly	↑	- allows to realize zero-emission in urban regions
Fuel cell	- status early research phase - no forecast for automotive application	↑	- status: F & E phase - no short or medium-term application

Fig. 5. Status of alternative drive concepts

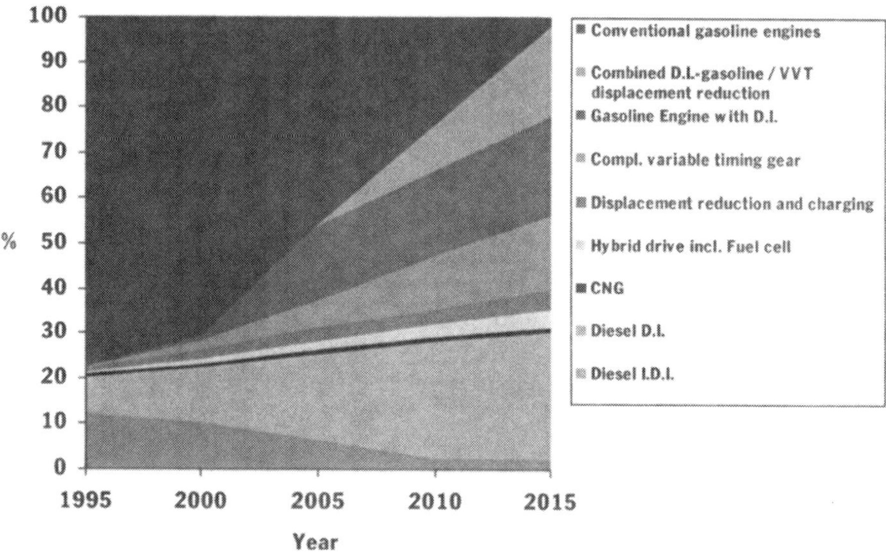

Fig. 6. Evolution of and forecast on future powerplant-systems

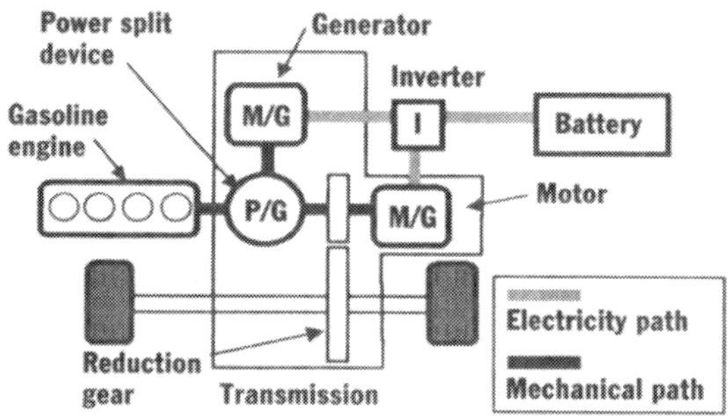

Fig. 7. Toyota hybrid system (THS) [29]

To date, however, mass production of this type of drive system has been hampered by its excessively high costs.

4.6.2
Fuel Cell

The fuel cell (Fig. 8) was invented as early as in 1839 by the Englishman William Robert Grove. It then was called the "galvanic gas battery".

The chemical reaction between fuel and air produces electricity with no mechanical energy involved. Hydrogen (H_2) combines with oxygen (O_2) producing water and electric current.

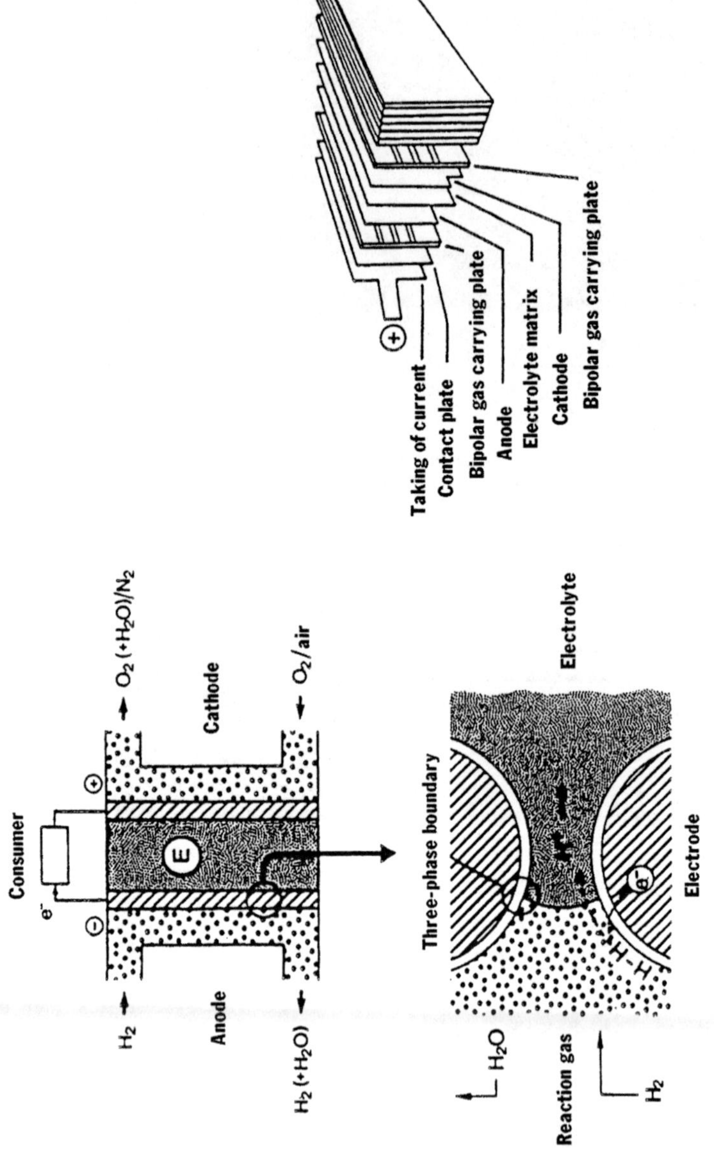

Fig. 8. Structures of a single fuel cell and fuel-cell stack

The core of each fuel cell is an electrolyte, that is a platinum-covered porous plastic foil. While hydrogen is fed to the anode, the platinum acts as a catalyst triggering the ionization of H_2. At the cathode-end of the electrolyte foil, the platinum material incites the atmospheric oxygen to absorb electrons. The positive H_2 ions and negative O_2 ions combine to form water (H_2O). Between the cathode and anode, an electric potential of approx. 0.6 to 1.2 V builds up. For this electric energy to be used, several fuel cells must be connected with each other, forming what is called a "stack". An electric motor for automotive application would need hundreds of individual cells (more than 200).

The first mention of fuel cells as potential alternative automotive drive units was made in the early 1960s. It took about 30 years for more intensive research to be finally started. Due to their theoretically high efficiency (Fig. 9) and low pollutant emissions, fuel cells are among the most promising alternative energy sources of the future.

Of the various potential fuel cell configurations, only two are currently being investigated in the automotive industry (Fig. 10).

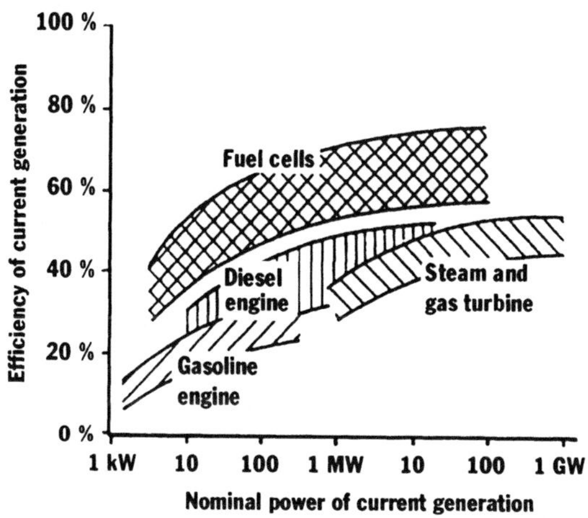

Fig. 9. Efficiency levels of electric-current production

PEM (PEFC) - Proton Exchange Membrane
SOFC - Solid Oxide Fuel Cell

	AFC	PEFC (PEM)	PAFC	MCFC	SOFC
Charge transfer	OH	H^+	H^+	CO_3^{-2}	O^{2-}
Electrolyte	Caustic potash solution	Polymer membrane	Concentrated phosphoric acid	Molten salt	Stabilised ceramic oxide electrolyte
Temperature	80 – 120° C	80° C	200° C	650° C	800 – 1000° C
Fuel/oxidising agent	H_2 and O_2 / air	H_2 and O_2 / air	Natural gas/ biogas and O_2/air	Natural gas/ biogas and O_2/air	Natural gas/ biogas and O_2/air
Particularities		Water management		CO_2 recirculation	Targets: Internal reforming temperature reduction

Fig. 10. Fuel cell types and their characteristics

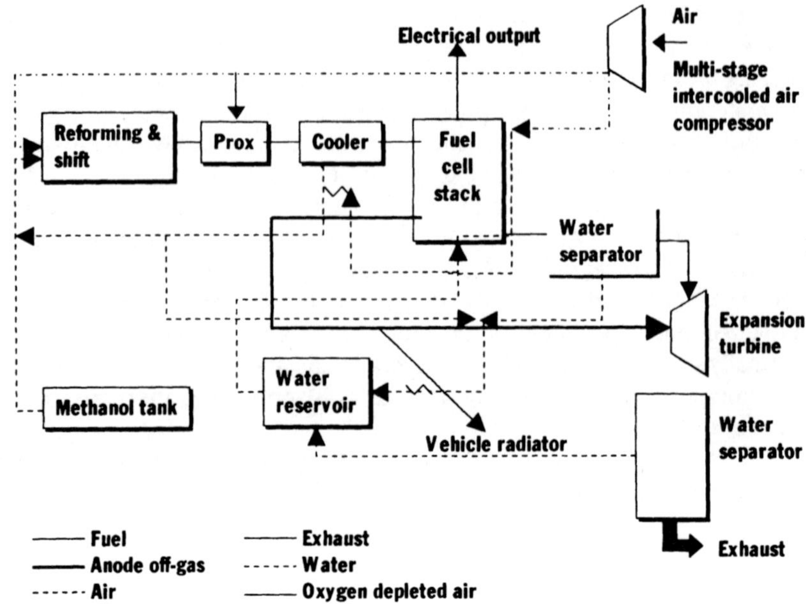

Fig. 11. The fuel cell system

With an operating temperature of approx. 80°C, PEM numbers among the so-called low-temperature fuel cells, whereas SOFC is what is called a high-temperature fuel cell with operating temperatures ranging between 800 and 1000°C.

For the fuel cell to function properly, various secondary units are required besides the cathode, anode and electrolytes (Fig. 11). These secondary units are needed to provide the fuel cell with air and fuel and to cool and control it. The efficiency of the fuel cell improves if it is fed with pressurized hydrogen and air (1.5 to 3 bars).

Less than 10% of the heat produced by a fuel cell is dissipated through the exhaust gas. For an assumed 40% overall efficiency, approx. 50% of the heat supplied must be dissipated through cooling. The required radiators and cooling power are 4 to 5 times bigger than with an Otto or Diesel engine.

The exhaust gas of the fuel cell (water vapor) is cooled down in an exhaust gas condenser. Part of the steam is used to humidify the fuel-cell membrane: These membranes function properly only if moistened.

The ideal fuel for fuel-cell operation is hydrogen produced from water by means of solar energy and solar cells (Fig. 12). However, it will probably not be possible to implement the technology required to create such ideal conditions before the middle of this century.

Today, hydrogen is produced from mineral oil or natural gas with resulting CO_2 emissions. Since the generation, transportation, storage, filling up and on-board storage of hydrogen causes certain problems, alternative fuels such as

- methanol,
- natural gas,

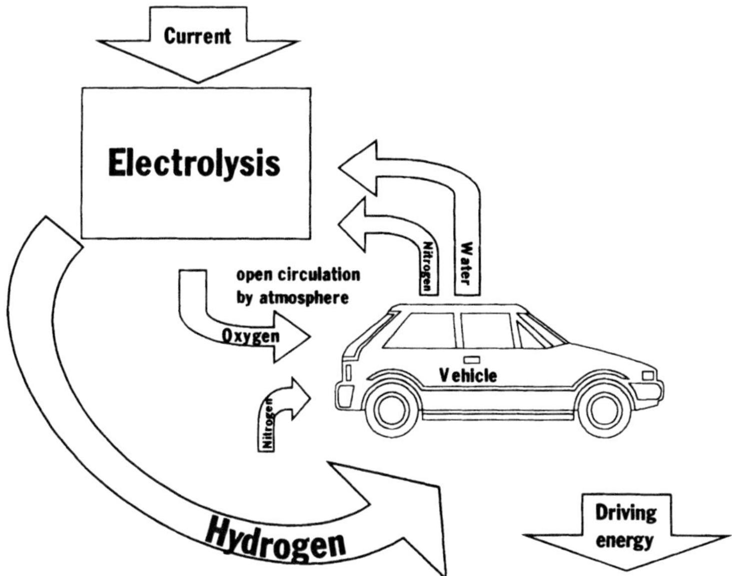

Fig. 12. Closed circulations, ideal automotive drive system e.g.: hydrogen

- gasoline,
- ethanol,

are also being examined for their capability to serve as hydrogen sources.

From these fuels, hydrogen and CO_2 can be obtained on-board in a so-called reformer. The required reforming temperatures range between 500 and 800°C. Reformers reach this temperature in 45 minutes to 2 hours.

To coat the fuel cells and reformer, about 50 g of platinum per car will be needed. This is about 10 times the amount required for a three-way catalyst.

To properly seal all the components of a fuel cell, about 1 km of gaskets is needed.

At the time being, apart from technical problems, the commercial launch of the fuel-cell technology is also hampered by the high manufacturing costs of fuel-cell-equipped cars and the enormous expenditure for building up the infrastructure required for fuel supply.

Costs for the generation of 1 kW of power by means of fuel cells are said to vary between 5,000 and 30,000 Euro (Fig. 13) compared with 25 to 40 Euro for the same amount of power from a gasoline or Diesel engine.

To be able to compete with internal combustion piston engines, the fuel-cell price for 1 kW of power should be lowered to 50 to 80 Euro – a target which might be reached by the end of this decade.

Several car manufacturers have announced their intention of launching their first standard fuel-cell-powered cars as of the year 2004. In Germany, more than 200,000 fuel-cell cars are expected to exist in 2010.

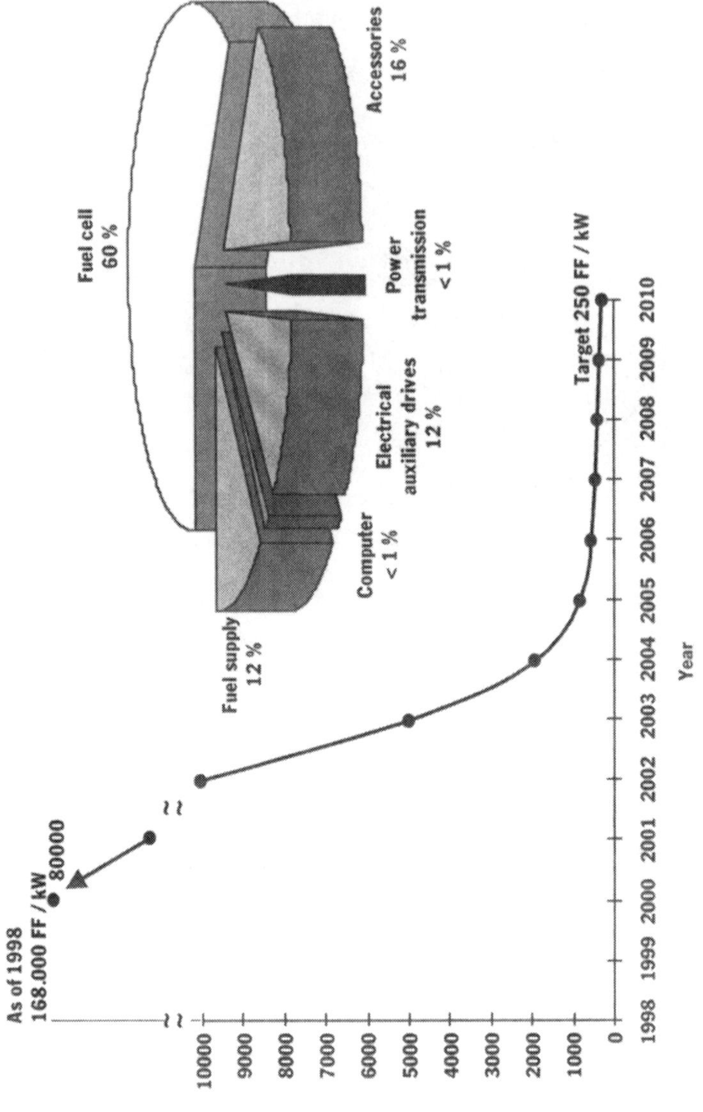

Fig. 13. Evolution of fuel-cell costs

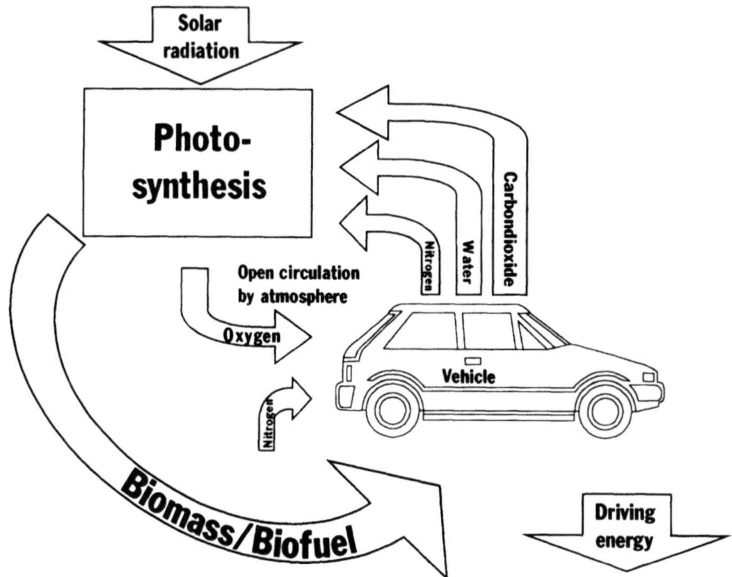

Fig. 14. Closed circulations, biomass as an energy source, e.g., biomass

4.6.3
Powerplants Using Alternative Fuels

Parallel to the search for alternative drive systems, engineers are also looking for alternative fuels. All those potential fuels are basically suited for conventional piston engines as well. When these fuels are put on the market, the conventional piston engines will also be adapted to suit their particular properties.

Of special importance for future applications are fuels made from so-called regenerative raw materials (biomass) (Fig. 14).

Via photosynthesis, solar energy is indirectly employed for the production of fuels: using the CO_2 and H_2O contained in the atmosphere, biomass is produced through photosynthesis. Biomass is an excellent raw material for the creation of various fuels such as alcohols and methyl esters. Power is then obtained by burning such biological fuels. The resulting products of combustion – that is CO_2 and H_2O – remain in the closed cycle of biomass production via photosynthesis. This method does not add to the increase of the atmospheric CO_2.

At the present time, this type of energy production is not yet a useful alternative, either technically or economically. But it is expected by many experts to gain increasing importance in the future.

4.7 References

1. Bertodo R, Smith P (1978) Prime Movers for Future Lift Trucks. SAE-Paper 780752
2. Eaton Corporation, 1973
3. Schorn N, Dürnholz M, et al (1993) Ist der Zweitaktmotor eine Alternative zum Viertaktmotor? VDI-Bericht 1066, Dresden
4. Fuhrmann E (1976) Entwicklungstendenzen beim Bau von Fahrzeugmotoren. Porsche AG
5. Wilson DG (1978) Alternative Automobile Engines. Scientific America 239:1
6. Should we have a new engine? An automobile power systems evaluation, vol I. Jet Propulsion Laboratory. California Institute of Technology, 1975
7. Nacamura H, Motoyama H, et al (1991) Passenger Car Engines for 21st century. SAE-Paper 911908
8. Beeck A, Joos F, et al (1999) Fortschrittliche Gasturbinen Technologie. BWK 51, 5/6
9. Amann CA (1987) How shall we power tomorrow's automobile? Automotive Engine Alternatives. Plenum Press, New York
10. Richey AE (1989) The Automotive Stirling Engine Programme. Final Report. SAE P-230
11. Downs D (1978) The Socially Acceptable Powerplants. Proceedings of Inst Mech Engineers 192
12. Buschmann G, Clemens H, et al (2000) Zero Emission Engine - Der Dampfmotor mit isothermer Expansion. MTZ 61:5
13. Pocrnja A (1982) Energieerzeugung aus Wärme. Erdöl und Kohle 35:5
14. Quissek F (1997) Fahrzeugantriebe der Zukunft - aus Sicht der VW-Forschung. Eurotax Nr. 1
15. Bentley JM, Teagan WP, et al (1997) Global Warming Impacts of Vehicle Propulsion Options. Arthur D. Little
16. Rügen macht e-mobil. VDI-Nachrichten Magazin 9/93
17. Japanese makers exhibit EV-efforts. Automotive News, 14 Dec. 1994
18. Battery price is key hurdle for development of electric. Automotive News, 23 December, 1996
19. Kobe G (1996) Battery of the Future. Automotive Industries 176:9
20. Schwungrad bringt Autos schnell auf Touren. VDI-Nachrichten, Nr 14, 2001
21. Cole AC (1997) Electric and hybrid Vehicles - a report on the SAE-Symposium. Automotive Engineer, September
22. Toyota's Hybrid ready for California. Automotive Industries, May 1997
23. Calif. Turns ZEV mandate upside down. Automotive news, 29.01.2001
24. Tachtler J, Dietsch T, et al (2000) Fuel Cell Auxiliary Power Unit - Innovation for the Electric Supply of Passenger Cars? SAE-Paper 2000-01-374
25. DaimlerChrysler, Umweltbericht 2000
26. Frank MH, Welter, DL, et al (2000) PEM Fuel Cell System Solutions for Transportation. SAE-Paper 2000-01-0373
27. Zizelman J, Botti J, et al (2000) Solid-oxide fuel cell auxiliary power unit: a paradigm shift in electric supply for transportation. Automotive engineering. September 2000
28. Schon in der zweiten Generation. Automobil Revue, 2000
29. Harada J (2001) Development of a new hybrid vehicle. Auto Technology 1

The Handbook of Environmental Chemistry Vol. 3, Part T (2003): 107–173
DOI 10.1007/b11990

Means of Transportation and Their Effect on the Environment

Hans Peter Lenz[1] · Stefan Prüller[2] · Dušan Gruden[3]

[1,2] Technical University of Viena, Getreidemarkt 9, 1060 Wien, Austria
[3] Dr. Ing. h.c. F. Porsche Aktiengesellschaft, Porschestraße, 71287 Weissach, Germany
[1] *E-mail: proflenz@ivk.tuwien.ac.at*
[2] *E-mail: stefan.prueller@ivk.tuwien.ac.at*
[3] *E-mail: dusan.gruden.@porsche.de*

Whereas the general desire for mobility has resulted in a constantly increasing vehicle population, the concern about the potential threat to the ecosystem posed by the life-style of present-day society is also growing.

This report gives a survey of the most recent findings on the contribution of road traffic to air pollution which have been obtained through continuous scientific research on the one hand, and compares motor vehicle emissions with pollution from other sources which have a global, regional or local impact.

This report also analyzes air quality data in different countries and relates these to the relevant legal provisions which had been introduced with the aim of protecting the environment and human health.

The analysis and evaluation of emissions is based on a detailed global investigation of emission components subject to the legally stipulated maximum permissible values such as carbon monoxide (CO), non-methane hydrocarbons (NMHC), nitrogen oxides (NO_x) and particulate matter (PM) on the one hand, and greenhouse gases, such as carbon dioxide, (CO_2), methane (CH_4), halogenated hydrocarbons (HHC) and dinitrogen monoxide (N_2O).

A comparison of natural and anthropogenic global emissions shows that carbon monoxide and nitrogen oxide emissions are roughly of the same order of magnitude. As regards carbon dioxide, non-methane hydrocarbons, dinitrogen monoxide and particulate matter, natural emission levels are significantly higher than anthropogenic ones, while methane and halogenated hydrocarbon emission are mainly caused by anthropogenic sources.

Carbon dioxide, methane, dinitrogen monoxide (nitrous gas) and halogenated hydrocarbons as well as ozone and water vapor account for a share of 0.5 to 1.5 per cent of the anthropogenic greenhouse effect. Of the aforementioned components, water vapor contributes the largest share to the overall greenhouse effect, with carbon dioxide ranking second. Carbon dioxide emissions from natural sources account for 96.5 per cent of total global carbon dioxide emissions thus by far surpassing emissions from man-made sources. The share of total global passenger and commercial vehicle traffic in global carbon dioxide emissions makes up approx. 0.5 per cent.

The evaluation of regional and local emissions and the calculation of vehicle emissions in Germany using the data base developed by the Department of Combustion Engines and Automotive Engineering of Vienna University has revealed a significant reduction of exhaust gas components for which limit values have been imposed by legislator.

- **Carbon monoxide:** in the EU the contribution of traffic to anthropogenic CO emissions currently makes up roughly 62 per cent. In Germany, emissions from passenger cars at present account for approx. 40 per cent. In the past three decades, anthropogenic CO emissions in Germany were brought down by more than 70 per cent. The reduction of emissions by approx. 50 per cent was confirmed by measurements of immission concentrations at roadside measuring stations. Thus immission concentrations are more than 80 per cent below the current clean air standard.
- **Non-methane hydrocarbons:** In the EU, road traffic currently accounts for some 29 per cent of total emissions. In Germany, anthropogenic NMHC emissions were reduced by approx. 42 per cent in the period from 1990 to 1998. Immission concentrations measured at

roadside stations were lowered to one quarter of the original level within a period of nine years.

- **Nitrogen oxides:** Within the EU, road traffic at present accounts for roughly 42 per cent of total emissions. In the course of the past decade, Germany succeeded in cutting in half total anthropogenic NO_x emissions. This also shows the trend towards declining immission concentrations. All measuring stations reported NO_2 emissions which were clearly below the currently valid clean air standard which is defined at 80 µg/m^3 NO_2 in the air.
- **Particulate matter:** In the EU, road traffic currently accounts for approx. 30 per cent of particulate matter (PM_{10}) emissions. Germany succeeded in lowering total anthropogenic PM emissions by as much as 80 per cent in the course of the past decade. All measuring stations report immissions far below the limit value of 150 µg/m^3.

The successful efforts made by the business community and the automotive industry to lower emissions significantly should, however, go hand in hand with other measures, taking into account that consumers' driving styles and vehicle use patterns as well as appropriate traffic planning also play a crucial role in bringing down emission levels.

One of the most urgent present-day priorities is the reduction of noises emission resulting from human activities. The noise produced by road traffic is the predominant element in overall noise emissions in crowded areas.

Since 1966 the permissible noise level for passenger cars has been reduced from 86 dB(A) to 74 dB(A). The gradual reduction of the noise emissions from the engine and exhaust system have resulted in an increase of the contribution made by the tire-road combination.

Today, the average lifetime of passenger cars is expected to be about 10 to 12 years. Once this period is expired, the question arises what to do with the used car, which is frequently also called "scrap car". The automotive industry have been confronted with the end of life vehicles problem for many years. Nowadays, more than 75 per cent of the weight of passenger car are recycled. Thus, automobiles constitute highly complex economic objects characterized by an unusually high recycling rate. For decades, the recycling system applied by the waste industries world-wide has been entirely based on the principle of economic market success. After years of intensive discussions an end of life vehicle directive was put into force by the European Union in September 2000.

The main points of this directive are the following:

- Taking back of used cars.
- Limitation of heavy metals, such as lead, cadmium, mercury and chromium(VI) compounds.
- Determination of the recycling rates.
- Authorization of the recycling standards or new vehicles during type approval.
- Availability of part/dismantling handbooks.
- Utilization of recycled material.

The current 75 per cent recovery rate is to be increased to 85 per cent by 2006 and to 95 per cent as of 2015. Thereafter, 5 per cent only of the so-called fluff is allowed to be dumped. The world-wide growing shortage of not-renewable resources and steeply decreasing landfill capacities have led to the conclusion that the manufacturer's responsibility for their products must be considerably extended. The term "sustainable development" describes a development which satisfies the requirements of the present without compromising the ability of future generations to met their own needs. At the time being, automotive industry is concentrating its efforts on doing life cycle assessment (LCA), i.e., drawing up ecological balances which cover the entire life time of an automobile.

Keywords. Natural and anthropogenic sources, Global, Regional and local emissions, Regulated and unregulated emissions, Greenhouse gases, Ozone formation potential, Immissions, Passenger cars, Commercial vehicles, Traffic noise, Car recycling, End of life vehicles, Life cycle assessment

1 Emission and Air Quality

Hans Peter Lenz · Stefan Prüller

This chapter takes stock of the current problems of air pollution resulting from road traffic. The public at large has been led to believe that air pollution is almost entirely caused by road traffic. This attitude will be critically analyzed for its validity.

As pointed out in Chapter 1, man's desire for mobility has led to a continuous rise of the number of vehicles on our roads.

Germany serves as an example to illustrate past trends in vehicle population figures and to make forecasts about future developments.

The breakdown of the vehicle population by categories is calculated on the basis of the number of newly registered vehicles in a year and their probable service life. The development of the passenger car population in Germany is shown in Fig. 1: For reasons of clarity, the passenger car population is not broken down by engine capacity. Figure 2 gives a survey of the development of the commercial vehicle population, broken down by vehicle types.

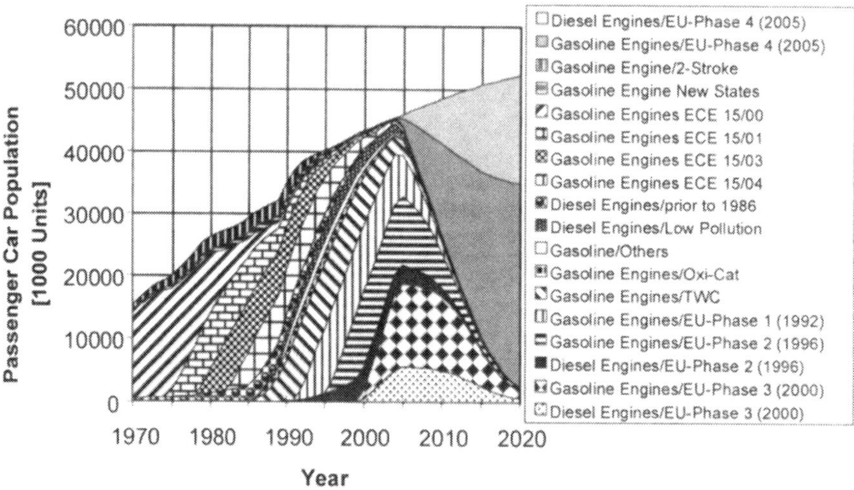

Fig. 1. Development of the passenger car population in Germany and expected trends

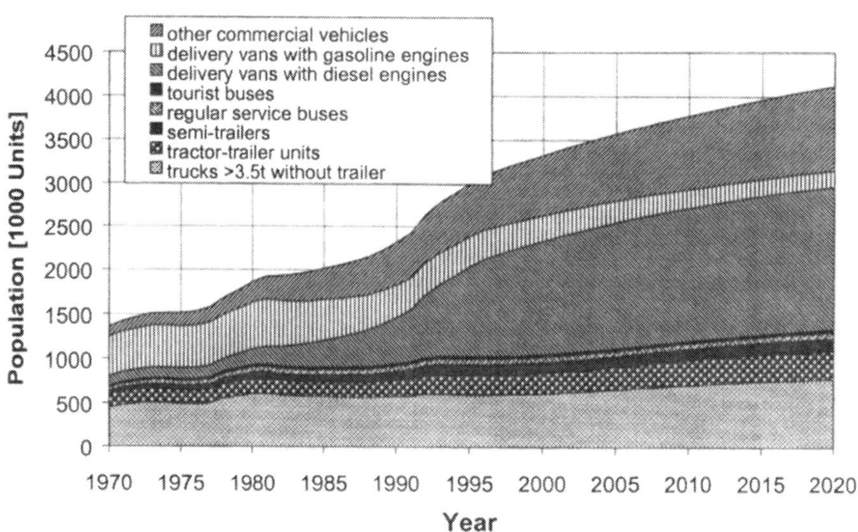

Fig. 2. Development of the commercial vehicle population in Germany and expected trends

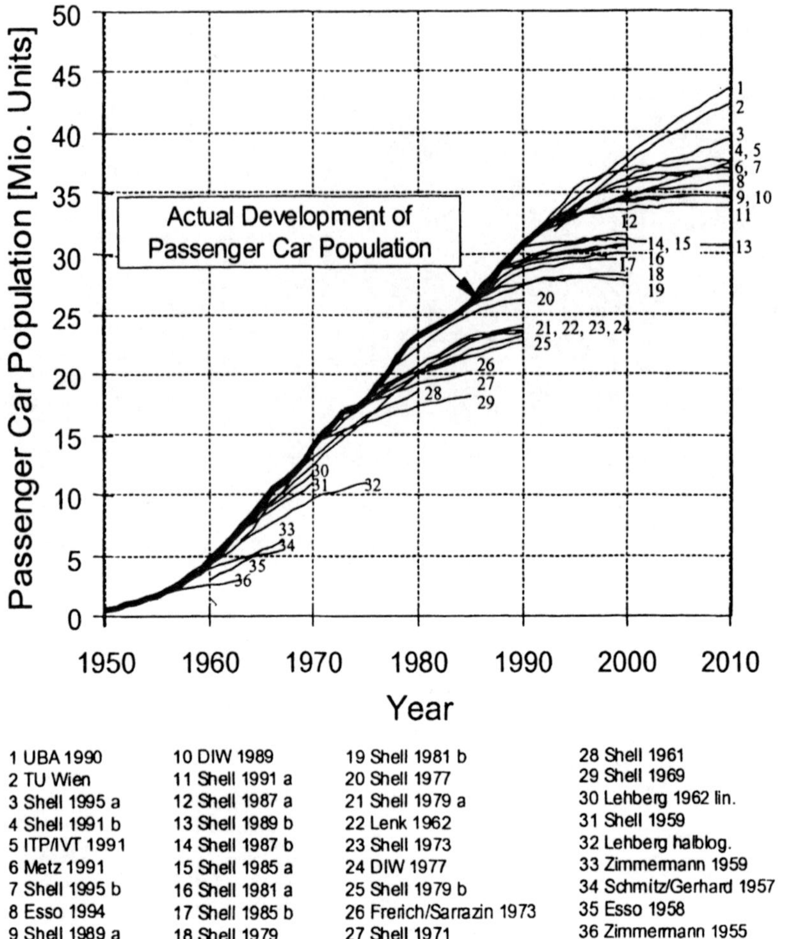

Fig. 3. Comparison of forecast passenger car population trends and real car population figures for the period 1950–2010 in the states of the former GDR [1]

As can be seen from Fig. 3 in which forecasts of the vehicle population trends made at different points of time are plotted, it is normally extremely difficult to estimate the future development of road traffic. To date, the real growth of the vehicle population has surpassed all forecasts.

1.1
Exhaust Gas Components Caused by Motor Traffic

The study of emissions from road traffic started more than half a century ago and has evolved further in the course of its history. The first reports on the influence of motorized road traffic were published in California in the 1940s. In this US state, the specific characteristics of the climate gave rise to "summer smog": under the impact of strong ultraviolet radiation at high temperatures, the interaction between nitrogen oxides and hydrocarbons causes a chemical reaction thus generating "photooxidants" (see chapter "Meassures for the Reduction of Exhaust Gas Emissions").

In Europe, road traffic caught public attention in the 1960s because of its carbon monoxide emissions which clearly have a harmful effect on human health. In response to this situation, limit values were fixed for the unburnt constituents of vehicle exhaust gases, such as carbon monoxide and hydrocarbons. In view of the trend towards rising levels of trace gases which are released into the air in combustion processes and their movement over long distances, the need arose in the 1970s and 1980s for the imposition of more stringent limit values for nitrogen oxides and unburnt hydrocarbons.

In an effort to combat this problem the Unites States introduced maximum permissible values for exhaust gasses from motor vehicles in the early 1960s followed by Japan in the mid-1960s, and by Europe in the 1970s. Limit values for stationary sources of air pollutants, such as power plants and boiler systems were not introduced until later.

In the industrialized countries, various measures were taken to reduce carbon monoxide emissions to safe levels as they had been shown to cause damage to human health. The drastic reduction of maximum permissible values for nitrogen oxide and hydrocarbon emissions in the early 1980s in the US and Japan and in the late 1980s in Central European countries, such as Austria, also led to considerably lower emissions of these trace gases from passenger cars and power plants in all the countries concerned.

Maximum permissible values for particulate emissions from Diesel-engine powered passenger cars were introduced in the US in the early 1980s and in Europe in the mid-1980s.

1.1.1
Limited and Not-Limited Exhaust Gas Emissions

Exhaust gas emissions subject to maximum permissible values are those containing CO, HC and NO_x components. Substances, such as lead, sulfur, benzene and other fuel components are also subject to quantity restrictions, but these limit values refer to their maximum permissible ratios in fuels, therefore theses substances are not referred to as limited exhaust gas components.

All other environmentally relevant substances required for the operation of motor vehicles are grouped together under the term "unlimited or unregulated exhaust gas components." In the US, for example, the Clean Air Act stipulates that all passenger car manufacturers must issue an "environmental impact certificate" evidencing the harmlessness of all potentially pollutant components of exhaust gases. There are neither clearly defined limit values to be observed, nor have analysis methods been indicated for issuing such a certificate.

In the early 1990s it became obvious that other exhaust gases from combustion processes which are harmless for human health do, however, affect the atmosphere. This phenomenon which became known as the "greenhouse effect" has in the meantime prompted scientists to focus their attention on carbon dioxide emissions.

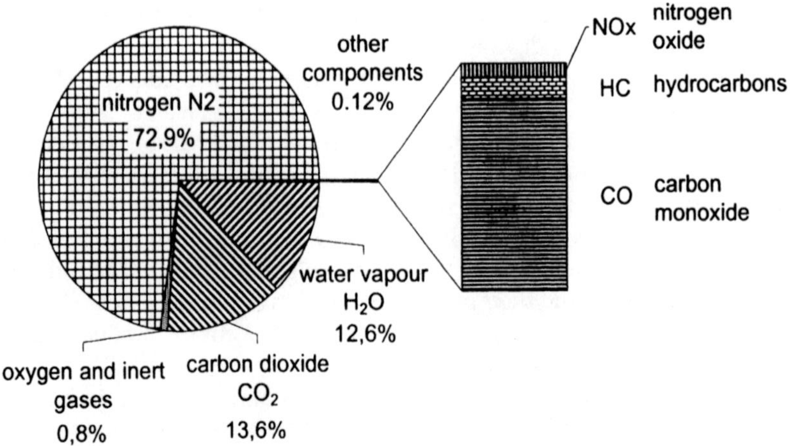

Fig. 4. Composition of the exhaust gases of a standard passenger car with catalyst determined through exhaust gas tests

Before dealing with the individual components of exhaust gas emissions in detail, we would draw your attention to Fig. 4 which illustrates the order of magnitude of pollutant emissions from modern mid-size cars with controlled three-way catalysts. Passenger car emissions consist primarily of the harmless components nitrogen 72.9%, carbon dioxide 13.6%, water vapor 12.6% and 0.8% innocuous gases, such as oxygen and inert gases forming part of the ambient air. The remaining components are typical reaction products from the incomplete combustion of fuels or products resulting from side-reactions of the initial substances. Besides carbon dioxide and water vapors which are the natural and final products of combustion, in vehicles with controlled three-way catalysts, only approximately one millionth part of the exhaust gases at the exhaust manifold are real pollutants. The same order of magnitude applies to Diesel-engine-powered vehicles.

1.2 Origin and Life-Cycle of Pollutants

The substances contained in the exhaust gases of vehicles can have an impact on the environment in the form of immissions. The specific immission situation in a given area is subject to a highly complex mechanism of interactions which are briefly described below. Emissions can be divided into two different categories, depending on the source:

- natural sources and
- anthropogenic sources.

Natural sources of trace substances in the atmosphere are the fauna, the flora, the ocean surfaces, volcanoes and the land surface as a whole, as well as atmospheric events such as lightning. Anthropogenic sources comprise all human activities, such as the generation of energy, industry and the trades, transport and traffic, domestic heating, waste incineration, dumping and agriculture and, to be precise, even human respiration.

A complex sequence of physical and chemical processes takes place between the time of emission of a substance and its deposition on various kinds of surfaces. These processes explain why the concentration of such substances in the air always varies both with location and time. The trajectory of a substance through the atmosphere is characterized by the stages described below:

emission – transport – transformation – immission – deposition.

Once they are released by their sources, emissions mix with the ambient air, are diluted to varying degrees depending on the weather conditions and are subsequently transported over longer or shorter distances by air streams. During their transport in the atmosphere, trace substances are transformed through physical and chemical processes.

Physical processes cause an exchange of substances among airborne particles (aerosol particles), cloud droplets, fog droplets and raindrops and the ambient gas phase. Chemical processes transform the mixture of trace gases through homogeneous and heterogeneous reactions. Homogeneous chemical reactions occur when only gaseous reaction partners are involved. Heterogeneous chemical reactions occur on the surfaces of aerosol particles, cloud droplets, fog droplets and raindrops. In addition, reactions also take place inside cloud or fog droplets and raindrops.

The chemical transformation of trace substances in the atmosphere yields a number of products such as carbon dioxide and water which cause no harm to human health. However, these chemical transformations also give rise to a number of secondary pollutants. The "trigger" for many of these processes is the irradiation of sun light.

Immissions are defined as the local concentrations of trace substances in the ambient air. These are influenced by the transport of locally emitted substances, their physical or chemical transformation and deposition. As all of these processes are subject to variations, immissions are also highly variable over time. In many cases they follow a distinct daily or annual pattern.

1.2.1
Long-Life Components with Global Impact

Global and local influences on the atmosphere due to exhaust gas components are analyzed and the impact of globally active emissions on the depletion of the protective ozone layer in the stratosphere or the greenhouse effect is assessed. In principle, a distinction is made between long- and short-life components in the atmosphere. Components such as carbon dioxide, methane, halogenated hydrocarbons and nitrous oxide which remain in the atmosphere for several years or even centuries are well intermixed with, and distributed evenly throughout, the atmosphere. Hence a few measuring stations scattered over the globe will suffice in order to determine the mean concentration of theses gases in the troposphere. These components influence the entire atmosphere. It is presumed that they influence the world's climate, contribute to the greenhouse effect and deplete the protective ozone layer in the stratosphere.

1.2.2
Short-Life Components with Local Impact

Short-life components, such as carbon monoxide, nitrogen monoxide, nitrogen dioxide and various hydrocarbons have a lifetime of between only a few minutes and several months. Their concentrations in the atmosphere vary widely and depend, to a great extent, on the geographic location of their sources. Theses components directly influence air quality in the immediate surroundings of the source. The short-life components will normally be transformed within a short time into water-soluble end products which will be washed out by precipitation. This definition of short-life components is, however, valid only under certain circumstances. When these components are blown up to high altitudes by the wind or by air streams (air craft) or when they are transported over long distances, they may have a regional or even global effect on air quality.

1.3
The Impact of Exhaust Gas Components

In principle, pollutants and trace substances can be analyzed separately on the basis of such criteria as nuisance value, toxicity for human beings, toxicity for the environment, and relevance for the climate.

1.3.1
Classical Analysis of Toxicity

1.3.1.1
Carbon Dioxide (CO_2)

CO_2 is the stable and natural end product from the combustion of biogenic and fossil fuels or from the aerobic decomposition of organic substances. Carbon dioxide is a colorless, odorless gas which remains almost totally inert in the atmosphere. CO_2 molecules do not decompose in the atmosphere, but are bound by plants on the surface of the earth through the absorption of energy (photosynthesis) or stored in the oceans. CO_2 is not toxic, but due to its high density which exceeds that of the air, it gathers in the lower sections of closed, sealed rooms. If it is the only gas present in such rooms it causes suffocation.

1.3.1.2
Carbon Monoxide (CO)

CO is a toxic, colorless and odorless gas, which is generated through incomplete combustion due to a lack of oxygen. When inhaled for half an hour, a mere 0.3% of CO in the air can lead to death by suffocation. As CO is converted into CO_2 (carbon dioxide) within a few hours in the presence of sufficient amounts of oxygen, this gas only has a local effect on air quality. Thanks to modern catalyst technology the threat to human health posed by car emissions has today been reduced to an absolute minimum.

1.3.1.3
Hydrocarbons (HC)

HC are generated through the incomplete combustion of fossil or biogenic sources of energy as well as through the evaporation of solvents and cleaning agents or petrol. Hydrocarbons are mixtures of various organic compounds, which sometimes smell foul and pose a threat of varying degrees to human health.

1.3.1.4
Nitrogen Oxides (NO_x)

Mixtures of nitrogen monoxide and nitrogen dioxide – $\Sigma(NO, NO_2)$ – are generally described as nitrogen oxides, shortly "NO_x".

Nitrogen monoxide is a colorless gas which does not cause irritations and which slowly oxidizes with ambient oxygen to form nitrogen dioxide (NO_2). Nitrogen dioxide is a brown, irritant gas which has the capacity to split dual compounds into alkenes. In particular, nitrogen monoxide is formed in cells after injuries and acts as a messenger substance.

1.3.1.5
Particulate Matter (PM)

The effect of particulates on human beings has been documented by numerous studies which are partly contradictory. Of all particulate emissions caused by vehicles due to the abrasion of road surfaces, tires, brake and clutch linings, wear of the exhaust manifold section and particulate matter generated during internal combustion processes of engines, it is the latter category of emissions on which the public debate is focused. The potentially toxicological effect of these particulates which are generated in most diverse processes results from the combination of specific chemical properties and the generally damaging effect that pollutant particles may have when they reach the lungs. The capacity to reach the lungs which is determined by particulate size, is one of the most important criteria for assessing the potential dangers of particulates.

1.3.2
Ozone Formation Potential

The interaction of various components of exhaust emissions and sunlight leads to a further chemical decomposition of the latter, thus generating, among other substances, ozone. This is why for quite some time now exhaust emissions analyses have also focused on their ozone formation potential.

High ozone concentrations are undesirable because they cause irritations in the respiratory tract and penetrate plants thus damaging them. The exhaust gases of internal combustion engines contain only tiny traces of ozone (O_3). However, chemical processes in the air triggered by solar irradiation give rise to so-called ozone-precursor substances which, in turn, lead to the formation of ozone. These ozone-precursor substances include volatile organic gases, such as hydrocarbons (HC), nitrogen oxides (NO_x) and carbon monoxide (CO).

In the relevant literature, the ozone formation potential is described by means of so-called MIR factors (maximum incremental reactivity), which takes into account the particular photochemical response of the individual hydrocarbons. By adding the MIR factors and by multiplying the sum thus obtained with the individual hydrocarbon emissions, the "ozone formation potential" (OFP) is determined.

$$OFP = \sum [NMOG_i \cdot MIR_i] \qquad (a)$$

- MIR ("Maximum Incremental Reactivity") [g O_3/g NMOG],
- $NMOG_i$ ("Non-Methane Organic Gases") [g/km].

For calculating the ozone formation potential of a vehicle, however, the relevant technical literature refers to the specific reactivity factor (SR), which indicates the potential in grams of O_3 per gram of NMOG. The ozone formation potential (indicated in g O_3/km) is determined by multiplying:

- SR (the specific reactivity) [g O_3/g NMOG] with
- $NMOG_i$ (the non-methane organic gases content) in g/km.

$$OFP = SR \cdot \sum [NMOG_i] \qquad (b)$$

1.3.3 *Greenhouse Gases*

In the mid-1980s, attention of the general public was drawn to trace gases which, in principle, have no harmful effect on human health:

- Carbon dioxide (CO_2),
- Methane (CH_4),
- Dinitrogen monoxide (nitrous gas or N_2O) and
- fully and partly halogenated hydrocarbons (HCH).

These trace gases lead to a change in the natural concentration of substances in the earth's atmosphere thus contributing to the greenhouse effect.

On the one hand, the capacity of CO_2 to trigger the greenhouse effect represents a vital property as without it the earth would have an average surface temperature of −18 °C and hence would be uninhabitable. On the other hand, this specific capacity has been the subject of a critical study of CO_2 because the CO_2 concentration has been steadily rising by 0.5% per annum since the middle of the 20th century. The resulting temperature increase which is ascribed to emissions of trace gases through human activities is commonly referred to as the "anthropogenic greenhouse effect". In an effort to bring this greenhouse effect under control, international conferences were convened which resolved that the concentration of these trace gases as well as that of sulfur hexafluoride was to be reduced.

In order to be able to assess the relative significance of the individual greenhouse gases, so-called CO_2 equivalents are applied [2]. Firstly, however, these CO_2

equivalents are not uniformly defined and not very well founded scientifically due to the complexity of the processes involved, and secondly, the order of magnitude of these values at least is not a controversial issue. When comparing the CO_2 equivalents, the varying lifetimes of individual compounds in the atmosphere must be taken into account. Hence the CO_2 equivalents depend very much on the reference period assumed for the greenhouse effect.

The CO_2 equivalent based on a 20 year reference period constitutes an indicator for short-term changes, whereas the CO_2 equivalent based on a reference period of 100 years is an indicator for long-term changes (Table 1).

Table 1

Component	Concentration in the atmosphere	Lifetime (in years)	CO_2-equivalent (reference period 20 year)	CO_2-equivalent (reference period 100 years)
H_2O	2 ppm - 2%	0.01–2	?	?
CO_2	369 ppm [2000]	50–200	1	1
N_2O	316 ppb [2000]	120–150	270–290	290–320
O_3 (tropospheric)	20 ppb [1993]	0.1–0.2	2000	2000
CH_4	1753 ppb [2000]	7–12.3	56-63	21–24
SF_6	4 ppt [1999]	3200	16500	24900
CKW				
CCl_4	97 ppt [2000]	42–50	1900	1300
CH_3CCl_3	45 ppt [2000]	6	350	100
FCKW				
11	266 ppt [1999]	50–60	4500-5000	3500–4000
12	545 ppt [2000]	102–130	7100–7900	7300–8500
113	81 ppt [2000]	85–90	4500–5000	4200–5000
114	20 ppt [1992]	200–300	6000–6900	6900–9300
115	<10 ppt [1992]	400–1700	5500–6200	6900
HFCKW				
22	119 ppt [1995]	13.3–15	4100–4300	1500–1700
123	not available at present	1.4–1.6	300–310	85–93
124	not available at present	5.9–6.6	1500	430–480
141b	13 ppt [2000]	8–9.4	1500–1600	440–630
142b	12 ppt [2000]	19–19.5	3700–4200	1600–2000
HFKW				
125	not available at present	28–36	4600–4700	2500–2800
134a	14 ppt [2000]	14–16	3200–3400	1200–1300
143a	not available at present	41–48.3	4500–5000	2900–3800
152a	not available at present	1.5–1.7	460–510	140
Halone				
1211	3 ppt [1996]	20	?	?
1301	2 ppt [1996]	65–110	5800	5800

ppm: parts per million; ppb: parts per billion; ppt: parts per trillion.

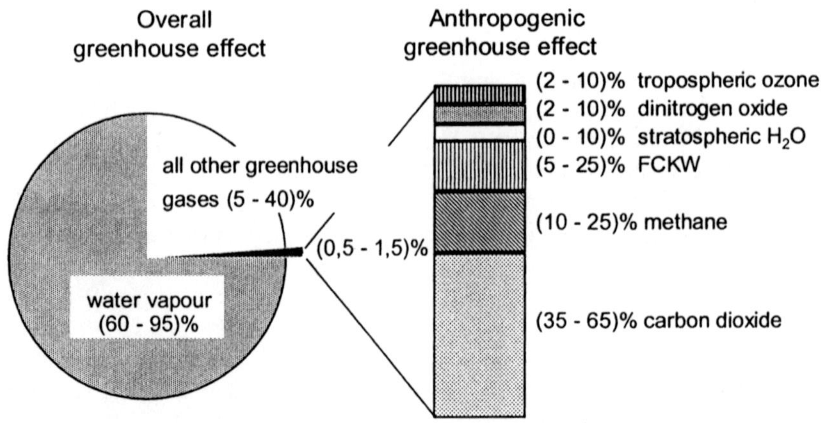

Fig. 5. Share of greenhouse gases in the anthropogenic temperature increases as compared to the shares of natural greenhouse [4, 7–15]

Estimates of the contribution of individual gases to the greenhouse effect vary significantly in the relevant technical literature. Water vapor constitutes by far the most important natural greenhouse gas (Fig. 5).

As can be seen in Fig. 5 on the right, opinions are divided as regards the contribution of the individual greenhouse gases to the additional anthropogenic greenhouse effect which makes up between 0.5% and 1.5% of the overall greenhouse effect. However, in all scientific publications carbon dioxide from anthropogenic sources is considered as the main contributor to the anthropogenic greenhouse effect.

1.4 Global Emissions of Greenhouse Gases by Groups of Emission Sources

1.4.1 *Carbon Dioxide (CO_2)*

The ratio between the emissions from one source and those from natural or other anthropogenic sources serves as an important indicator of the degree to which their impact on the environment can be reduced. Before dealing with differences between emissions caused by passenger transport and goods traffic, we should like to present a survey of estimated global CO_2 emissions.

It is primarily the combustion of fossil fuels which represents a major source of anthropogenic CO_2 emissions. The natural sources include the decomposition of organic materials in the soil through the action of microbes, the weathering of rocks and the release of CO_2 by the oceans. The CO_2 emissions caused by breathing, disintegration and the exchange of substances between the oceans and the atmosphere are one order of magnitude above those from all anthropogenic sources taken together, which means that emissions from natural sources by far exceed those caused by man (Fig. 6).

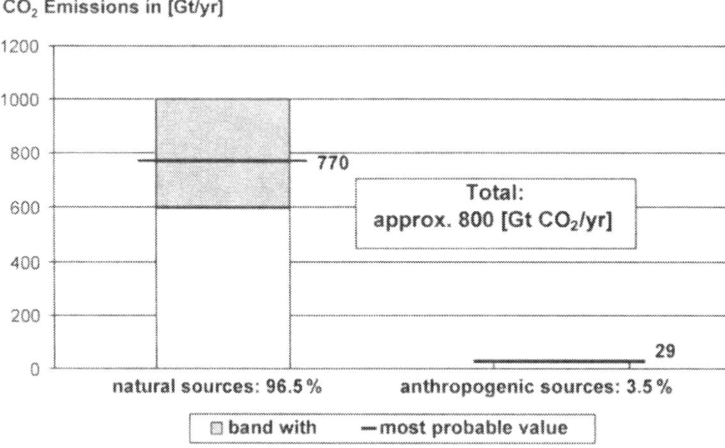

Fig. 6. Breakdown of global carbon dioxide emissions per annum (reference year 2000) [7, 15–23, 81, 82, 86]

Despite all efforts made by many industrialized countries, global anthropogenic CO_2 emissions will continue to rise. Countries in Asia and Oceania, as well as Central and Latin American will record the sharpest increase in CO_2 emissions over the next few years because their rapid economic growth will be accompanied by rising energy consumption rates. The shares of North America and Europe in global CO_2 emissions will decline further in the future (Figs. 7 and 8).

Fig. 9 gives a survey of the percentage shares of global natural and anthropogenic CO_2 emissions by sources. Road traffic makes up a mere 11.5% of global anthropogenic emissions with passenger cars contributing only 5.5% to global man-made emissions. Power plants are the largest sources of emissions, followed by domestic heating and industry.

Fig. 10 shows the trend of global anthropogenic carbon dioxide emissions over time. It is worth noting that passenger cars and road traffic make up a relatively small share of total CO_2 emissions.

1.4.2
Methane (CH_4)

Figure 11 illustrates the contributions of different sources to global methane emissions. Anthropogenic methane emissions which are mainly caused by the growing of rice in wet zones and the breeding of ruminants, exceed those from natural sources. Passenger cars and commercial vehicles, which account for 0.3% and 0.2% of total anthropogenic methane emissions respectively, contribute only insignificantly to these.

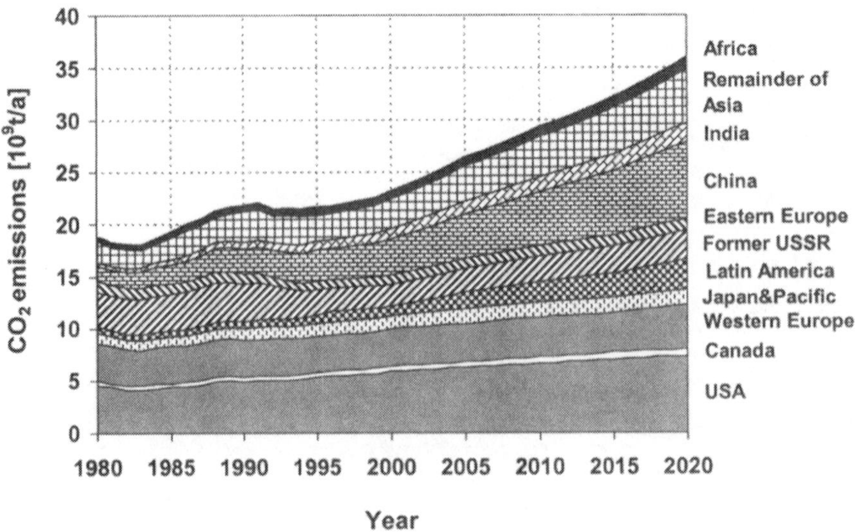

Fig. 7. Trends and forecasts of global CO_2 emissions from combustion of fossil fuels 1980–2020 [16, 17, 24–29, 81, 82, 86]

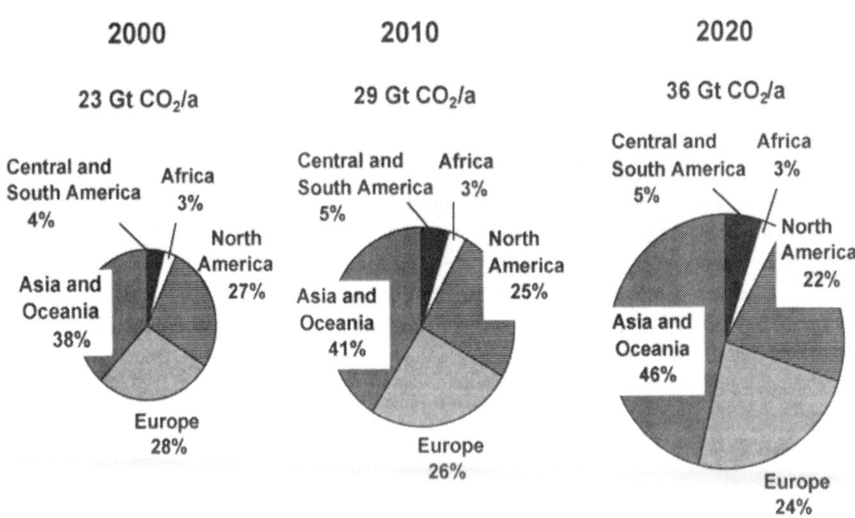

Fig. 8. Trends and forecasts for the contribution of individual continents to carbon dioxide (CO_2) emissions from fossil fuel burning [16, 17, 24–29, 81, 82, 86]

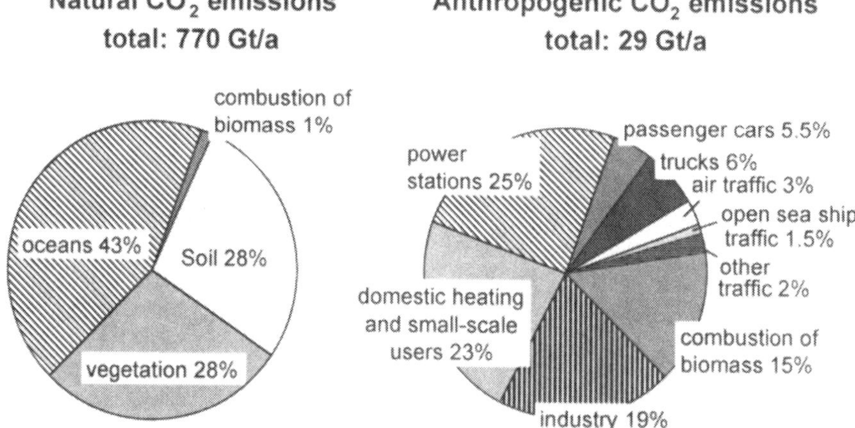

Fig. 9. Breakdown by source of the most probable values for global carbon dioxide emissions (reference year 2000) [7, 15–23, 29, 81, 82, 86]

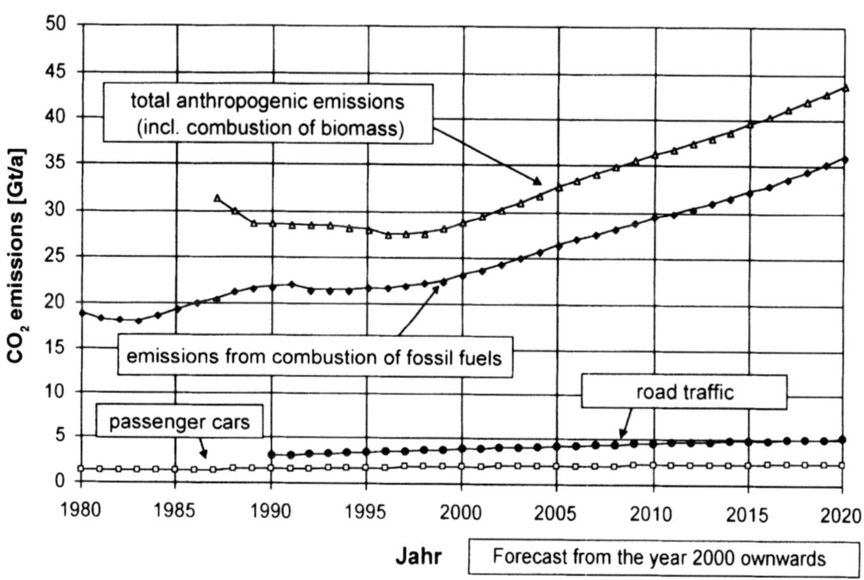

Fig. 10. Development of global anthropogenic carbon dioxide (CO_2) emissions over time (forecast from the year 2000 onwards) [7, 17, 21, 29–36, 81, 82, 86]

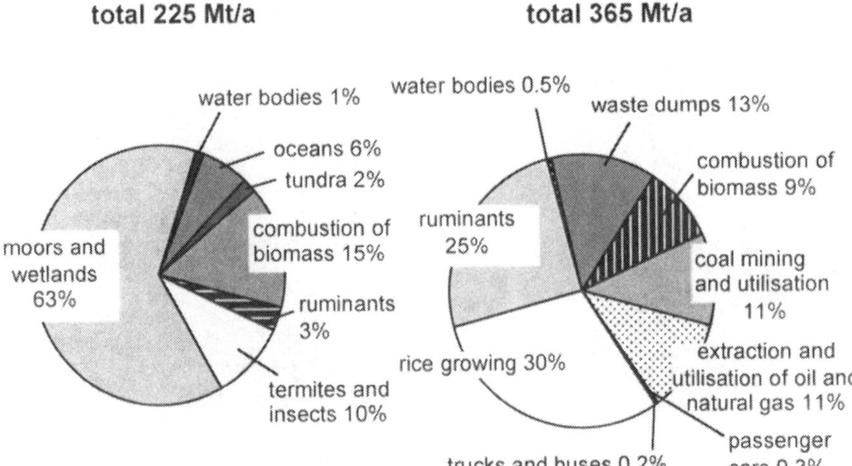

Fig. 11. Breakdown by source of the most probable values for methane (CH_4) emissions (reference year 2000) [18, 31, 34, 35, 37–47, 59, 86, 87, 88]

1.4.3
Halogenated Hydrocarbons (HCH)

Organic compounds containing such elements as fluoride, chloride, bromide, iodine, and belonging to the group of salt-forming substances (halogens) are defined as halogenated hydrocarbons [48]. With the exception of methyl chloride (CH_3Cl) which is released by the oceans and during the combustion of biomass, and partly also of methyl bromide (CH_3Br) released by marine algae, halogenated hydrocarbons are emitted exclusively by anthropogenic sources. They serve as solvents, filling agents for refrigeration units and heat pumps, as a dielectric in transformers, propellants, fire extinguishing and cleaning agents, as pesticides and medical drugs.

In contrast to the procedure applied throughout this study, no detailed quantitative values can be indicated for the individual groups of emission sources as the available data are not sufficient for this purpose. Since 1991 the automotive industry has supported efforts aimed at the rapid discontinuation of the use of chlorofluorocarbons (CFCs). In the production of foamed parts, for example, the automotive industry changed over to CFC-free foaming agents, and for filling vehicle air conditioning systems manufacturers switched to the halogenated fluorocarbon 134a, which is kinder to the environment. The reduction of CFC emissions is reflected by the lower immission levels measured, as is shown for CFC-11 (Fig. 12). As a result of the long retention time of gases in the atmosphere, the concentration of immissions did not decline immediately once emissions were reduced, but has remained rather constant for a while and it will now start to decline in the next few years.

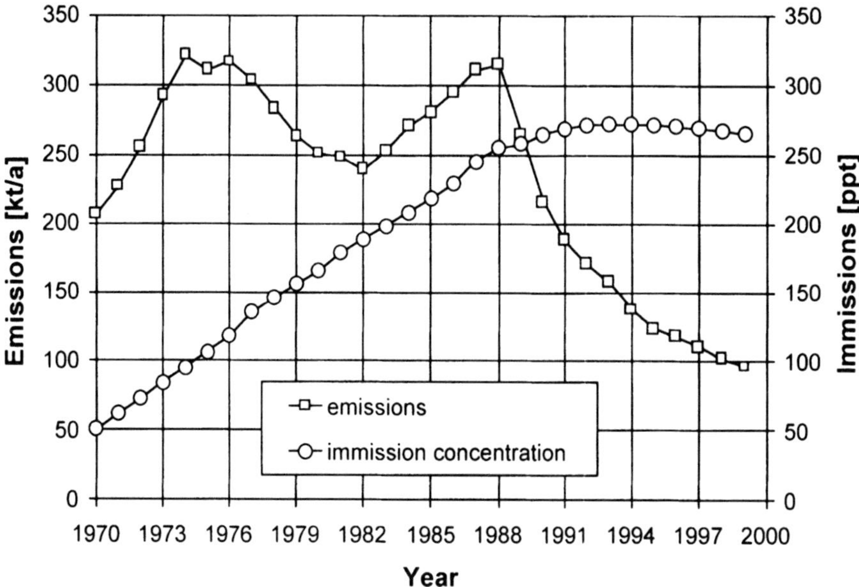

Fig. 12. Development of global CFC 11 emissions and immission concentrations over time [49, 50]

1.4.4
Dinitrogen Monoxide (N_2O)

This nitrogen compound which is also referred to as nitrous gas belongs to the group of nitrogen oxides. Dinitrogen monoxide (nitrous gas) is a colorless, odorless gas which dissolves easily in water. Not until as late as 1938 was it possible to prove the existence of nitrous gas in the atmosphere by measuring its infrared absorptions at 7.57 μm and 7.77 μm.

Owing to its long retention time in the atmosphere, N_2O is very homogeneously distributed in the global troposphere. Although there are only minor annual fluctuations, no regular pattern of yearly concentrations of tropospheric N_2O has been observed.

The shares of the different N_2O sources in global emissions are graphically represented in Fig. 13.

Oceans represent the main sources of natural emissions, followed by tropical forests. Agriculture constitutes the main source of anthropogenic N_2O emissions.

Anthropogenic N_2O emissions make up approximately 50% of natural emissions, with the utilization of fertilizers being mainly responsible for anthropogenic nitrous gas emissions, followed by fire clearing of tropical forests and by industry.

Excessive quantities of fertilizers which are not washed out normally act as a source of N_2O emissions as these are decomposed through denitrification. The general rule is that all combustion processes may act as sources of N_2O emissions. Furthermore, N_2O escapes during its preparation and use and is also generated by waste recycling.

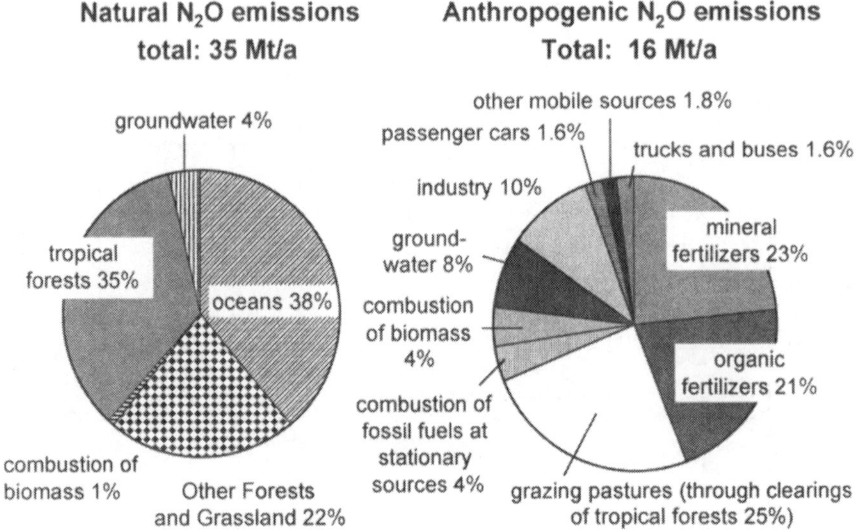

Fig. 13. Breakdown of global nitrous gas (N_2O) emissions by sources (most probable values) (reference year 2000) [18, 51–56, 84]

The production and use of nitrous gas for anesthesia has been included in the emission category "industry". Passenger cars and commercial vehicles contribute 1.6% each to the global emissions of anthropogenic nitrous gas and thus play an only insignificant role.

Although road traffic contributes only insignificantly to nitrous gas emissions, increased attention has recently been drawn to the growing number of passenger cars with gasoline engines and controlled three-way catalysts, which emit nitrous gas as an undesirable by-product of NO_x conversion. Given today's state of the art, N_2O formation mechanisms operating in the three-way catalyst are largely known. It is to be expected that improved catalyst technologies and corresponding control measures for engine operation will result in a reduction of nitrous gas emission levels by gasoline-engine powered passenger cars [57].

1.5 Emissions and Immission – A Comparison for Limited Pollutants Between the EU and Germany

1.5.1 *Carbon Monoxide (CO)*

1.5.1.1 European Union

As regards CO emissions in the European Union the greatest uncertainties exist with regard to the contribution of small-scale users and households. The burn-

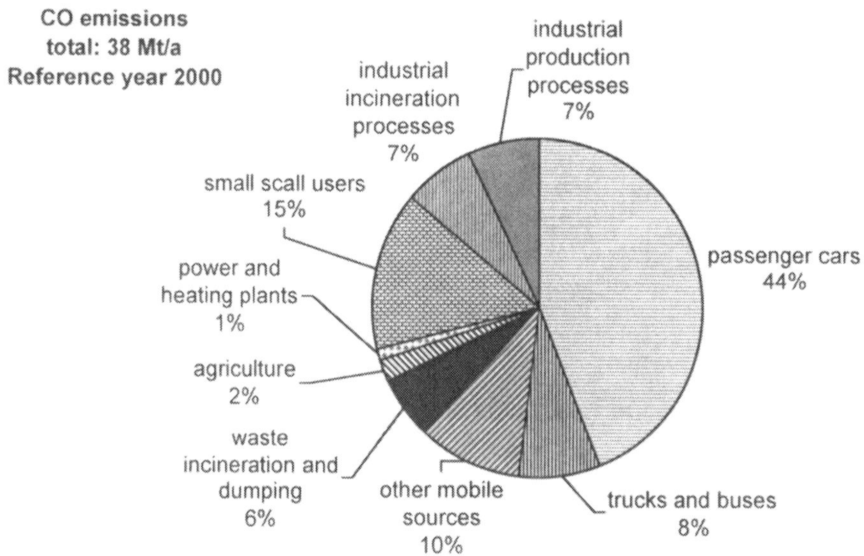

Fig. 14. Contribution of different sources to anthropogenic carbon monoxide emissions (CO) in Europe (EU 15) [58, 80]

ing of wood accounts for a certain share of total CO emissions. Emission factors for small-scale incineration plants, for example, especially when these are wood-fired, vary by two orders of magnitude and even the data on the quantities of fuel wood consumed are anything but uniform. Figure 14 illustrates the contribution of different emission sources to total CO emissions in the EU.

1.5.1.2
Germany

Since 1972, carbon monoxide emissions have declined continuously. This decline can be ascribed both to statutory provisions governing exhaust gas emissions from road traffic and the conversion of small incineration plants to liquid and gaseous fuels (Fig. 15).

Passenger car traffic continues to be the largest source of carbon monoxide emissions in Germany, accounting for roughly 40% of total CO emissions. However, emissions from road traffic will decline further as vehicles without catalysts will be replaced by cars with controlled three-way catalysts and low-pollutant Diesel engines. Investigations conducted by the Vienna University of Technology showed that the contribution of off-road vehicles to total emissions is three times higher than had originally been assumed. Accordingly, off-road vehicles in Germany accounted for 15% of total carbon monoxide emissions in 2000.

Emissions from off-road vehicles come from, e.g., lawnmowers, construction machinery, railway vehicles, combustion-engine powered portable machines, inland navigation and emissions from military vehicles.

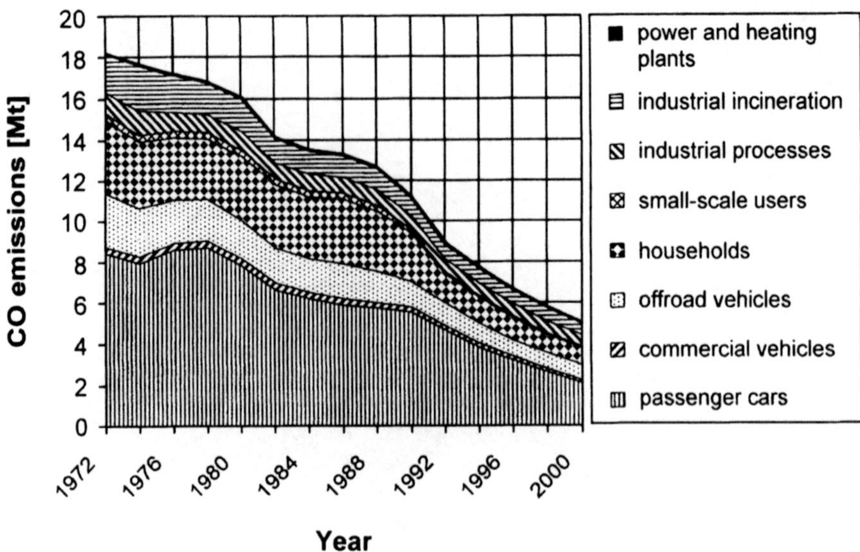

Fig. 15. Trend of anthropogenic carbon monoxide emissions in Germany for the period from 1970–2000 [16, 60, 80]

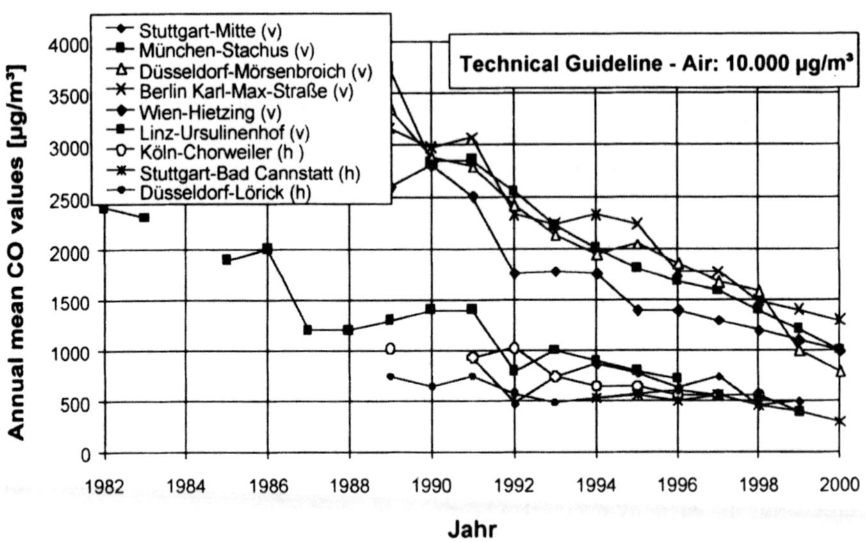

Fig. 16. Trend of mean yearly CO concentrations in the atmosphere at roadside measuring stations (v) and urban background stations (h) in Germany and Austria [61–63, 72, 78, 79]

Figure 16 illustrates carbon monoxide concentrations in the atmosphere at roadside measuring stations and urban background stations in Germany and Austria. The immission concentrations measured at the roadside stations reflect a marked reduction of emissions from road traffic. In the period under review, immission concentrations were lowered by roughly 50% and hence are currently more than 80% below the limit value set forth in the Technical Guideline for Air Quality. The reduction of emissions from domestic heating systems is mirrored by the immission concentrations measured at urban background stations.

The European Union intends to impose the limit value of 10 mg/m³ (maximum mean value within a time-span of eight hours on one day) which was fixed by the WHO, from the year 2005 onwards [84].

1.5.2
Non-Methane Hydrocarbon Emissions (NMHC)

The term "non-methane hydrocarbon emissions" comprises a wide variety of compounds of different constituents. It generally denotes all higher-ordered hydrocarbon compounds other than those containing methane.

1.5.2.1
The European Union

Non-methane hydrocarbon emissions from anthropogenic sources follow the general trend, which showed a peak in 1989 and have declined since then. In the reference year 2000, the largest quantities of man-made NMHC emissions in the European Union resulted from the use of solvents (32%), followed by exhaust gas emissions from passenger cars (13%) (see Fig. 17).

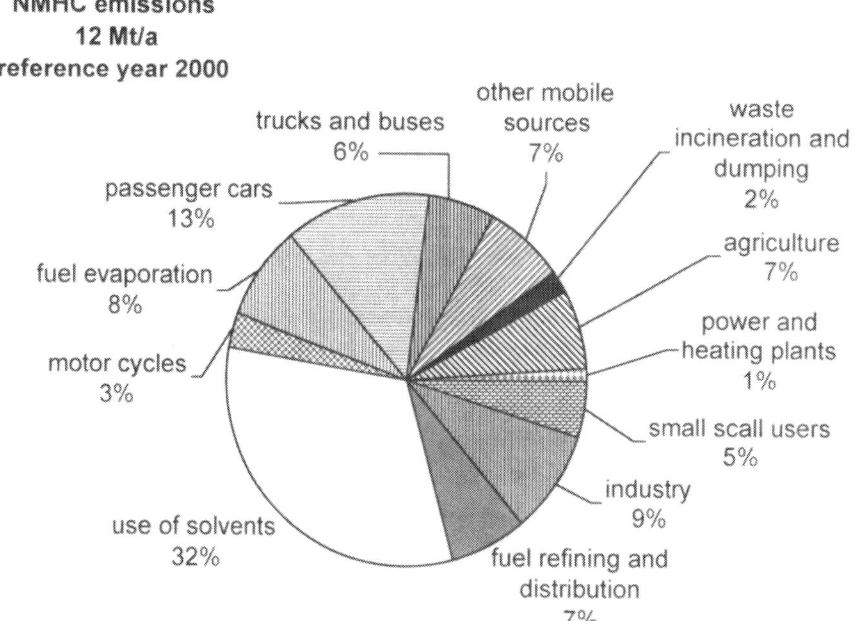

Fig. 17. Contribution of different sources to yearly emissions of volatile non-methane hydrocarbons (NMHC) in Europe (EU 15) [18, 58, 64, 65, 83]

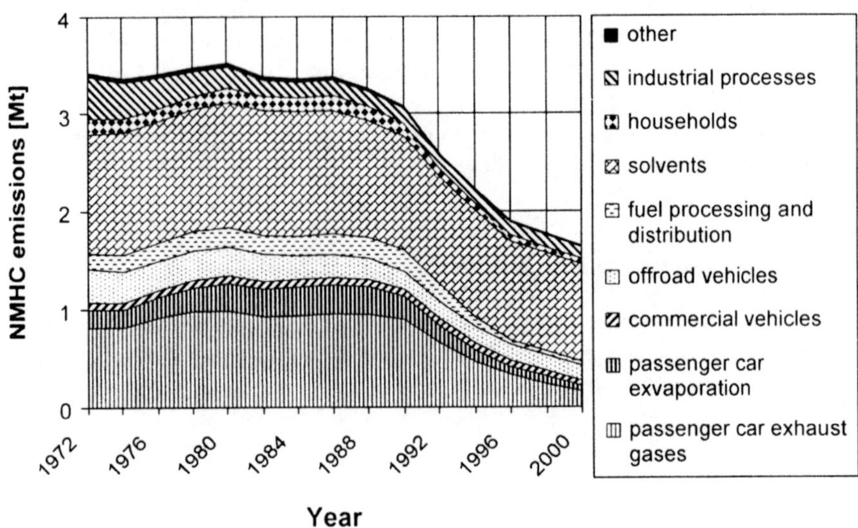

Fig. 18. Development of anthropogenic non-methane hydrocarbon emissions in Germany in the period from 1970 to 2000 [16, 60, 80]

1.5.2.2 *Germany*

Germany has in the meantime met the obligation it assumed when it signed the UNECE Protocol to bring down emissions of methane-free volatile organic compounds by 30% below the 1988 level by the year 1999. The 42% reduction of these emissions in the period from 1990 to 1998 was mainly achieved in road traffic through statutory exhaust as limit values and the replacement of vehicles with 2-stroke engines by vehicles with state-of-the-art drive technology in the new German states. Statutory provisions imposing limit values on emissions from fuel distribution also resulted in a lowering of overall emissions. Figure 18 illustrates the development over time of anthropogenic non-methane hydrocarbon emissions in Germany.

No limit or reference values were fixed for the mixture of non-methane hydrocarbons for assessing air quality. Since the introduction of the lambda-controlled catalyst for gasoline-engine passenger cars a significant improvement of air quality has been observed at the road measuring stations. Within a period of nine years, the concentration of immissions at roadside measuring stations could be brought down to one-quarter of the concentrations originally observed (Fig. 19).

1.5.3 *Nitrogen Oxide (NO_x)*

Nitrogen oxides is the generic term for nine different nitrogen compounds. Of these, only nitrogen monoxide (NO) and nitrogen dioxide (NO_2) which are

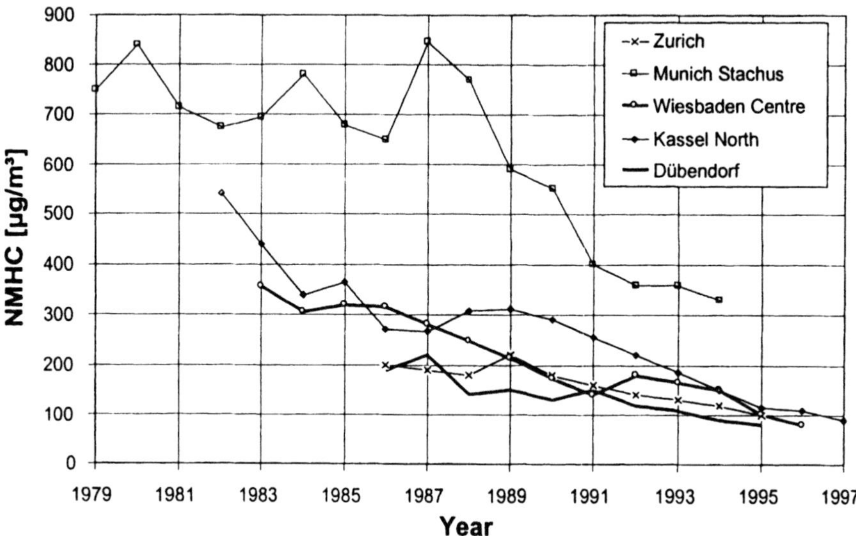

Fig. 19. Development of mean annual values of on-methane hydrocarbon (NMHC) emissions at measuring stations close to urban traffic in Germany and Switzerland [62, 66, 67]

relatively stable at normal temperatures are of practical significance. In practice, the term nitrogen oxides is generally applied to total nitrogen monoxide and nitrogen dioxide emissions.

1.5.3.1
European Union

In Europe, nitrogen oxide emissions peaked in 1989. Within the European Union, traffic accounts for approximately 42% of total nitrogen oxide emissions (Fig. 20), and the share of agriculture is estimated to make up 25% (calculated in accordance with [69]). However, this estimate is based on a number of uncertainties and could, in reality, amount to as much as 35% [18].

1.5.3.2
Germany

Figure 21 illustrates the development of anthropogenic sources of nitrogen oxide emissions in Germany.
Data on agriculture, as indicated in Fig. 20, are not available in Germany.
Under the UNECE Protocol on the Reduction of Nitrogen Oxide Emissions, Germany committed itself to bring down these emissions to the 1987 level by the year 1994. Within this period, emissions were reduced by roughly 40%. The additional commitment voluntarily made by Germany and the 11 EEC Member States, going beyond this Protocol (reduction of NO_x emissions from the 1986 level by 30% by the year 1998) resulted in a reduction of emissions by as much as 50%.

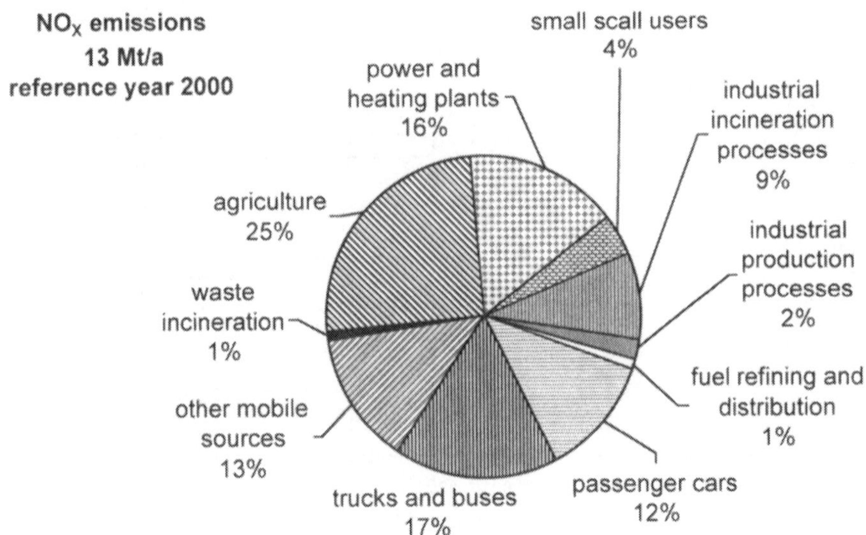

Fig. 20. Shares of different sources of anthropogenic nitrogen oxide (NO_x) emissions within the EU (EU 15) [16, 18, 58, 64, 65, 68, 69]

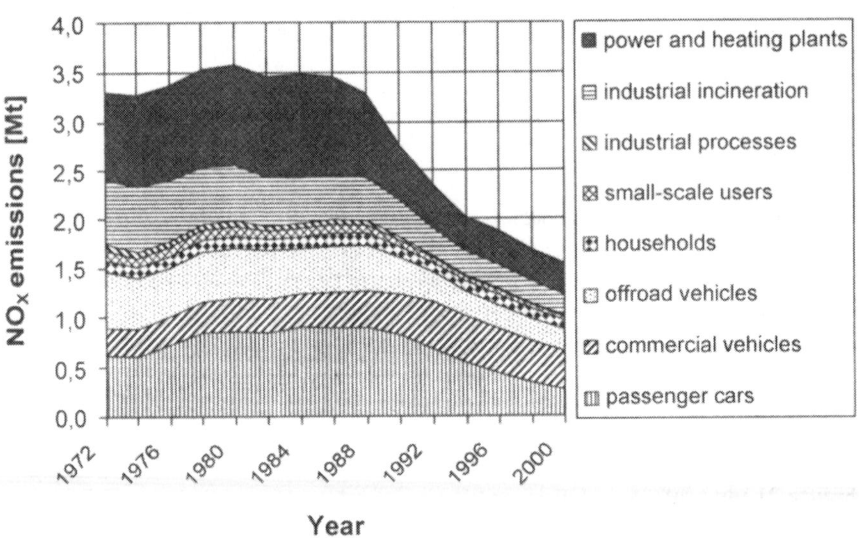

Fig. 21. Development of anthropogenic nitrogen oxide emissions in Germany [16, 60, 80]

Fig. 22 gives a survey of the development of nitrogen oxide concentrations in the air, expressed as mean values annually calculated from the readings at selected measuring stations in Germany and Austria. Immission concentrations have followed a downtrend both at the urban background stations and roadside stations.

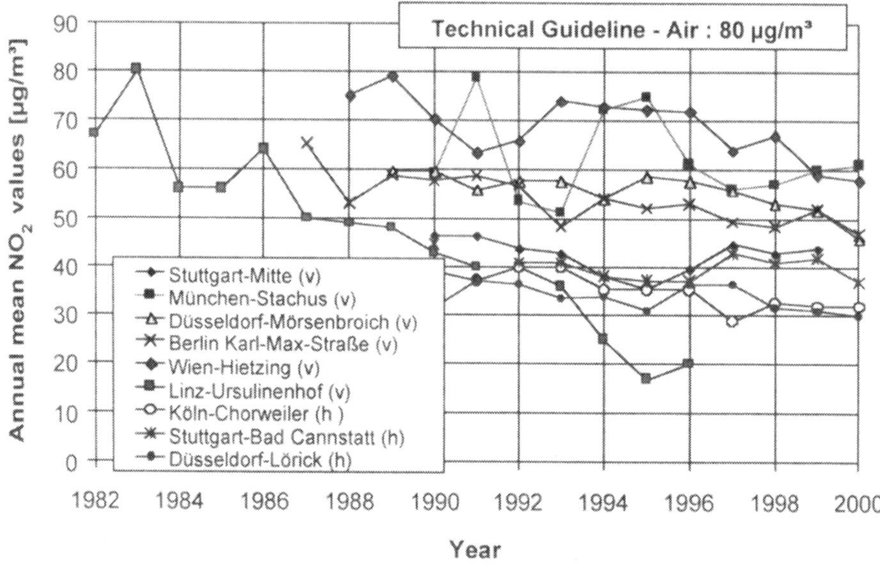

Fig. 22. Trend of mean yearly NO_2 concentrations in the atmosphere at roadside (v) and urban background stations (h) in Germany and Austria [61–63, 72, 78, 79]

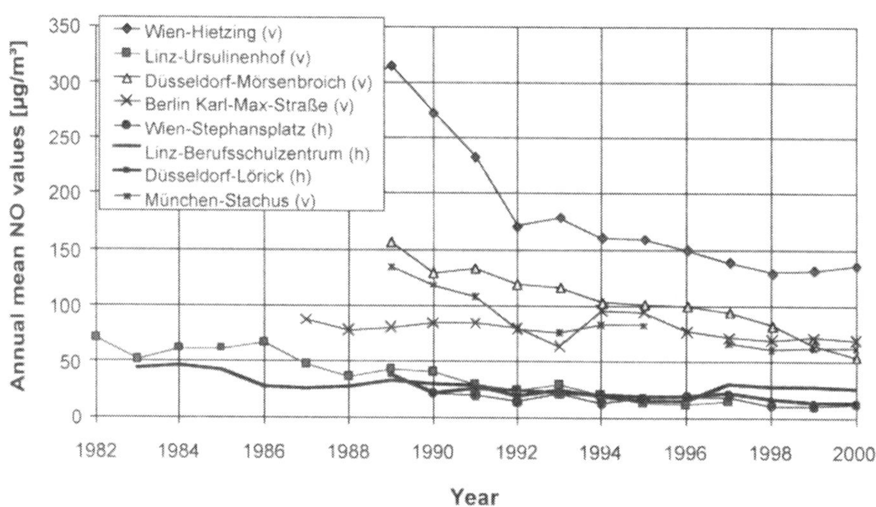

Fig. 23. Trend of mean yearly NO concentrations in the atmosphere at roadside (v) and urban background stations in Germany and Austria [61–63, 72, 78, 79]

However, concentrations below the guideline value of 40 µg/m³ fixed by the World Health Organization WHO as the threshold for the protection of human health are recorded at only few measuring stations, but all measuring stations considered reported values which were significantly lower than the limit value of 80 µg/m³ specified in the Technical Guideline-Air.

Figure 23 shows the curves of nitrogen monoxide concentrations for different years. With this component, a significant downtrend of immission concentrations can also be observed.

1.5.4
Particulates (PM)

The terms "airborne dust" or "particulate matter" (PM) refer to airborne particles that may either be generated or emitted directly by natural or anthropogenic sources or that are produced indirectly by gaseous precursor substances and may also be of natural or anthropogenic origin.

1.5.4.1
European Union

Available literature did not allow the drawing up of an inventory of total airborne emissions in Europe. In accordance with [70] PM_{10}-, $PM_{2.5}$- and $PM_{0.1}$-emissions in the European Union were estimated as can be seen from Fig. 24–26. As combustion processes produce primarily small particles, the emission balance shows relatively high percentages of emissions from such sources. As the handling of bulk materials results primarily in the release of particulates that are greater than 10 µm this source is negligible for the PM_{10}-emission balance and the balances for smaller particles. Emissions from agriculture cannot be reliably assessed and may therefore have been over-estimated in the figure.

Particulate emission inventories reveal, in particular, that the contribution of road traffic to emissions increases in relative terms as particulate size diminishes. In practice this means that traffic accounts for a share of 21% of total PM_{10}-emissions, and for a share of 50% of $PM_{0.1}$-emissions.

1.5.4.2
Germany

Figure 27 gives a survey of the development of anthropogenic dust emissions in Germany. Emissions from passenger car and commercial vehicle traffic also include emissions resulting from the wear on tires, brake linings and road surfaces.

The decline in airborne dust emissions by approximately 80% in the period from 1990 to 2000 can be attributed primarily to efforts made to bring down emission levels in the "new" German states, where a large number of outdated incineration facilities and industrial plants were closed and where the efficiency of existing firing and dedusting units in power stations and remote heating plants was improved within a short time. Conversion from solid to low-emission liquid

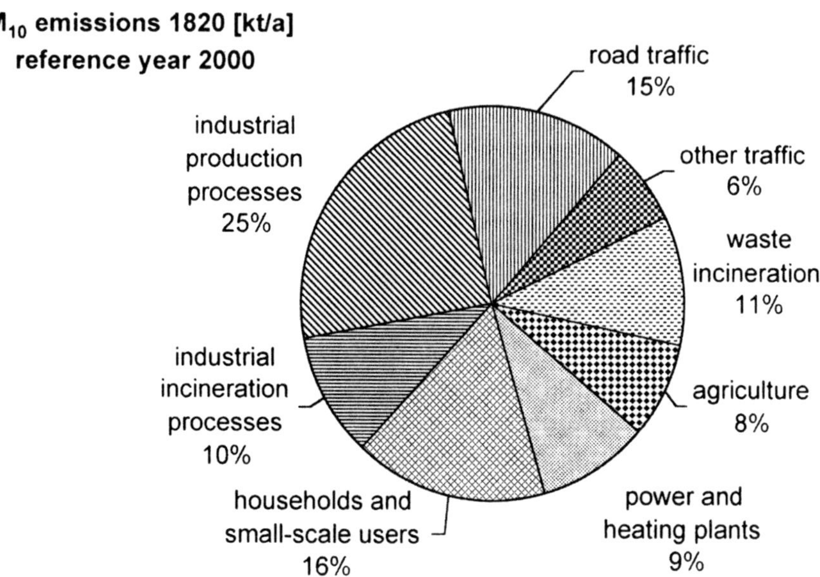

Fig. 24. Anthropogenic PM_{10} emissions in the European Union broken down by different sources [35, 60, 70, 71]

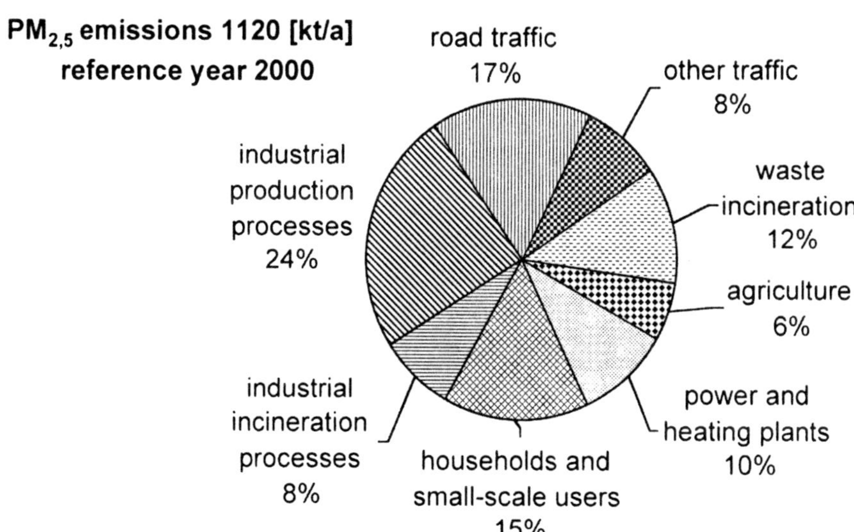

Fig. 25. Anthropogenic $PM_{2.5}$ emissions in the European Union, broken down by different sources [35, 60, 70, 71]

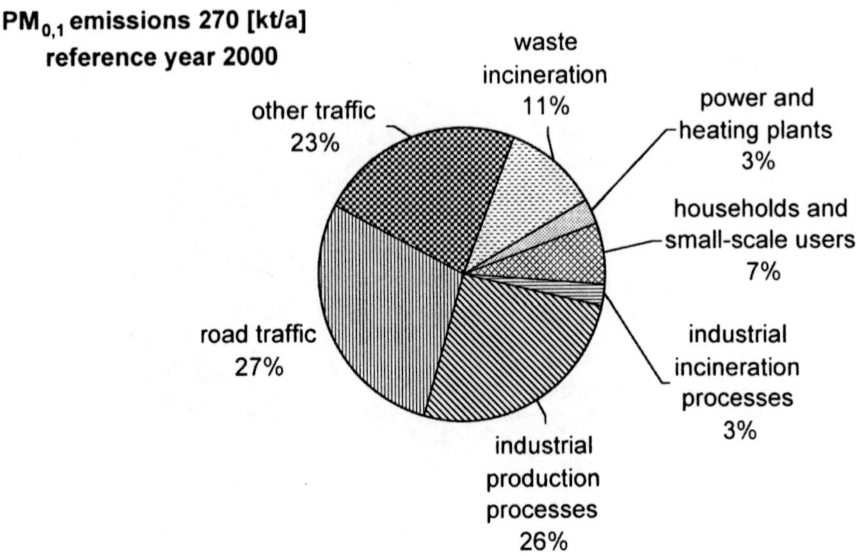

Fig. 26. Anthropogenic $PM_{0.1}$ emissions in the European Union, broken down by source [35, 60, 70, 71]

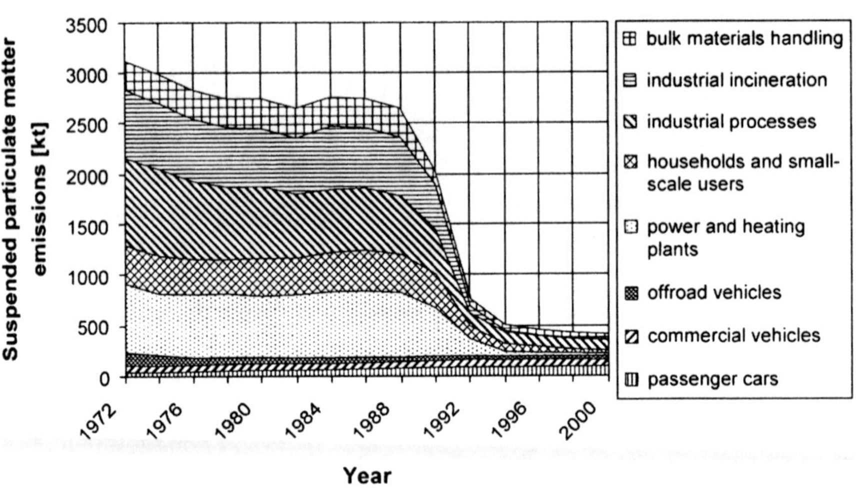

Fig. 27. Trend for suspended particulate matter emissions in Germany (including particulate matter resulting from wear in road traffic) [16, 60, 80]

and gaseous fuels, especially in small incineration plants, also had a positive effect on the development of airborne dust emissions.

Figure 28 illustrates dust emissions at selected roadside stations and urban background stations in Germany and in Austria. The emissions monitored at all of these stations remained significantly below the limit value of 150 μg/m³ defined in the Technical Guideline for Air Quality.

The European Union intends to impose the limit value of 40 μg/m³ by the year 2005 and 20 μg/m³ by the year 2010 (mean yearly values) for PM_{10} which were fixed by the WHO [85].

With regard to road traffic, this study focuses mainly on particulate emissions from Diesel-engine powered passenger cars and commercial vehicles. In everyday usage, these emissions are wrongly referred to as soot emissions although soot represents only one component of the particulate matter.

In technically correct language, soot refers to the elementary carbon components of airborne dust particles. Figure 29 gives a survey of the trend of soot immission concentrations in the air at different roadside measuring stations in two German cities. For reference, the limit value set forth in the 23rd Regulation to the Federal Act on Protection against Immissions, which has been in force since July 1, 1998, is also indicated. At all measuring stations, the yearly mean values have been shown to follow a downward trend; since 1998, they have remained below the currently applicable maximum permissible value.

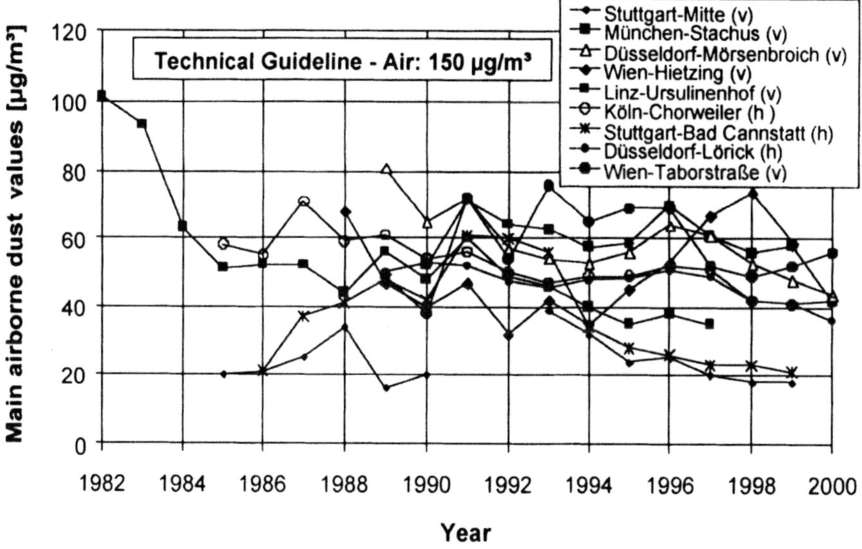

Fig. 28. Trend of mean yearly airborne dust concentration at roadside measuring stations (v) and urban background measuring stations (h) in Germany and in Austria [61–63, 72, 78, 79]

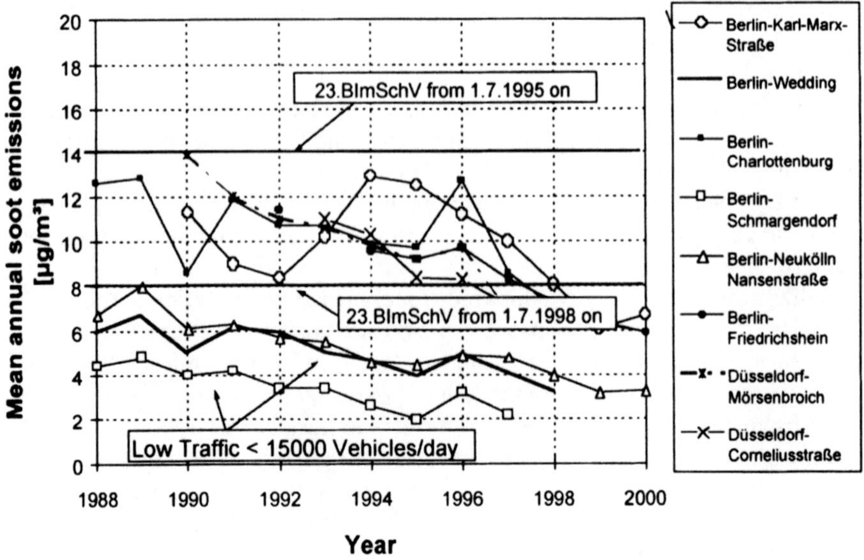

Fig. 29. Trend of soot immission concentrations in the air at roadside measuring stations in Germany [78, 79]

1.6 Environmental Impacts of Passenger Car Traffic

In practice, the emission behavior of passenger cars and commercial vehicles is not only determined by specific vehicle characteristics which partly result from legal requirements, but is also influenced to a significant extent by the driving style of the vehicle user.

A number of parameters need to be taken into account when analyzing emissions from passenger cars and commercial vehicles. These parameters include not only legislative measures relating to maximum permissible emission values of vehicles, fuel quality, or the inspection of vehicles in use but also vehicle user habits.

The main problem associated with the calculation of emissions, which in essence, consists in determining vehicle populations and total kilometers driven and in correlating these data with emission factors, is the proper selection of these input parameters and the forecasting of future developments on this basis.

1.6.1 *Influence of Service Time on Emissions from Passenger Cars*

Thanks to the database of the Department of Internal Combustion Engines and Automotive Engineering at the Technical University of Vienna [57] it was possible to calculate the correlation between exhaust gas emissions and the year of construction and mileage driven of vehicles in use. Figs. 30–32 show the inter-

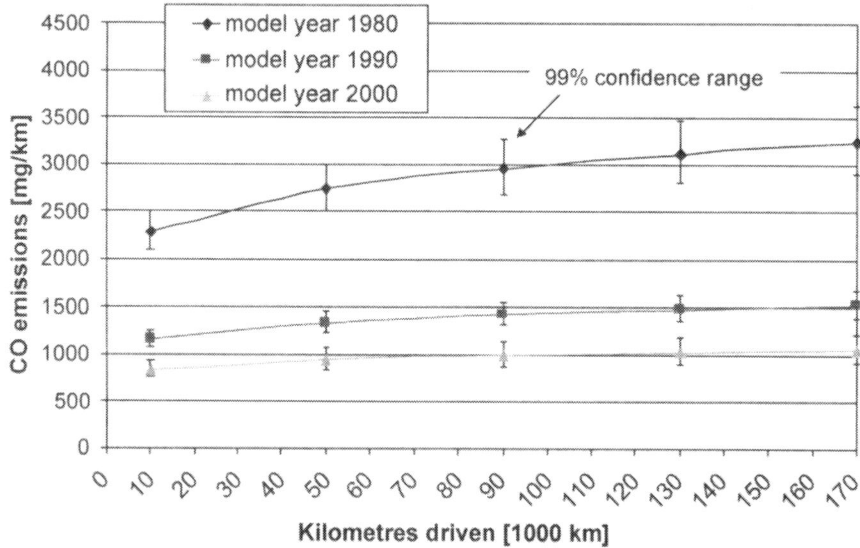

Fig. 30. Impact of year of construction and kilometres driven on CO emissions of gasoline-engine-powered passenger cars with controlled three-way catalysts in the applied ignition test 75 cycle

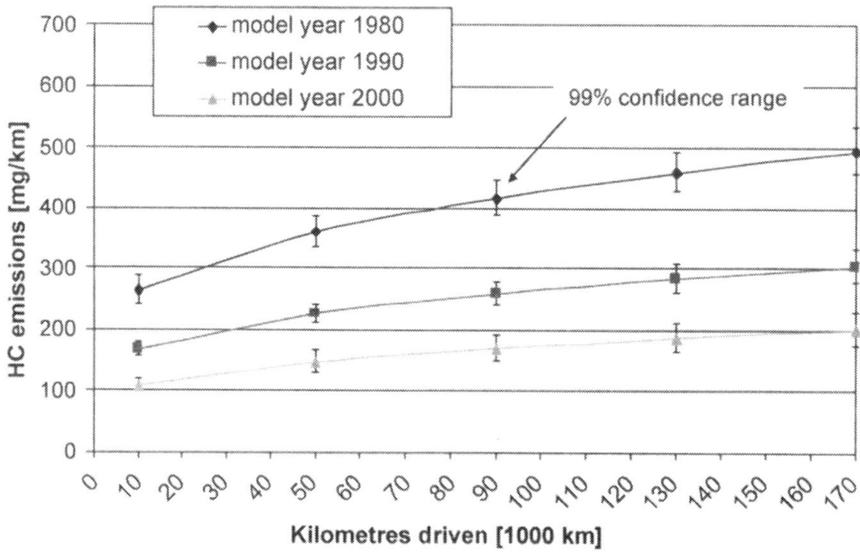

Fig. 31. Impact of year of construction and kilometres driven on HC emissions of gasoline-engine-powered passenger cars with controlled three-way catalysts in the applied ignition test 75 cycle

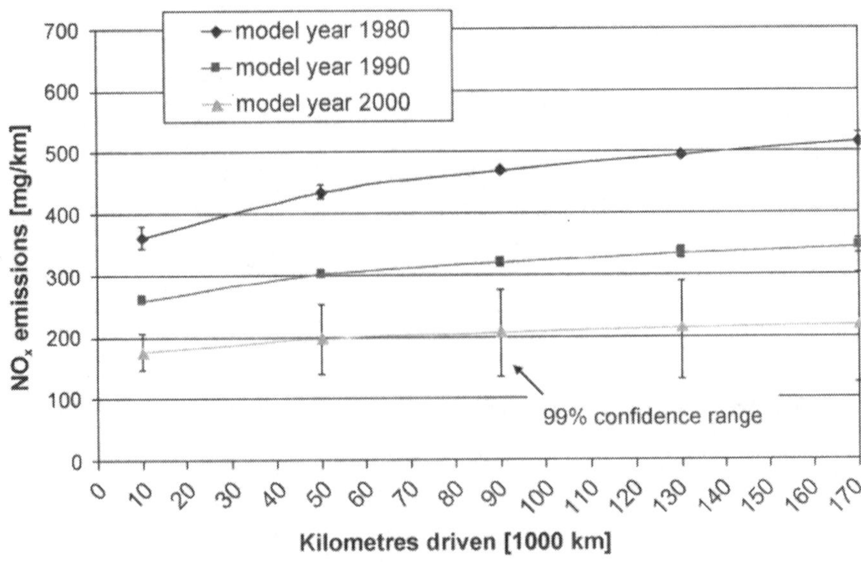

Fig. 32. Impact of year of construction and kilometres driven on NO_x emissions of gasoline-engine-powered passenger cars with controlled three-way catalysts in the applied ignition test 75 cycle

actions between these parameters and their impact on carbon monoxide, hydrocarbon and nitrogen oxide emissions. As can be seen, emissions increase with longer service life due to wear on vehicle parts. It is, however, worth noting that thanks to modern vehicle technology it is now possible to counteract wear-induced higher emissions to some extent.

As Diesel-engine powered passenger cars do not show a marked increase in emissions with longer service life, this category of vehicles has not been considered separately.

For the functions shown in Figs. 30–32, confidence ranges were calculated for assessing the quality of available data (emission values). When a large number of data are available, the confidence ranges are small, but if only a few widely scattered readings exist, the confidence ranges widen accordingly.

1.6.2
Passenger Car Emissions in Germany

With the aid of the computer program developed by the Vienna University of Technology, emissions from passenger cars in both the old and the new states of Germany in the period from 1970 to 2020 have been determined for 44 separate passenger car categories, broken down by the relevant emission regulations and by engine capacity. Investigations conducted by Shell [73] and Esso [74] as well as the ifo Institute for Economic Research [75] were used as a basis for calculating trends.

The trends for nitrogen oxide (NO_x), hydrocarbon (HC) and carbon monoxide (CO) emissions from passenger cars in Germany are illustrated in Figs. 33–35. In theses Figures, the abbreviation Otto/Ukat denotes gasoline-engine powered vehicles with unregulated catalysts and Otto/Gkat refers to vehicles with gasoline engines with controlled three-way catalysts prior to the "Euro 1" phase. The shares in total emissions of the different vehicle categories as defined in the relevant legal provisions are indicated by a vertical line along the abscissa.

As can be seen from Figs. 33–35, all three emission components (CO, HC and NO_x) will be continuously reduced beyond the year 2000 and will subsequently remain constant at a very low level, even though the total number of kilometers driven will go up. By comparing the level recorded for 2020 with the year 1987, when emission curves started to decline, the effectiveness of regulatory emission limits becomes obvious.

The introduction of ever more stringent emission limits has resulted in a reduction in specific particulate emissions over the past decade. Fig. 36 describes the extent to which maximum permissible exhaust emission values constitute an instrument for reducing exhaust gas emissions from motor traffic, against the background of a constantly growing vehicle population.

Figure 37 illustrates fuel consumption patterns by vehicle category as defined by the relevant legal provisions in Germany. As can be seen from this graph, fuel consumption in all vehicle categories will go down further thanks to high fuel efficiency technologies, despite ever rising vehicle population figures.

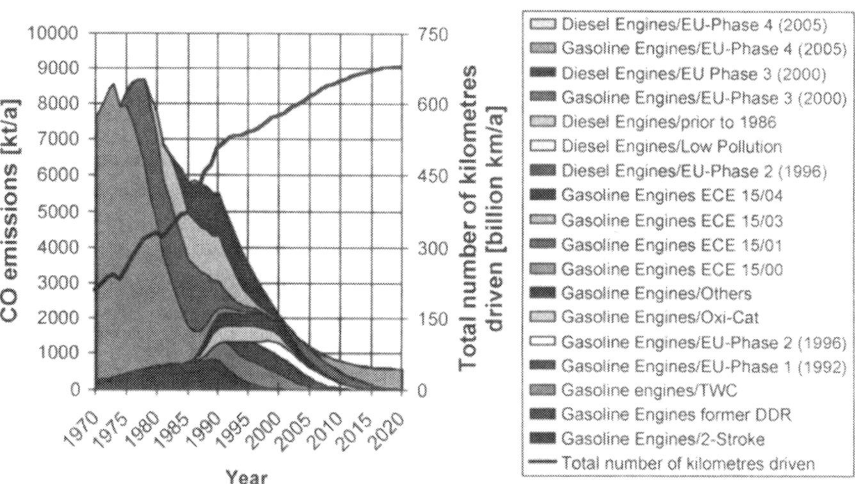

Fig. 33. Trend of carbon monoxide (CO) emissions from passenger cars broken down by vehicle category in Germany

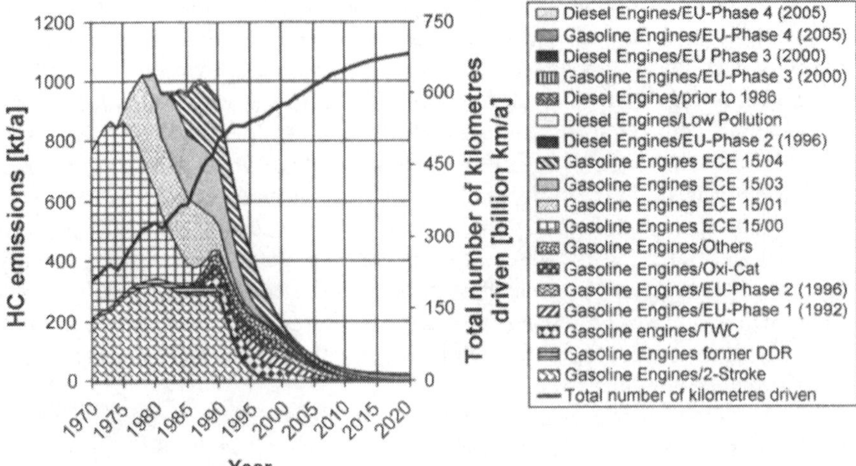

Fig. 34. Trend of hydrocarbon (HC) emissions from passenger cars broken down by vehicle category in Germany

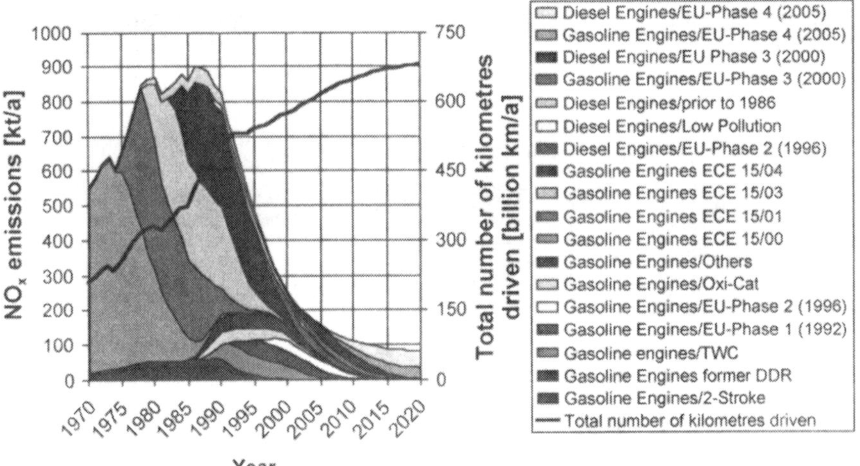

Fig. 35. Trend of nitrogen (NO_x) emissions from passenger cars broken down by vehicle category in Germany

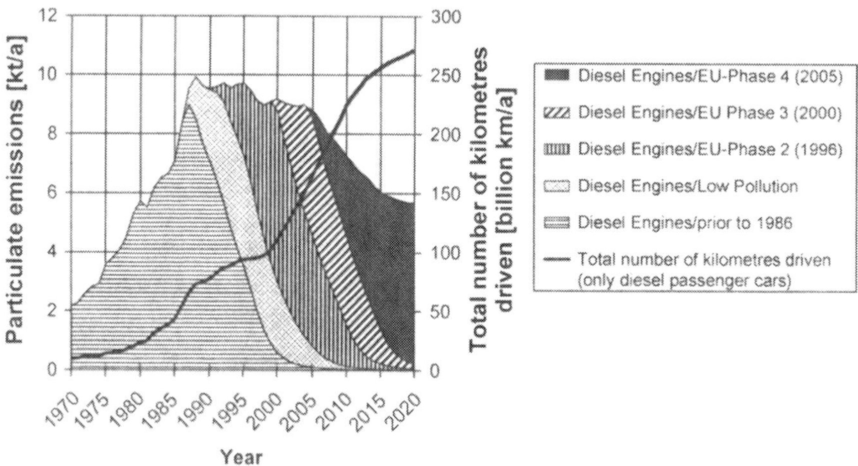

Fig. 36. Trend of particulate (PM) emissions from passenger cars broken down by vehicle category in Germany

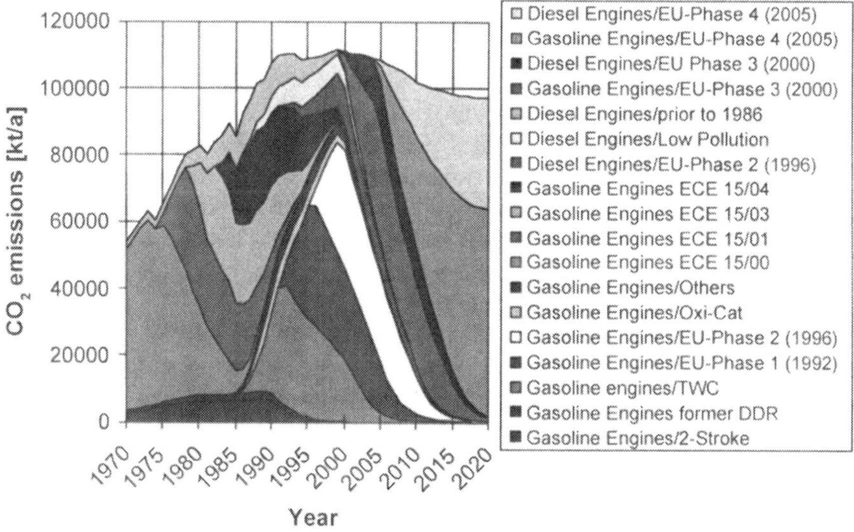

Fig. 37. Past, present and future trends in fuel consumption from passenger cars, broken down by vehicle category in Germany

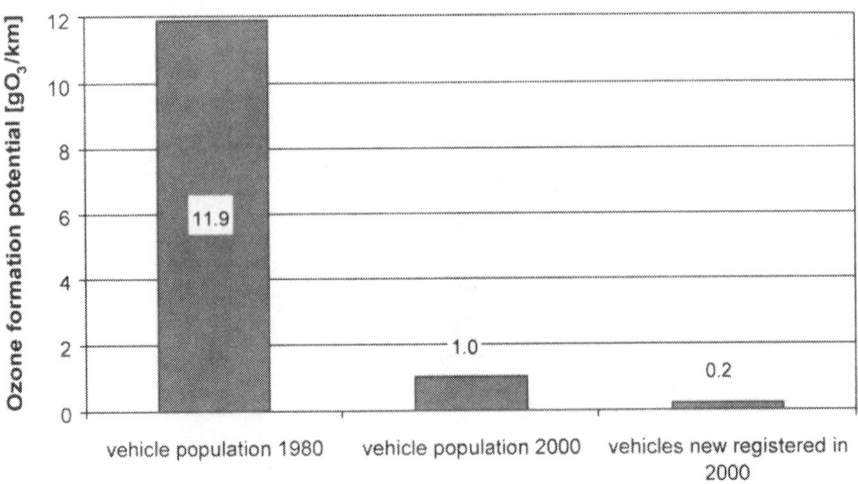

Fig. 38. The development of the ozone formation potential of the passenger car population and newly registered cars in Germany

1.6.3
Ozone Formation Potential of Passenger Car Emissions in Germany

Figure 38 depicts the ozone formation potential in g O_3/km for the vehicle population in 1980, 2000 and vehicles newly registered in 2000 ("Euro 3" vehicles). The sharp decline in ozone formation potential for the vehicle population in 2000 as against the reference year of 1980 is clearly visible. The ozone formation potential of newly registered vehicles is approximately 80% below that of the vehicle population of 2000, which means that with the partial replacement of the passenger car population by model 2000 vehicles, the ozone formation potential will continue to fall significantly in the future.

Figure 39 gives an indication of the ozone formation potential of the passenger car population in Germany, broken down into vehicles equipped with catalysts and vehicles without catalysts. Considerations to take such measures as imposing driving bans or speed restrictions on vehicles which are not equipped with state-of-the-art exhaust gas treatment systems and do not meet limit values, do not produce the desired effect of reducing ozone concentrations in the air during phases of early ozone warning. The ozone formation potential of the vehicles without catalytic converters in use is steadily and significantly going down every year so that such measures would fail to produce the desired result.

1.6.4
Greenhouse Gas Formation Potential of Passenger Car Emissions in Germany

Figure 40 shows a comparison of greenhouse gas emissions for newly registered vehicles constructed in the years 1975 and 2000. The computed emission levels

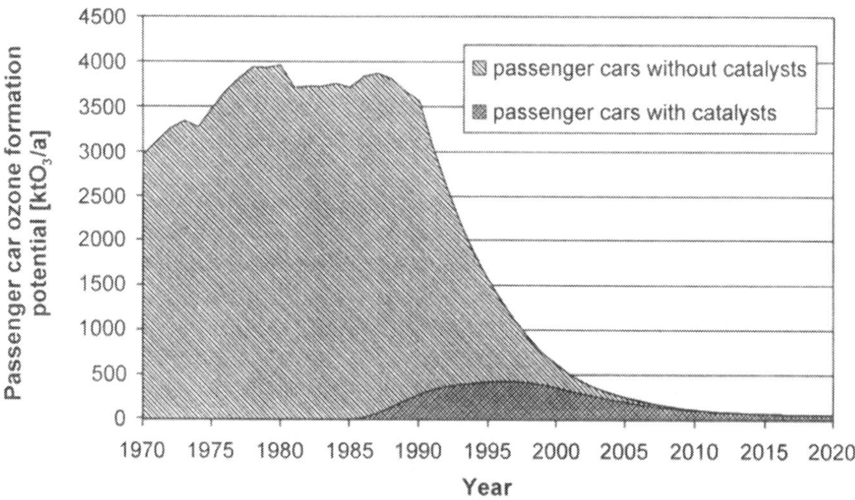

Fig. 39. Development of the passenger car ozone formation potential of vehicles equipped with catalysts and without catalysts in Germany

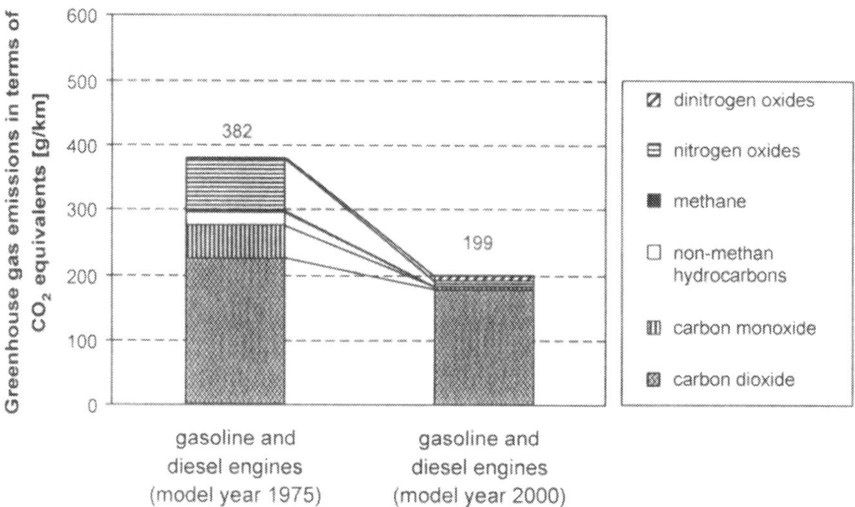

Fig. 40. Comparison of the greenhouse gas formation potential as CO_2 equivalents for passenger cars constructed in the years 1975 and 2000

represent mean values of all gasoline engine and Diesel engine powered vehicles newly registered in Germany in these two years.

CO_2 emissions of newly registered vehicles will decline further in the future as the Association of German Automobile Makers (VDA) has committed itself to lowering the overall fuel consumption of newly registered vehicles by 25% below the year 1990 level by the year 2005.

Thanks to improvements in engine technology it was possible to bring down greenhouse gas emissions from 382 g/km to 199 g/km. The lowering of carbon monoxide, non-methane hydrocarbon and nitrogen oxide emissions accounted for the lion's share of this reduction. As can be seen from the Figs., emissions of the greenhouse gases dinitrogen monoxide N_2O and methane CH_4 are negligible in modern passenger cars. With the drastic reduction of the emission levels of these exhaust gas components, their relevance for greenhouse gas formation has declined.

With the replacement of passenger cars in the existing population by model 2000 vehicles (Euro 3 vehicles), greenhouse gas emissions can be lowered from 230 g/km to 199 g/km, as is illustrated in Fig. 41. As the components carbon monoxide, hydrocarbons and nitrogen oxides, all of which are subject to limit values, represent only a small share of greenhouse gas emissions, no significantly lower level will be attained with the introduction of standards in accordance with "Euro 4".

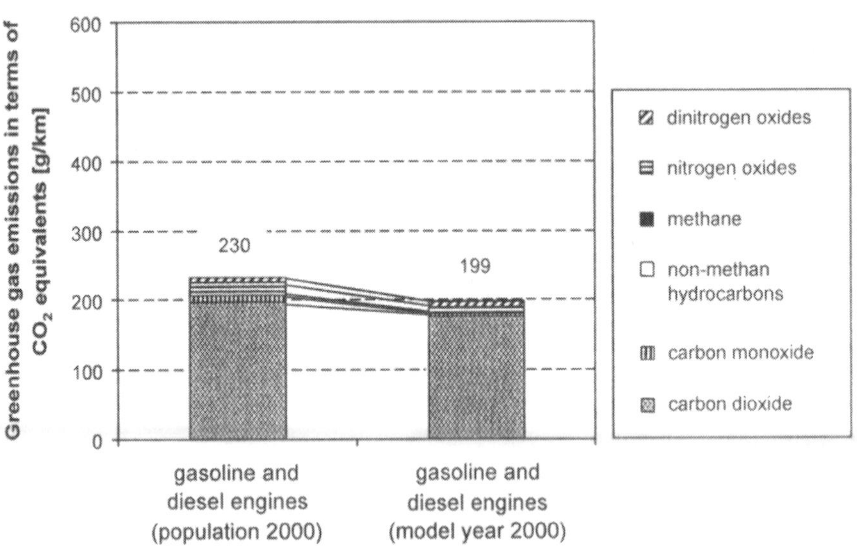

Fig. 41. Comparison of greenhouse gas emissions as CO_2 equivalents between the existing car population and passenger cars newly registered in 2000

1.7
Environmental Impacts of Commercial Vehicle Traffic

1.7.1
Emissions from Commercial Vehicles in Germany

The question as to whether current or future exhaust gas limit values will constitute an effective tool for reducing exhaust gas emissions from commercial vehicles in the face of constantly growing vehicle populations will be analyzed on the basis of carbon monoxide, hydrocarbon, nitrogen oxide and particulate emissions.

Figures 42, 44, 46 and 48 give a breakdown of emissions by vehicle category in accordance with the individual legislation stages. In Figs. 43, 45, 47 49 emissions are listed for each vehicle type.

Figure 50 illustrates past and present CO_2 emissions from commercial vehicle traffic and future trends.

1.7.2
Ozone Formation Potential of Commercial Vehicle Emissions in Germany

Figure 51 shows the ozone formation potential in g O_3/km for the 1980 and 2000 vehicle populations, as well as that for vehicles newly registered in 2000 (Euro 3-vehicles). This graph reveals the sharp decline in ozone formation potential for the 2000 vehicle population from the 1980 level. The ozone formation potential of newly registered vehicles is approximately 60% below that of the 2000 vehicle population which means that the ozone formation potential will drop further as vehicles constructed in 2000 replace the existing commercial vehicle population.

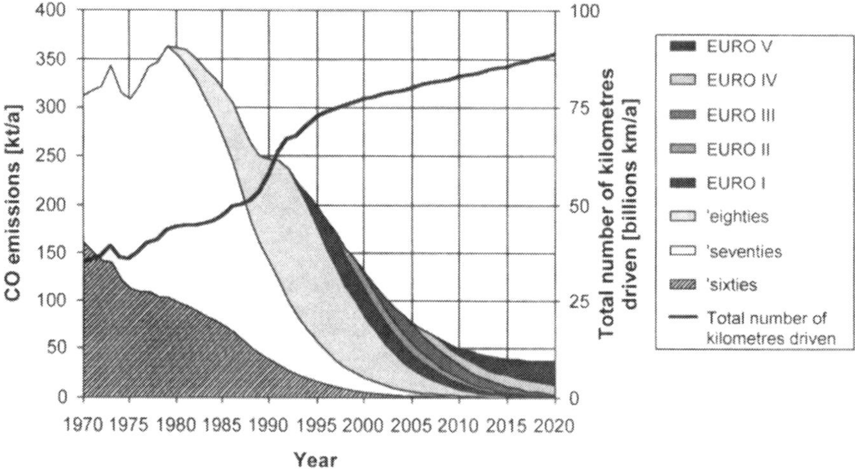

Fig. 42. Development of carbon monoxide (CO) emissions from commercial vehicles by vehicle category in Germany

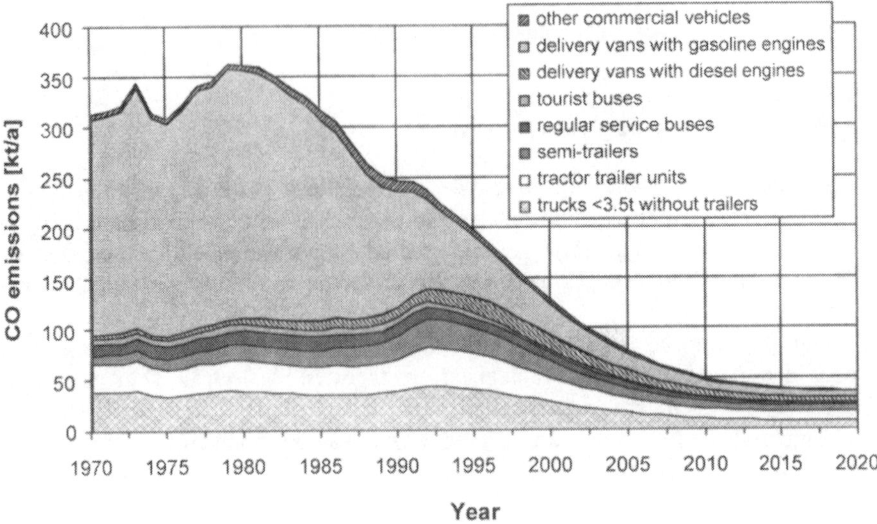

Fig. 43. Trend in carbon monoxide emissions from commercial vehicles by vehicle category in Germany

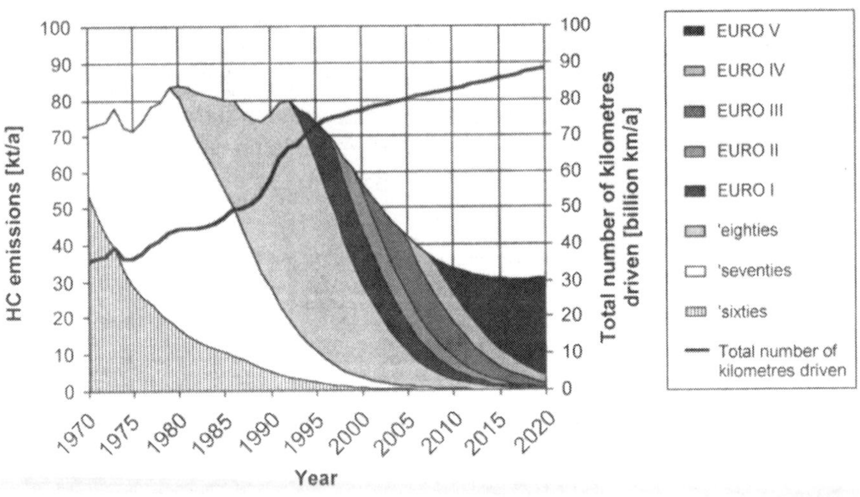

Fig. 44. Development of hydrocarbon (HC) emissions from commercial vehicles by vehicle category in Germany

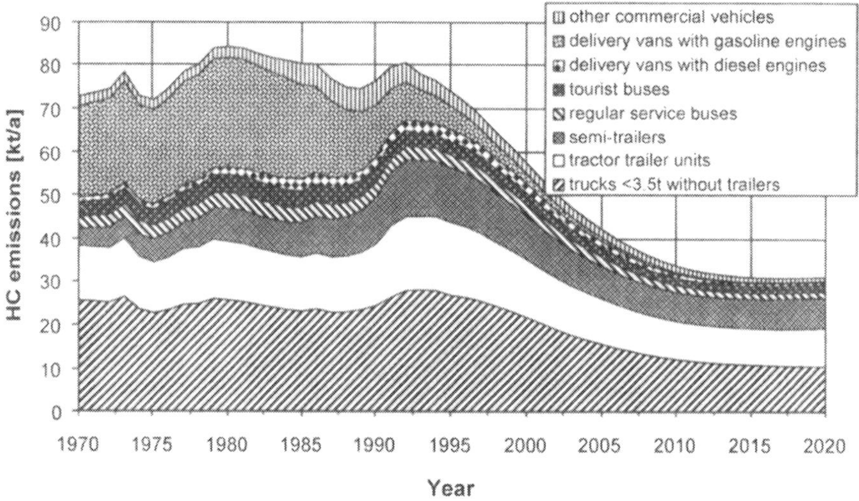

Fig. 45. Development of hydrocarbon (HC) emission from commercial vehicles broken down by vehicle category in Germany

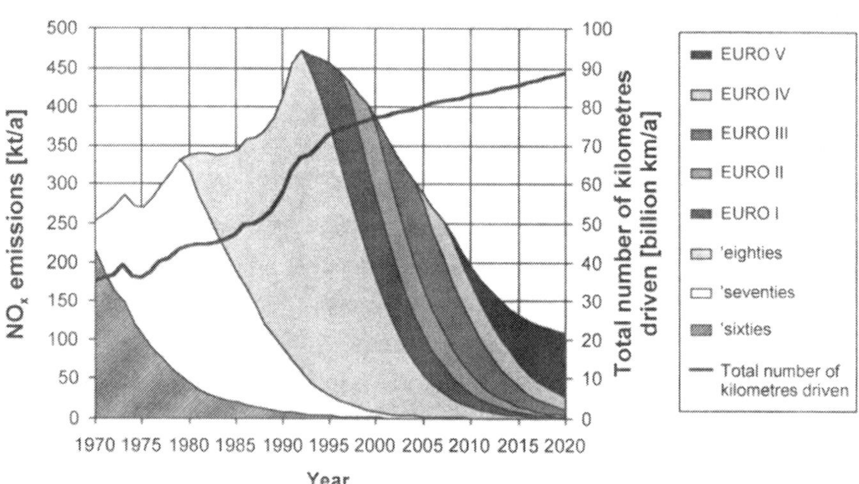

Fig. 46. Trends in nitrogen (NO_x) emissions from commercial vehicles by vehicle category in Germany

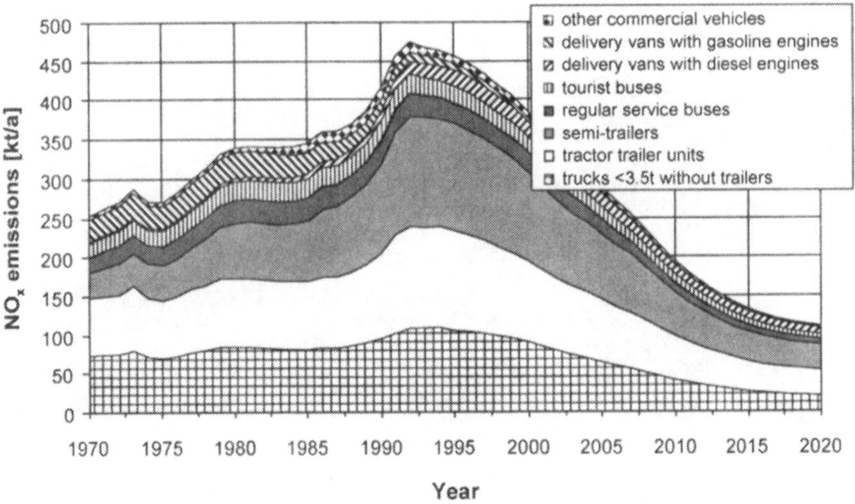

Fig. 47. Trends in nitrogen (NO_x) emissions from commercial vehicles broken down by vehicle category in Germany

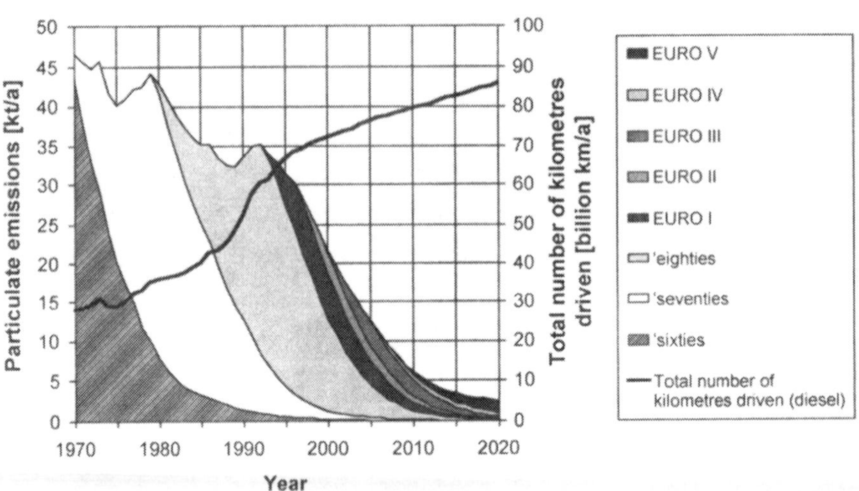

Fig. 48. Trends in particulate (PM) emissions from commercial vehicles broken down by vehicle category in Germany

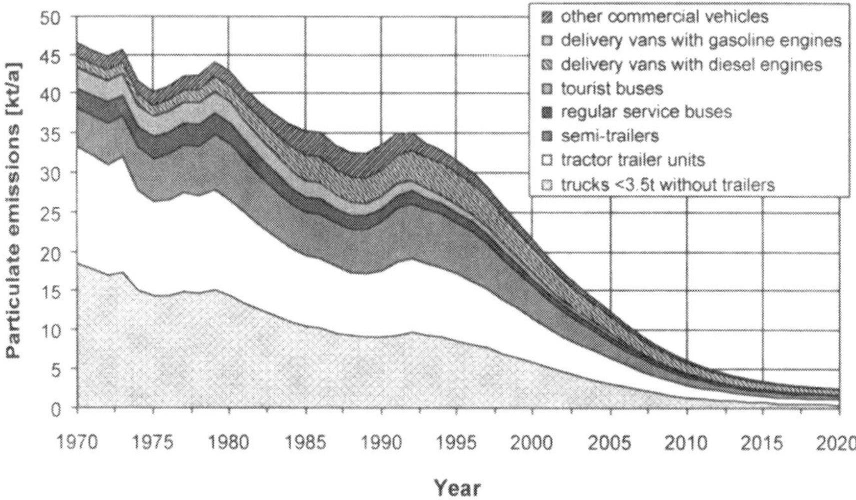

Fig. 49. Trends in particulate (PM) emissions from commercial vehicles broken down by vehicle category in Germany

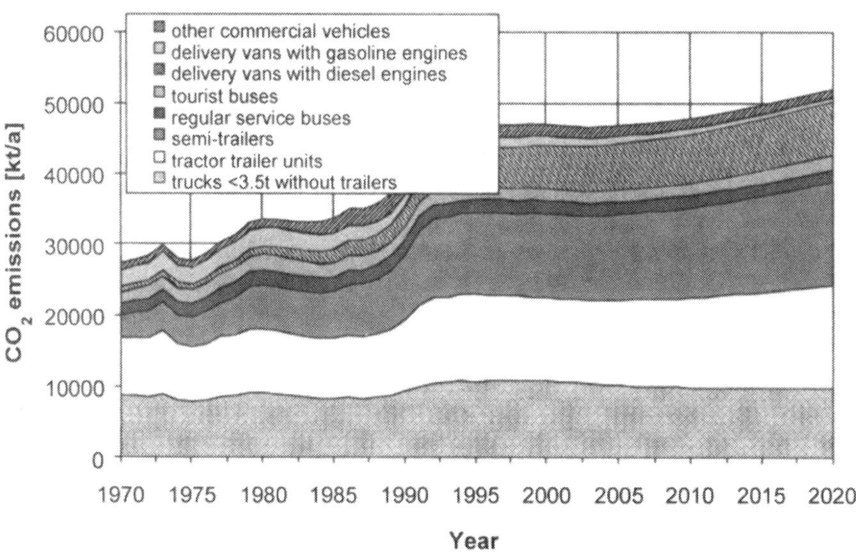

Fig. 50. Trends in CO_2 emissions from commercial vehicles broken down by vehicle category in Germany

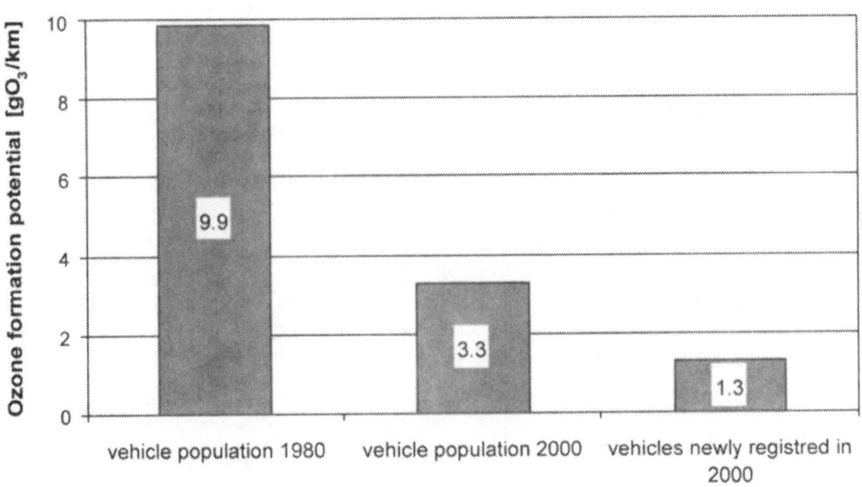

Fig. 51. Comparison of ozone formation potentials of the commercial vehicle population and newly registered vehicles in Germany

The development in ozone formation potential clearly reflects the technological progress made in vehicle design. Fig. 52 illustrates the influence of technological advances resulting in a reduction of HC emissions on the ozone formation potential, which has shown a significant reduction since the year 1980.

1.7.3
Greenhouse Gas Formation Potential of Commercial Vehicle Emissions in Germany

Figure 53 shows a comparison of the greenhouse gas formation potential of newly registered commercial vehicles in 1975 and 2000. It can be seen that emissions contributing to the greenhouse effect were brought down from 1127 g/km in 1975 to 730 g/km in 2000.

As can be seen from Fig. 54, a reduction of emissions from 859 g/km to 730 g/km can be achieved by the substitution of the old commercial vehicle population by new model 2000 vehicles (Euro 3 vehicles). This reduction will come entirely from lower nitrogen oxide emissions. Therefore, a further decline in greenhouse gas emissions can be expected from the introduction of the Euro 4 legislation.

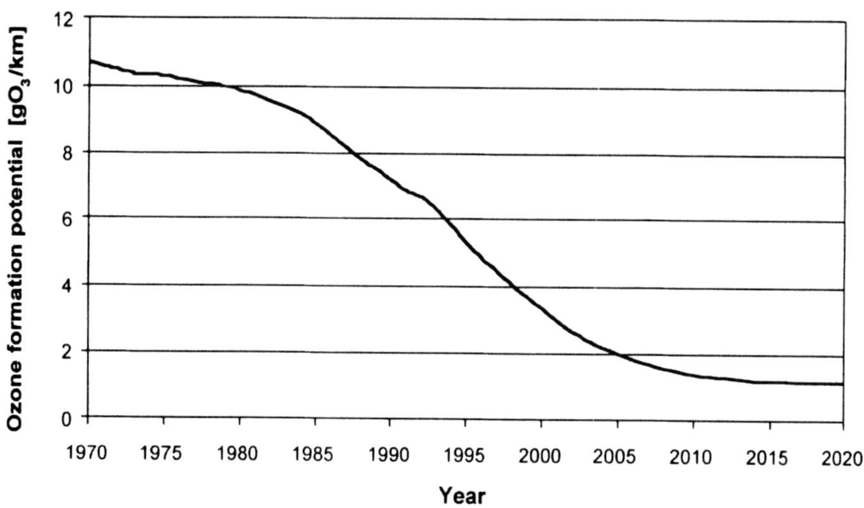

Fig. 52. Trend in ozone formation potential of the commercial vehicle fleet in Germany

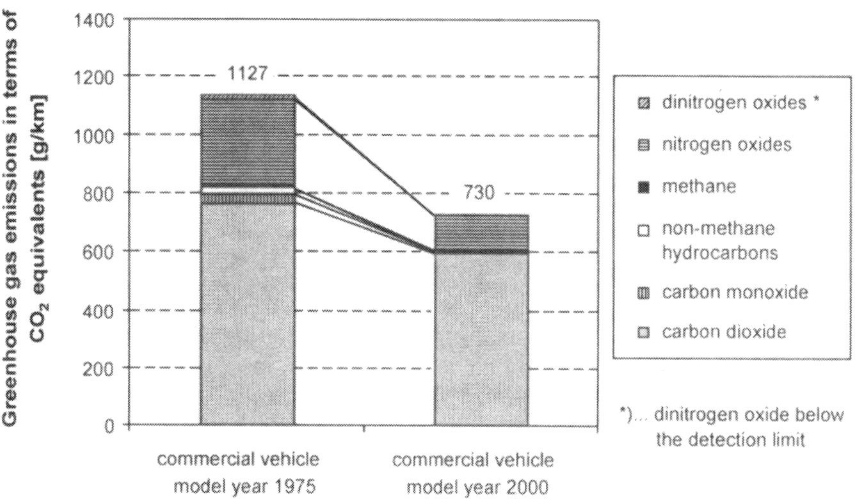

Fig. 53. Comparison of the greenhouse gas formation potential as CO_2 equivalents of commercial vehicles constructed in 1975 and 2000

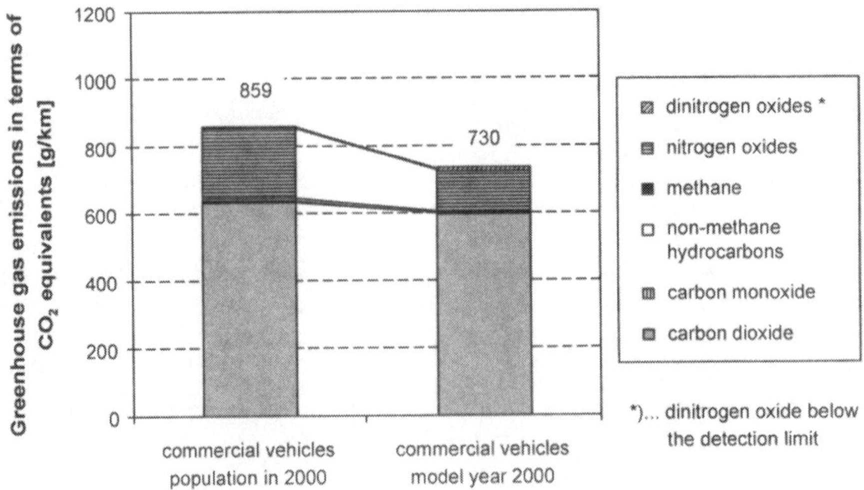

Fig. 54. Comparison of greenhouse gas emissions as CO_2 equivalents of the commercial vehicle population and newly registered vehicles in 2000

1.8 References

1. Plaßmann E, Waldeyer H, Brosthaus J (1993) Minderung der Luftbelastung und des Energiebedarfs durch die Verkehrsmittel im Vergleich, Verkehr in der Mitte Europas – Deutscher Ingenieurtag 1993. VDI-Berichte 1041, VDI, Düsseldorf
2. Kuhn M (1990) Klimaänderungen: Treibhauseffekt und Ozon. Hochschulschriftenreihe Forschung, Band 1, Kulturverlag Thaur/Tirol, pp 157
3. Wellburn A (1997) Luftverschmutzung und Klimaänderung – Auswirkung auf Flora, Fauna und Mensch. Springer, Berlin, Heidelberg, New York, ISBN 3-540-61831-7
4. Heintz A, Reinhardt G (1991) Chemie und Umwelt. 2. Auflage, Vieweg & Sohn Verlagsgesellschaft Braunschweig Wiesbaden, ISBN 3-528-16349-6
5. NN (1990) Treibhauseffekt, Ursachen, Konsequenzen, Strategien. Umweltbundesamt Wien. Monographien Bd. 23, Wien
6. NN (1997) CDIAC: Carbon Dioxide Information Analysis Center, Messwerte auf Datenträger. Oak Ridge
7. Lenz H P, Kohoutek P, Pischinger R, Hausberger S (1995) Beeinflußungsmöglichkeiten des motorisierten Straßenverkehrs auf die CO_2-Emissionen. Int. Kongreß der AVL List GmbH, Graz 24./25. August 1995
8. NN (1992) Deutscher Bundestag – 12. Wahlperiode: Erster Bericht der Enquete-Kommission "Schutz der Erdatmosphäre" zum Thema Klimaänderung gefährdet globale Entwicklung, Zukunft sichern – Jetzt handeln. Drucksache 12/2400
9. Hansen JE, Lebedeff S (1988) Global Surface Air Temperatures: Update through 1987. Geophys Res Lett 15:323–326
10. Ramanathan V, Cicerone RJ, Singh B, Kiehl JT (1985) Trace Gas Trends and their Potential Role in Climate Change. Geophys Res Lett 90:5547–5566
11. N.N. (1990) Supplementary figures from the Policymakers Summary of IPCC Working Group 1. Intergovernmental Panel on Climate
12. Lesch K-H, Cerveny M, Leitner A, Berger B (1990)Treibhauseffekt Ursachen, Konsequenzen, Strategien. Umweltbundesamt Wien, Monographien, Bd. 23

13. Stahl W, Berner U (1997) Die Rolle der BGR in der Klimadiskussion, gehalten am 20.08.1997 in der Bundesanstalt für Geowissenschaften und Rohstoffe BGR, Hannover
14. Rhode H (1990) A Comparison of the Contributions of Various Gases to the Greenhouse Effect. Science Reports 248:1217–1219
15. MacKenzie JJ, Walsh MP (1990) Driving Forces: Motor Vehicle Trends and their Implications for Global Warming, Energy Strategies, and Transportation Planning. WRI World Resources Institute, Washington, USA
16. N.N. (2001) Daten zur Umwelt, Der Zustand der Umwelt in Deutschland 2000. Umweltbundesamt Berlin
17. NN (1993) Organization For Economic Co-Operation and Development: OECD Environmental Data. Compendium
18. Krapfenbauer A, Wriessnig K (1995) Anthropogene Umweltbelastung – Die Rolle der Landbewirtschaftung. Sonderdruck – Die Bodenkultur. Journal für landwirtschaftliche Forschung 46:269ff
19. Bolle HJ (1991) Treibhauseffekt. Vortrag bei der Österreichischen Gesellschaft für Erdölwissenschaft, Schwechat
20. Heimann M, Maier-Reimer E (1996) On the Relations between the Oceanic Uptake of CO_2 and its Carbon Isotopes. Global Biogeochem Cycles 10:89–110
21. Walsh MP (1994) Motor Vehicle Pollution Control: A Global Overview. XXV FISITA Congress, 17–21 October, Beijing, SAE Technical Paper 945112
22. Woodwell GM (1996) Das Kohlendioxid-Problem in Atmosphäre, Klima, Umwelt, 2. Auflage, Spektrum Akademischer Verlag
23. Faber M, Jöst F, Proops J, Wagenals G (1996) Wirtschaftliche Aspekte des Kohlendioxid-Problems, in Atmosphäre, Klima, Umwelt. 2. Auflage, Spektrum Akademischer Verlag
24. NN (1997) World Energy Outlook, Head Economic Analysis Division. OECD/IEA, Paris
25. NN (1998) World Energy Outlook, Head Economic Analysis Division. OECD/IEA, Paris
26. NN (1995) Organization For Economic Co-Operation and Development: OECD Environmental Data. Compendium
27. NN (1997) Organization For Economic Co-Operation and Development: OECD Environmental Data. Compendium
28. NN (1999) World Energy Outlook, Head Economic Analysis Division. OECD/IEA, Paris
29. Metz N (1999) Verteilung der weltweiten CO_2-Emissionen von 1970 bis 2010. 20. Internationales Wiener Motorensymposium, Wien
30. NN (1995) World Resources 1994/1995 – A Guide to the Global Environment, United Nations Environment Programme. World Resources Institute
31. NN (1997) World Resources 1996/1997 – A Guide to the Global Environment, United Nations Environment Programme. World Resources Institute
32. NN (1994) World Energy Outlook, Head Economic Analysis Division. OECD/IEA, Paris
33. NN (1997) International Energy Outlook, Energy Information Administration (EIA). Washington
34. NN (1995) Environmental Data, Compendium 1995. OECD Organization for Economic Cooperation and Development, Paris
35. NN (1997) Environmental Data, Compendium 1997. OECD Organization for Economic Cooperation and Development, Paris
36. Teufel D, Bauer P, Braunfeld S, Kilian G, Wagner T (1995) Folgen einer globalen Motorisierung. Umwelt- und Prognose-Institut Heidelberg e.V., UPI-Bericht Nr. 35
37. Streit B (1994) Lexikon Ökotoxikologie. Zweite, aktualisierte und erweiterte Auflage, ISBN 3-527-30053-8, VCH, Weinheim
38. Wei Min Hao, Ward DE (1993) Methane Production from Global Biomass Burning. J Geophys Res 98:20,657–20,661
39. Crutzen PJ (1991) Methane's sinks and sources. Nature 350:380–381
40. Hein R, Crutzen PJ, Heimann M (1997) An Inverse Modeling Approach to Investigate the Global Atmospheric Methane Cycle. Global Biogeochem Cycles 11:43–76
41. Müller J-F, Brasseur G (1995) Images: a three-dimensional chemical transport model of the global troposphere. J Geophys Res 100:16,445–16,490

42. Anastasi C, Simpson VJ (1993) Future Methane Emissions from Animals. J Geophys Res 98:7181–7186
43. Mosier AR, Schimel DS (1991) Influence of Agricultural Nitrogen on Atmospheric Methane and Nitrous Oxide. Chemistry and Industry
44. Batjes NH, Bridges EM (1994) Potential emissions of radiatively active gases from soil to atmosphere with special reference to methane: development of a global database. J Geophys Res 99:16,479–16,489
45. Anastasi C, Dowding M, Simpson VJ (1996) Future CH_4 emissions from rice production. J Geophys Res 101:89–110
46. Law KS, Nisbet EG (1996) Sensitivity of the CH_4 Growth Rate to Changes in CH4 Emissions from National Gas and Coal. J Geophys Res 101:14,387–14,397
47. Lassey KR, Lowe DC (1992) A source inventory of atmospheric methane in New Zealand and its global perspective. J Geophys Res 97:3751–3765
48. Orthofer R (1990) Flüchtige halogenierte Kohlenwasserstoffe: Eine Abschätzung der Emissionssituation in Österreich. Österreichisches Forschungszentrum Seibersdorf, OEFZS-4526, NU-123/90
49. Fisher DA, Midgley PM (1994) Uncertainties in the Calculation of Atmospheric Releases of Chlorofluorocarbons. J Geophys Res 99:16,643–16,650
50. N.N. (2001) Production, Sales and Atmospheric Release of Fluorocarbons through 1999. AFEAS, USA, Washington DC
51. Elkins JW (1989) State of the Research for Atmospheric Nitrous Oxide (N_2O) in 1989. Contribution for the Intergovernmental Panel on Climate Change (IPCC)
52. Bouwmann, et al. (1995) Uncertainties in the global source distribution of nitrous oxide. J Geophys Res 100:2785–2800
53. Potter CS, et al. (1996) Process modeling of controls on nitrogen trace gas emissions from soils worldwide. J Geophys Res 101:1361–1377
54. Khalil MAK, Rasmussen RA (1992) The global sources of nitrous oxide. J Geophys Res 97:14,651–14,660
55. Bange HW, Rapsomanikis S, Andreae MO (1996) Nitrous oxide in coastal waters. Global Biogeochem Cycles 10:167–207
56. NN (1991) Greenhouse Gas Emissions, the Energy Dimension. Heat of Publication Service, OECD, Paris
57. Kohoutek P (1996) Entwicklung einer Datenbank zur Berechnung von Abgasemissionen benzinbetriebener PKW unter besonderer Berücksichtigung der Kraftstoffzusammensetzung. VDI-Fortschritt-Berichte, Reihe 12, Nr. 275, Düsseldorf
58. NN (1996) CORINAIR 1996
59. Augenbraun H, Matthews E, Sarma D (1999) The global Methane cycle, Institute on climate and planets, http://icp.nasa.gov
60. Own calculations
61. Schermann G (1999) Monatsmittelwerte 1998, Luftmeßnetz der MA 22, personal communication
62. NN (2001) Immissionsmeßwerte Landesumweltamt Bayern, personal communication
64. NN (1999–2001) Diverse Monats und Jahresberichte der Landesanstalt für Umweltschutz Baden-Württemberg, Gesellschaft für Umweltmessungen und Umwelterhebungen (UMEG)
64. Berge E, Beck J, Larssen S, Moussiopoulos N, Pulles T (1997) Air Pollution in Europe 1997. EEA-European Environment Agency, ISBN 92-9167-059-6, Copenhagen
65. Simpson D, Guenther A, Hewitt C N, Steinbrecher R (1995)Biogenic Emissions in Europe, 1. Estimates and Uncertainties, J Geophysi Res 100:22,875–22,890
66. NN (1997) Luftbelastung 1996, Meßresultate des Nationalen Beobachtungsnetzes für Luftfremdstoffe (NABEL). Schriftenreihe Umwelt Nr. 286, Bundesamt für Umwelt, Wald und Landschaft, Bern
67. Balrutsch M, Hanewald K, Siegmund A, Stec-Lazaj W, Wunderlich W (1994) Lufthygienischer Jahresbericht 1994. Hessische Landesanstalt für Umwelt, 1994 und persönliche Mitteilung

68. Stohl A, Williams E, Wotawa G, Kromp-Kolb (1996) A European Inventory of Soil Nitric Oxide Emissions and the Effect of these Emissions on the Photochemical Formation of Ozone. Atmospheric Environment 30:3741-3755
69. Chameides WL, Kasibhatala RS, Yienger J, Levy H (1994) Growth of continental-scale metro-argo-plexes, regional ozone pollution and world food production. Science 264:74–76
70. Berdowski JJM, Mulder W, Veldt C, Visschedijk AJH, Zandveld PYJ (1997) Particulate Matter Emissions (PM10–PM2.5–PM 0.1) in Europe in 1990 and 1993. TNO-Report, TNO-MEP-R 96/472, Apeldoom
71. Harrison RM, et al. (1996) Airborne Particulate Matter in the United Kingdom. Third Report of the Quality of Urban Air Review Group, University of Birmingham, Institute of Public and Environmental Health
72. Redl M, Lorenz D (1999) Luftmeßnetz Jahresbericht 1998: Amt der OÖ Landesregierung Meßberichte
73. NN (1997) Motorisierung- Frauen geben Gas, Neue Techniken senken Verbrauch und Emissionen. Deutsche Shell AG, Shell PKW-Szenarien, Hamburg
74. NN (1995) Mobil bleiben, Umwelt schonen, Energieprognose 1994. Esso AG, Hamburg
75. Ratzenberger R, Hild R, Langmantel E (1995) Vorausschätzung der Verkehrsentwicklung in Deutschland bis zum Jahr 2010. ifo Institut für Wirtschaftsforschung Abt. Verkehr, München
76. NN (1995) Handbuch für Emissionsfaktoren des Straßenverkehrs. Version 1.1, Oktober 1995, Umweltbundsamt Berlin
77. NN (1999) Handbuch Emissionsfaktoren des Straßenverkehrs. Version 1.2, Januar 1999, Umweltbundsamt Berlin
78. NN (1993, 1994, 1995, 1996, 1997, 1998) Luftqualität in Nordrhein-Westfalen, TEMES- und LIMES-Jahresberichte, Landesanstalt für Immissionsschutz Nordrhein-Westfalen, Essen
79. NN (1998–2002) Luftgüte-Meßnetz BLUME, Senatsverwaltung für Stadtentwicklung, Umweltschutz und Technologie, diverse Jahresberichte Luftgütemeßdaten
80. NN (1999) The AOPII Emission Base Case. SENCO, Sustainable Environment Consultants Ltd
81. Zittel W, Treber M (2000) Analysis of BP Statistical Review of World Energy with respect to CO_2-Emissions. Germanwatch, Bonn
82. Sundt N (2000) US Study Foresees Big Rise in Global CO_2 Emissions; Growth dominated by increased fossil fuel use in less developed countries; Global change, www.globalchange.org
83. NN (2000) Transport and environment: Statistics for the Transport and Environment Reporting Mechanism.; Theme 8 Environment and Energy, edn 2000, Eurostat, ISBN 92-828-9330-8
84. NN (2000) Richtlinie 2000/69/EG des Europäischen Parlaments und des Rates vom 16. November 2000 über Grenzwerte für Benzol und Kohlenmonoxid in der Luft
85. NN (1993) Richtlinie 1999/30/EG des Rates vom 22. April 1999 über Grenzwerte für Schwefeldioxid, Stickstoffdioxid und Stickstoffoxide, Partikel und Blei in der Luft
86. NN (2001) Climate Change 2001, The Scientific Basis, Intergovernmental Panel of Climate Change (IPCC)
87. Collins WJ, Stevenson DS (2000) The European regional ozone distribution and its links with the global scale for the years 1992 and 2015, Atmospheric Environment 34:255–267
88. Cao M, Gregson K, Marshall S (1998) Global Emission from Wetlands and its sensitivity to climate change. Atmospheric Environment 32:3293

2
Noise Emission

Dušan Gruden

2.1
Introduction

There is hardly any place or any moment in our everyday life which is completely free of technical noise. Therefore, one of the most urging present-day priorities is the reduction of the noises resulting from human activities.

As the results of public surveys carried out in European cities show, the most annoying source of noise is road traffic.

In fact, since traffic zones and residential areas are so closely intertwined, the noise produced by motor cars is the predominant element in overall noise emissions.

The impact of sound on the hearing depends on the intensity and duration of the sound pressure. The sound pressure level is indicated as dB(A). Some noise intensity examples (sound pressure levels) are given in Fig. 1.

2.2
Legislation

The measures to be taken to reduce the noise levels in Germany are laid down in the "Technische Anleitung zum Schutz gegen Lärm" (Technical Noise Protection Instructions) or "TA Lärm" in short. These instructions indicate immission limits for areas directly adjacent to industrial plants which may not be exceeded. These immission values have been summarized in Table 1.

Automotive noise regulations are meant to reduce the noise emissions from motor cars and thus contain the noise emissions produced by road traffic within acceptable limits.

The European Union has its own traffic noise regulations. Since 1966, the permissible noise levels for passenger cars has been periodically lowered (Fig. 2). The basic regulation 70/157/EWG which defines the pass-by and standing noise measuring procedures as well as the corresponding limit values for type approval has been amended (latest status: 92/97/EWG) and complemented over the years. The current limit for passenger cars is 74 dB(A).

The noise levels are determined using a procedure which is precisely defined under ISO R-362. The 74 dB(A) limit specifies that 10 current passenger cars may

	dB(A)
Rustling of leaves, whispering	30 – 40
Conversation, office talk	50 – 60
Surf, water fall	60 – 70
Loud calling	65 – 75
Work noise in a factory	80 – 90

Fig. 1. Levels of noise from various sources

Table 1. Technical noise protection instructions (TA Lärm)

Immission limits according to "Technical Noise Protection Instructions" dB(A)
a) Areas where there are commercial or industrial facilities and apartments of plant owners/managers and surveillance/standby personnel (industrial areas)
b) Areas where there are mainly commercial facilities (commercial parks)
c) Areas where there are commercial facilities and apartments with neither commercial facilities nor apartments predominating (mixed areas, rural areas)
d) Areas where apartments are predominant (general residential areas)
e) Areas where there are apartments only (mere residential areas)
f) Health resorts, hospitals and nursing homes
g) Apartments structurally connected to an industrial facility

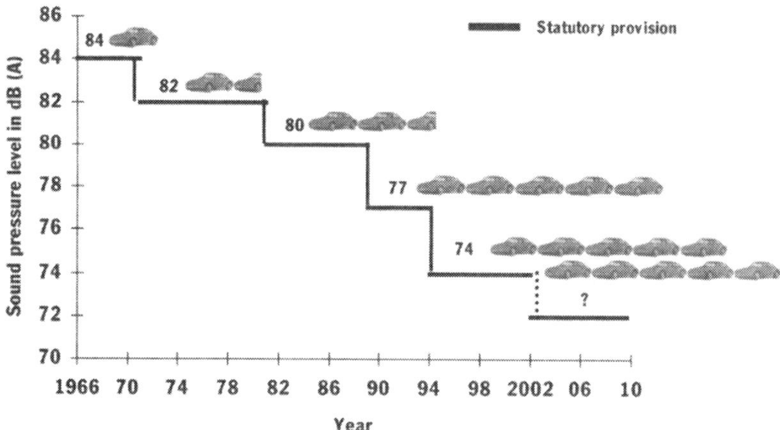

Fig. 2. Evolution of the statutory traffic noise limits

not emit more noise than one single car did at the time when the noise emission limitation was first introduced.

Basically, the measuring methods in the USA and Japan are similar to those of the European standard with the sound pressure levels being less severe, however.

According to the European 84/424/EWG standard, limit values also exist for commercial vehicles (trucks and busses) (Fig. 3).

The original commercial-vehicle noise level of 91 dB(A) was gradually reduced to the 1995 level of 80 dB(A) which means that one truck of model year 1974 emitted the same amount of noise as eight to twelve new trucks built in 2000.

Similar noise regulations also exist for motorcycles and mopeds as well as for other engine-powered machines.

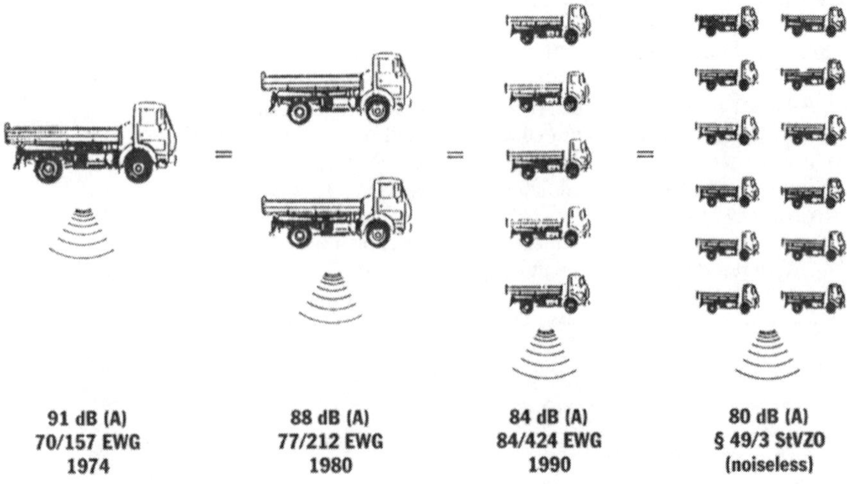

Fig. 3. Evolution of commercial-vehicle noise emissions since 1974

2.3 Sources of Noise Emissions

Automobiles include a multitude of sound sources such as, for example, the powertrain (engine, cooling-air fan, alternator, transmission, intake and exhaust noises), the tires and the aerodynamic noises all of which have most different acoustic effects (Figs. 4 and 5).

When the legislator stipulated automotive noise-emission limits 25 years ago, the "accelerated passage" test was chosen in order to simulate the worst-case traffic condition and achieve the greatest possible reduction of the partial-noise sources which prevailed at that time, i.e., the exhaust-gas system and the engine (Fig. 6).

According to the laws of logarithmic adding-up by which acoustics are governed, it is always necessary to reduce the loudest noise source in order to efficiently lower the overall noise level.

The lower and lower limit values and the gradual reduction of the noise emissions from the engine and exhaust system have necessarily resulted in an increase of the contribution made by the tire/road combination. At driving speeds of little more than 40 km/h already, the traffic noise is clearly dominated by the tire/road noise.

As far as commercial vehicles are concerned, the following factors were found to have the greatest noise reduction potential [6]:

- Tires (longitudinal and traction profiles): 3 to 5 dB(A),
- Road pavements (concrete, drain asphalt): 4 to 5 dB(A),
- Driving style,
- Road pavement and driving style combined: 10 dB(A).

The tire/road noise level can only be lowered if the road builders and tire manufacturers combine their efforts and jointly continue improving the tire/road system.

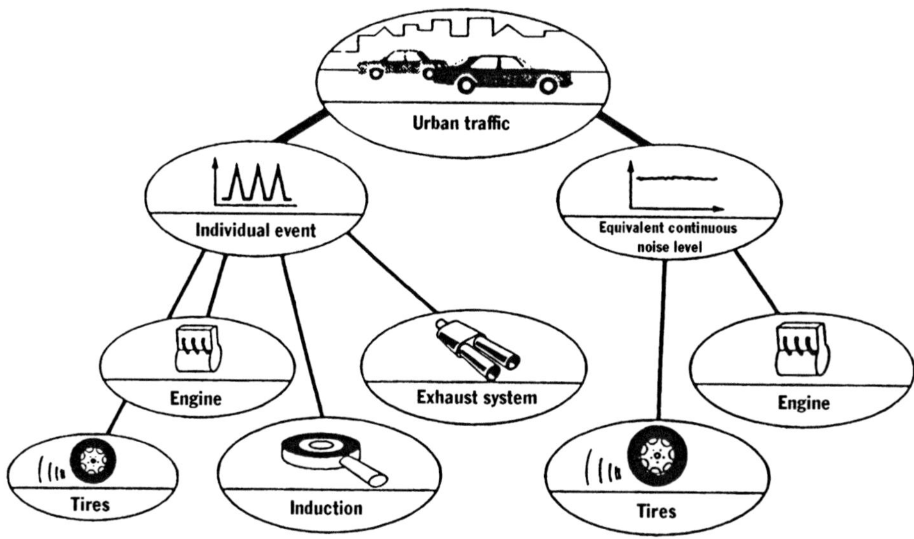

Fig. 4. Dominant passenger-car partial-noise sources in urban traffic

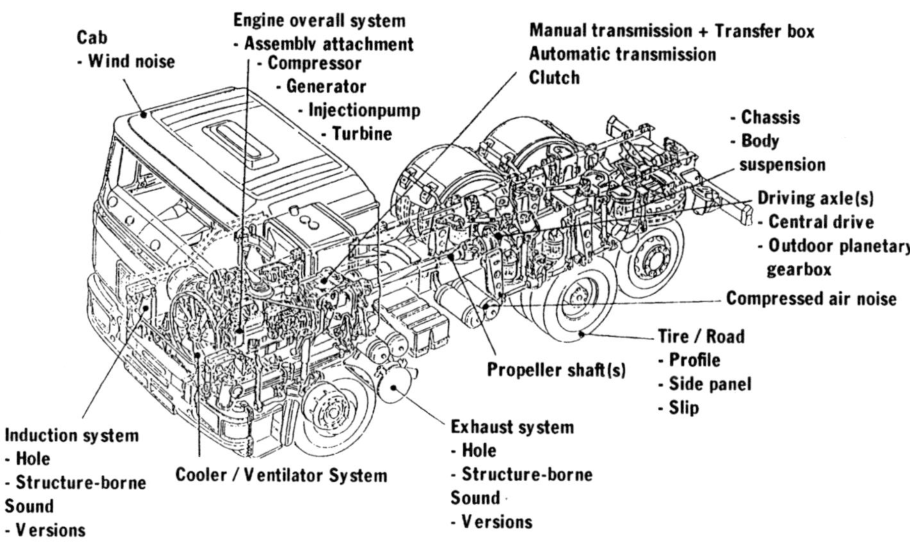

Fig. 5. Most important partial-noise sources of commercial vehicles (example: MAN) [3]

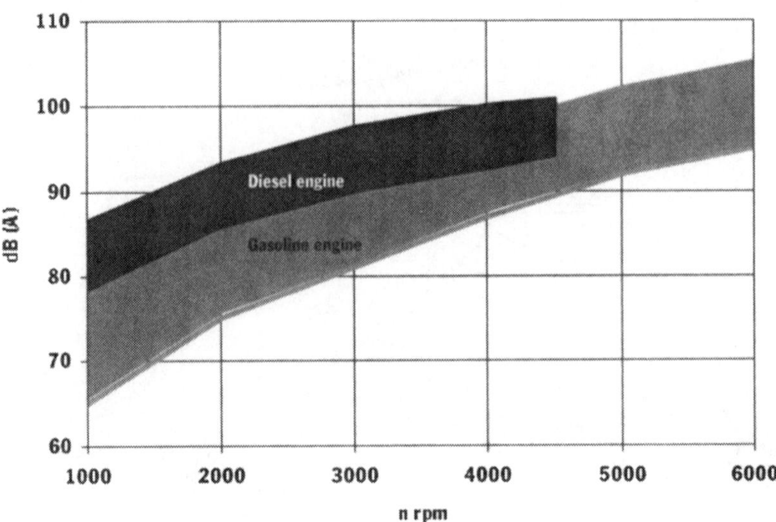

Fig. 6. Otto and Diesel engine combustion noises

2.4 References

1. Umweltschutz in Baden-Württemberg. Macht Lärm uns krank? Landesanstalt für Umweltschutz, Stuttgart, 1986
2. Kampf den lästigen Geräuschen. Umweltmagazin (1992) 12
3. Geib W (1996) Fahrzeugakustik, Komplexität, Zielkonflikte, Potentiale. Haus der Technik Essen
4. Lewis A (1997) The Science of Sound. Automotive Engineer, September
5. VDA/WdK-Informationsveranstaltung Straßenverkehrsgeräusche. Aschheim, 1999
6. Huber J (2000) Gesetzliche Geräuschvorschriften-Szenario mit Blick auf das Reifen-Fahrbahn-Geräusch. VDA-Technischer Kongreß, September
7. Pischinger S, Genuit K, et al (2000) Gestaltung des Geräusches von Verbrennungsmotoren zur Beeinflussung des Höreindrucks unter Berücksichtigung der Luft- und Körperschallübertragung. FVV-Forschungsvorhaben

3
Car Recycling – End of Life Vehicles

Dušan Gruden

3.1
Introduction

Today, the average life of a passenger car is expected to be about 10 to 12 years. Once this period has expired, the question arises what to do with the used car which is frequently also called "scrap car".

World-wide, 18 to 22 million cars a year reach the status of automotive scrap: up to 3.5 million in Germany, between 6.5 and 12 million at EU level, 9 to 10 million in the USA and about 1.1 million in Japan.

Automotive industry and economy in general have been confronted with the end of life vehicles problem for many years. Since the end of World War II techniques have been developed – first in the USA and then also in other countries – allowing the valuable substances to be extracted from a scrap car and recycled into the production process.

A motor car consists of roughly 10,000 parts and contains 40 different materials. The most frequently used valuable substances have been summarized in Fig. 1.

Besides the automobile, there few other products composed of such a great variety of materials and which have reached such a high degree of recyclability.

Nowadays, more than 75% of the weight of a passenger car are recycled. Thus, automobiles constitute highly complex economic objects characterized by an unusually high recycling rate.

Most of the metals, for example, are returned into the economic recycle process. There has been a market for iron and non-iron metals for many decades.

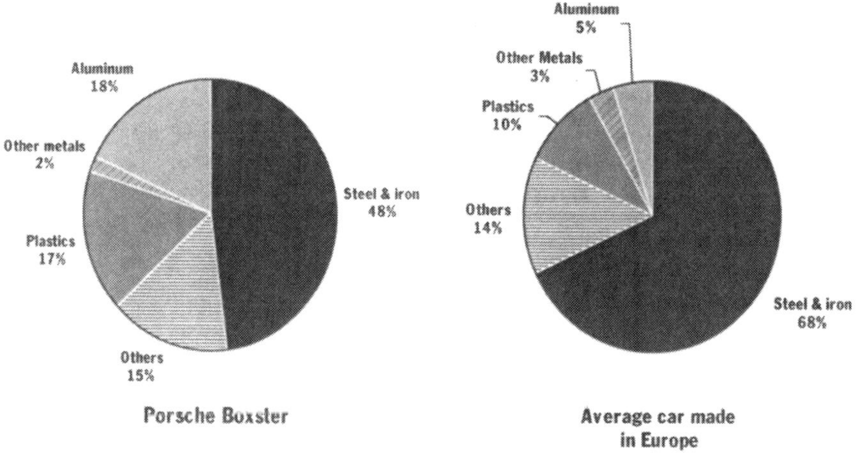

Fig. 1. Passenger-car material portions

As far as most of the plastic materials are concerned, however, the situation is quite different: At the time being, there is little demand for non-metallic materials extracted from used cars (this mainly applies to plastics, textiles and glass).

During the last few decades, a multitude of plastic materials have been introduced into automotive construction many of which are very difficult to recycle – if recyclable at all. Currently, as little as 5 to 10% of the plastic materials are actually recycled while the rest is dumped.

One of the most pressing motivations for industry to intensify the recyclability of its products is the so-called product responsibility. According to the Kreislaufwirtschaftsgesetz (KrWG or Law on Closed Economic Circuits) the person or firm who/which develops, produces, treats or sells a product must assume the responsibility for this very product. The afore-mentioned law is meant to preserve the existing natural resources and to guarantee the environmentally compatible disposal of waste. To begin with, waste should be avoided as far as this is practically feasible. Then, the amount of waste must be reduced and its noxiousness lowered. In the next approach, the waste materials must be recycled or used for energy production (energetic utilization). In the last resort only, waste should be disposed of. "Waste is raw material at the wrong place" – this is the motto upon which the closed-circuit law is based and which has given rise to a new kind of product responsibility: Products must be designed in such a way that waste is minimized during their production and utilization already. And product responsibility also includes the environment-friendly disposal of the product at the end of its life cycle.

The problem of used-car recycling mainly concerns passenger cars. As a rule, trucks or commercial cars are not scrapped. Thanks to their long service lives of 20 to 30 years they run through various utilizations and – once withdrawn from service – are often used as spare-part sources. The share of iron and non-iron metals in trucks is about 80 to 83%.

The complexity of the recycling problem and the intertwinement of the various enterprises involved require the close co-operation of all the economic areas concerned, i.e.:

- the automotive manufacturers and its suppliers,
- the recycling industry,
- the material manufacturers.

3.2 State of the Art of Used-Car Recycling

3.2.1 *Legal Boundary Conditions*

For decades, the recycling system applied by the waste industries world-wide has been entirely based on the principle of economic market success. However, the efforts of modern society to minimize the ecological consequences of human activities also include the measures to be taken at the end of a product life. Finally – after long years of intensive discussions between the legislator and the industry branches concerned – an end of life vehicle directive was put into force by the European Union in September 2000. The main points of this directive are the following:

- Taking back of used cars (Article 5.4).
- Limitation of heavy metals such as lead (Pb), cadmium (Cd), mercury (Hg) and chromium VI (CrVI) (Article 4.2).
- Determination of the recycling rates (Article 7.2).
  The current 75% recycling rate (Fig. 2) is to be increased to 85% by 2006 and to 95% as of 2015. Thereafter, 5% only of the so-called fluff is allowed to be dumped.
- Authorization of the recycling standards for new vehicles during type approval (Article 7.4).
- Availability of part/dismantling handbooks (Article 8).
  For each modern vehicle, there must be a draining/dismantling handbook with useful information on how to handle its individual parts and components.
- Utilization of recycled materials (Article 4.1).

According to the traditional recycling methods, any operative vehicle units and parts are reused and, additionally, the metallic materials are recovered.

Such units as engines, alternators, transmissions etc. are dismantled, repaired and sold. Such body elements as fenders, engine hoods, seats etc. can be reused without any major previous efforts required. The rest of the car is shredded and dumped.

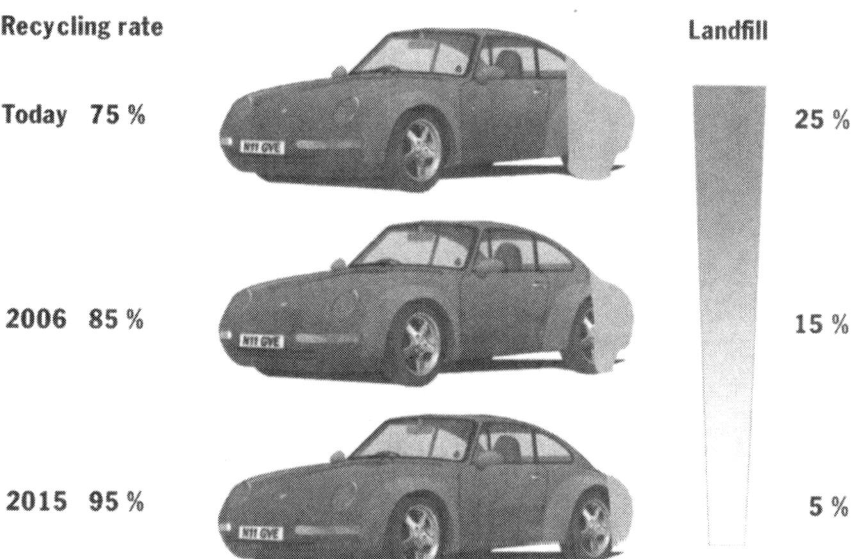

Fig. 2. Passenger-car recycling rates – European Used-Car Directive

Modern recycling systems for so-called ELVs – or End of Life Vehicles – use a different approach: the last owner hands over his used vehicle to an authorized and certified agency.

In the first phase of used-car recycling, all operating materials and fluids are drained in a recycling-conforming manner. This is done by means of draining systems which have to meet certain specifications.

After the vehicle has been drained, as many parts as possible have to be removed. This applies mainly if there are closed recycling circuits which help to noticeably reduce the amount of scrap. For many of the more recent cars dismantling handbooks exist which list the removable assemblies and the materials of which they are made. This allows the different materials to be separately collected in different boxes, closed containers and skeleton containers which is absolutely indispensable for an efficient, high-quality recycling. The separation and sorting technique is the absolute pivot of any waste processing. The better the segregation of the materials, the easier and more economic is the recycling process.

For the time being, the recycling of plastic materials is technically feasible only if certain boundary conditions are observed and if both segregation and cleanliness meet highest quality standards.

After all valuable and recyclable parts have been removed, the rest of the vehicle – which essentially consists of the naked body – is shredded.

During shredding (Fig. 3), what remains of the automobile is mechanically torn to small pieces and fed into segregated material flows.

The shredded fraction contains:

- Iron and steel (Fe content),
- Aluminium, zinc, lead, copper (non-ferrous metals),
- Non-metal materials (plastics, glass, paint, pieces of carpets, dirt) known as "fluff".

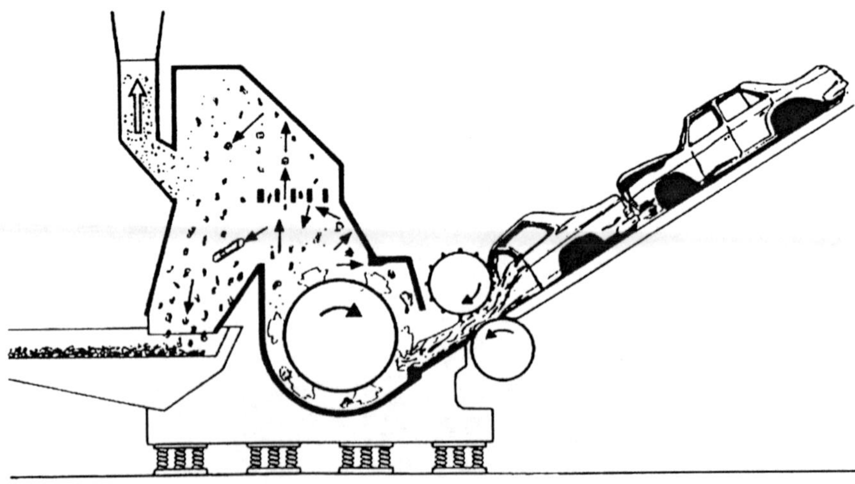

Fig. 3. Principle view of a material shredding process

3.2.2
Iron Fraction

After having been shredded, the metallic materials of the body structure are fed into the processing cycles for iron (steel mills) and non-ferrous metals (refineries). The recycling rate of metals is more than 95%. As a rule, metals can be re-utilized several times without suffering any quality losses. Thanks to these secondary raw materials, several hundred million tons of primary iron ore can be saved and do not need to be mined, processed, transported and smelted. In addition, the smelting of primary iron ore during steel production requires 4 times the energy needed for the melting of steel scrap. Thus, used-car recycling is an important means of saving energy, raw materials and processing costs.

3.2.3
Non-Ferrous Metals

10 to 20% percent of the weight of an average automobile consist of non-ferrous metals. The different alloys must be identified and segregated before being fed into the recycling process.

The recuperated materials are taken to their producers (e.g., the metallurgical plants or chemical industry) where they are processed into new materials.

More than 90% of the aluminium contained in used vehicles is recuperated. Secondary aluminium made out of scrap is of the same high quality as primary aluminium.

The recycling of aluminium from scrap requires 80 to 90% less energy than is needed for the production of primary aluminium from bauxite.

3.2.4
Fluff

Shredding does not only produce pure, high-quality metal scrap but also a certain amount of residues which have to be treated as special waste. The fluff contained in a passenger car is about 250 kg almost one third of which are plastic materials. These residues form the critical link in the used-car disposal chain.

Today, only part of such shredding residues as plastics, elastomeres, glass, textiles, paints and hardware remaining in the used-car body can be efficiently disposed of. Most frequently, the shredding residues are simply dumped. Currently, apart from the dismantling and draining of non-metallic assemblies, new methods are being examined to reduce the amount of such residues with special focus placed on the disposal of plastics.

The methods for plastics recycling can be broken up into three different categories: energetic recycling, chemical recycling and material recycling.

3.2.5
Energetic Recycling

Any recycling method is aimed at making optimum use of the inherent energetic and raw-material potential of the waste to be recycled. A major portion of the plastics and composite materials end up as shredding residues which are burnt in garbage incineration plants.

The thermo-energetic recycling of waste is a method which is particularly suited for the mixed and strongly contaminated residues of fluff.

Due to their high energy content, plastics lend themselves to thermal utilization whenever material recycling is not possible for either economic or ecological reasons. Thermal recycling mainly recommends itself, if the calorific value of the waste is >11 MJ/kg and the combustion efficiency is higher than 75%.

3.2.6
Chemical Recycling

The chemical recycling of plastic materials is a particularly interesting utilization method. It consists in chemically separating the plastics into low-molecular compounds (monomers) which can be returned into the manufacturing process. There are different types of chemical recycling, i.e.:

- pyrolysis,
- hydrolysis,
- gasification,
- cooking,
- hydrogenation.

In the course of these processes, the materials are decomposed into their constituents which are then used as raw materials for the manufacture of new products.

At the time being, intensive efforts are being made to further develop and improve the chemical recycling of plastic materials.

3.2.7
Material Recycling

Material recycling consists in reprocessing scrapped parts in closed material circuits for utilization in the production of new products.

As far as mixed-up plastic materials are concerned, however, material recycling is of limited use only because of the inherent quality problems: To obtain high-quality recycled materials we need pure, unmixed plastics waste.

3.3
Recycling-Friendly Design – Think of the End from the Start

About 80% of the ecological impacts of a product are "born" in the development phase or, in other words, have to be answered for by the development departments. The production concept laid down during development is binding upon all subsequent departments. It therefore is vital for development to precisely analyze the environmental impacts of the chosen production processes, to determine the waste caused by packing materials, for example, or the possibilities of environment-friendly disposal at the end of the product's life. Thus, development departments hold a key position with regard to environmental protection as they have to fulfill the basic requirements which will relate to the environmental impact of both production processes and products.

Automotive industry in Europe (automotive manufacturers and suppliers) is making intensive efforts to further increase the current recycling rate of 75 to 85% by 2006 and to 95% by the year 2015. This target can only be reached if such issues as environmental protection and the preservation of the world-wide raw-material sources are given due consideration during new-car development and design.

A designer's responsibility does not end with the completion of the functional, production-related and economic concept of his product. It rather should be considered as a "holistic responsibility" giving consideration to the entire life span of a product from its creation to the end of its lifetime including the final recycling and/or environmentally-friendly disposal.

For the recycling rate to be as high as possible and economically justifiable at that, the boundary demands of recyclability must be borne in mind from the very beginning of vehicle and component design.

As far as long-life products such as automobiles are concerned, the particular challenge which designers are faced with consists in having to take recycling-oriented decisions during early product design even though recycling will actually have to be done 10 or 20 years later – at a time when the recycling technologies might be quite different from those assumed today. Accordingly, designers must continuously remain informed about and further improve the latest methods for the design of recycling-friendly products.

The recyclability of an automobile heavily depends on the number and type of materials used and on the ease with which these materials can be identified and segregated.

The designers' main concern is to choose environment-friendly and easy-to-recycle substances and to reduce the number of materials used and to ensure their markings. Recyclability can be further promoted by intensifying the use of either recycled materials or materials containing high portions of recycled substances.

The desired effect of material recycling is to reduce the consumption of raw materials and energy. The overall amount of specific energy needed to recycle a material includes the so-called primary and secondary energies. The primary energy represents the energy consumed during the winning of raw materials and their subsequent processing. It also includes the transport energies required to do so.

Material recycling allows a major portion of this energy to be saved. The so-called secondary energy includes all the energies needed during the recycling process to produce the same product.

Table 1 lists the specific primary and secondary energies required for the primary production and/or the recycling of the most important automotive materials.

When selecting recycling-friendly materials, a series of additional criteria have to be kept in mind such as, for example, the regulations concerning the avoidance of hazardous materials and the marking of components.

Table 1. Primary and secondary energies for some important materials (MJ/kg)

Steel	40,000	18,100
Iron	34,000	24,000
Aluminium	190,000	26,700
Glass	30,000	13,000
Lead	41,100	8,000
Copper	100,000	45,000
Rubber	67,600	43,600
Polypropylene	74,300	42,300
Polyvinyl chloride	65,400	29,300
Polyester	95,800	50,000

The automotive industry has made a point of not using the following (hazardous) materials:

- cadmium,
- mercury,
- CFC,
- asbestos,
- lead,
- chromium(VI) compounds.

One of the main prerequisites for the recycling of plastics is the precise marking of the different materials without which it would not be possible to correctly separate the various substances and produce high-quality basic materials. Currently, all the plastics contained in a modern automobile and accounting for more than 100 g are marked individually. It is important for these materials to be precisely identified in order to allow them to be fed into the reutilization cycle.

Material identification is governed by:

- ISO 1043 symbols for plastics identification,
- ISO 11469 for plastics; material-specific identification and marking of plastic parts,
- VDA 260 marking of parts made from polymeric materials,
- SAE J1344.

In the foreseeable future, positive recycling inputs are expected to come from renewable natural materials such as flax, cotton, leather, coconut fibers etc.

3.4 Lifetime of Products

A recycling-friendly design must also make sure that parts can be easily repaired. Components, spare parts and replacement units must be designed and manufactured in a way so as to maximize their useful lives and to make them easy to remove and to be reused, as far as this is technically feasible. The best method of value preservation is to reprocess the parts and units and to use them as replacements. A reconditioned replacement engine, for example, costs about 40%

less than a completely new one. As a rule it can be said that by doubling the average life of a product the yearly amount of waste can be cut by half. Therefore, the target should be to create products which have a long life span, are easy to repair and can be recycled without problems.

While following the recommendations for a recycling-friendly design, it is most important, however, also to meet first of all the respective quality and safety standards. Under no circumstances must the quality and safety levels of a car be deteriorated by the use of recycled materials or parts.

3.5
Environmental Auditing, Life Cycle Assessment (LCA)

Environmental audits serve to identify and compare the desired/actual status of the materials used, the amounts of waste produced and the legal requirements to be observed in production. Based upon the results of this audit, ecological counter-measures are defined, if required. Among the suitable auditing instruments are the Environmental Management and Auditing System (EMAS) as well as ISO 14001.

Apart from the ecological impacts at the production sites, consideration must also be given to the environmental pollution caused by a product over its entire lifetime. In the automotive domain, life cycles are of particular importance because the environmental impacts of a vehicle during its utilization are far greater than during its production.

Drawing up a convincing ecological balance for a product as complex as an automobile is the most difficult phase in ecological auditing because it covers the entire life cycle from raw material extraction to final car disposal.

The world-wide growing shortage of not-renewable resources and the steeply decreasing landfill capacities have led to the conclusion that the manufacturers' responsibility for their products must be considerably extended.

All automotive manufacturers are making intensive efforts to further improve the ecological aspects of their vehicles using tools such as environmental management systems according to ISO 14001 to analyze the ecological impacts over the entire life-cycle of their products.

The term "sustainable development" describes a development which satisfies the requirements of the present without compromising the ability of future generations to meet their own needs.

At the time being, automotive industry is concentrating its efforts on doing Life Cycle Assessments (LCA), i.e., drawing up ecological balances which cover the entire life time of an automobile (ISO 14040).

LCA includes the whole range of activities from raw-material extraction to the various possibilities of utilization at the end of the life cycle, such as incineration, dumping, recycling, and reutilization also covers the various stages of production, transport, distribution, utilization and disposal (Fig. 4).

LCA is an instrument which has been primarily designed to internally optimize a product and thus to reduce its ecological impacts over its lifetime.

According to ISO 14040, the term "Life Cycle Assessment" is defined as "a systematic approach designed to collect and analyze the material and energetic in-

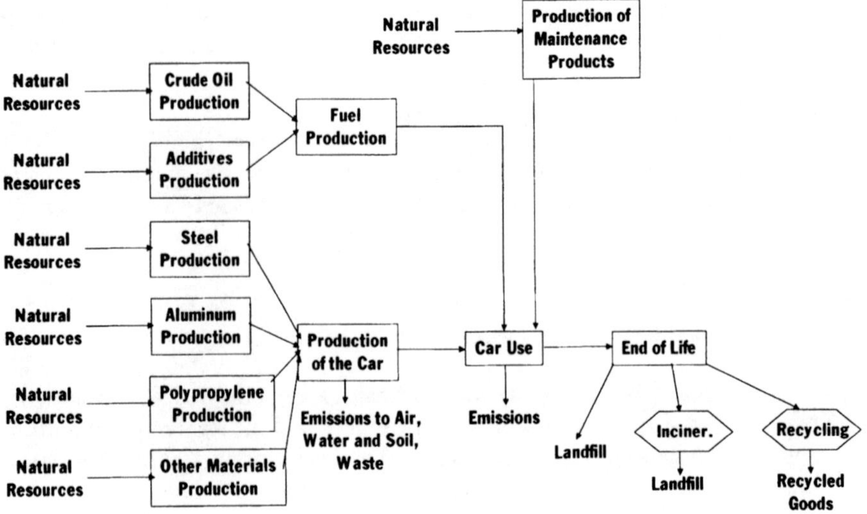

Fig. 4. Life cycle of a car, description of system boundaries [27]

puts and outputs of a system and the environmental impacts resulting therefrom during the execution of an activity or the utilization of a product over its entire life time."

Environmental auditing is a relatively new scientific discipline which is still under development. Currently, various auditing methods are available whose results can differ strongly from each other.

3.6 References

1. Life cycle analysis: Getting the picture on vehicle engineering alternatives. Automotive Engineering March, 1996
2. Recycling state of the art for scrapped automobiles. American Iron and Steel Institute, 1992
3. ICAR Report. Japanese Consortium on Automobile Recycling. The Japanese Research Institute, Tokyo 1993
4. Design for recyclability. Renault Standard 00-10-060/AA, 11/97
5. Weber R (1975) Recycling und Automobilbau – Gesichtspunkte für zukünftige Entwicklungen. Automobil Industrie
6. Brüdgam S, Barrenscheen J (1993) Demontagegerechte Produktgestaltung in der Automobilindustrie. ATZ/MTZ-Sonderheft, Fertigungstechnik
7. Verwertung und Entsorgung von Airbags und Gurtstraffern. Verband der Automobilindustrie (VDA), Frankfurt 1997
8. Grey D, Löhner L (1996) Abfallvermeidung und -verminderung im Maschinen- und Anlagenbau. Umwelt 26: Nr. 11/12
9. Stahl WR (1991) Längere Nutzungsdauer von Produkten – eine Strategie für die Zukunft? Genf
10. Märkte für ausgewählte Sekundärrohstoffe in der Bundesrepublik Deutschland. DIN-Wochenbericht 36/93
11. Jochum HG (1995) Recycling von Automobilkomponenten. Forums- Fachtagung, Stuttgart

12. Automobile Recycling. AAMA-American Automotive Manufacturers Association, Washington D.C. 1993
13. Schredderabfall: Verbrennen ist sinnvoll. Umweltschutz Nr. 12, 1996
14. Krüper M (1994) Umweltmanagement. DGMK-Haupttagung. DGMK-Tagungsband 9403
15. Murphy J (1998) Plastic Recycling – the need for an infrastructure. Automotive interiors international
16. Projekt zur Altautoverwertung der deutschen Automobilindustrie – Verwertung von Kunststoffen aus dem Automobil. VDA 1995
17. Entsorgung von Altfahrzeugen in Baden-Württemberg. Landesanstalt für Umweltschutz, Baden-Württemberg 1994
18. Beits W (1995) Umweltschutz beginnt beim Konstrukteur. Mobil Nr. 1
19. Design for Recycling. Automotive Engineering. August 1997
20. Schlotter U (1996) Werkstoffliche Verwertung von Kunststoffen aus Altfahrzeugen. Kunststoffe 86:8
21. Kramer B, Brüdgam S, et al (1996) Kunststoffrecycling: Aufgaben und Realisierung. DGMK-Tagungsband, Hamburg
22. Recycling, Sonderpublikation. Automobil Industrie 40 (1995), Mai
23. Fuhrmann E (1978) Eine neue Herausforderung für den Ingenieur. Eintrittsvorlesung, TU Wien
24. Fuhrmann E, Schäfer R (1975) Studie "Forschungsprojekt Langzeitauto". ATZ 77 4, 5
25. Optimising plastic parts for recycling. Automotive engineering, May 1996
26. Coppens C, Le Coq M, et al (2001) Evaluating and Improving the Recovery aptitude of an automobile function: the PSA Approach. Ingenieurs de l'Automobile. Janvier
27. International Automobile Recycling Congress. ICMAG Geneva, 2001

The Handbook of Environmental Chemistry Vol. 3, Part T (2003): 175–253
DOI 10.1007/b10463

Legislation for the Reduction of Exhaust Gas Emissions

Wolfgang Berg

Consultant, International Automotive Regulatory Affairs, Schilfweg 17, 70599 Stuttgart, Germany, *E-mail: dr.berg@berg-automotive.de*

About 40 years ago, the term "emission control legislation" opened a new chapter in the history of the automobile: conventional engine development criteria like, e.g., increase of performance, reduction of mass and volume and improvement of reliability and durability were amended by legislative requirements about the reduction of pollutant emissions from the vehicle's fuel system and exhaust gas.

These requirements have initiated extensive activities by the various involved industry branches, e.g., in the field of basic research about the formation of pollutant exhaust gas constituents during the combustion process and technical means for their reduction, about the question by which representative driving cycle the vehicle's operation conditions on the road can best be reproduced on chassis dynamometers as the basis for testing it's emissions and fuel economy under comparable laboratory conditions, and last not least in the field of measurement equipment and test methods for reliable and reproducible determination of the vehicle's emission characteristics.

The development of this "emission control legislation" took place at a highly dynamic pace and has not lost its impact on the involved industry since its beginning to date. On the contrary, it has constantly increased in comprehensiveness and complexity over the years and is still continuing to do so in spite of the fact that common efforts of auto-, mineral oil- and supplier industry have resulted in the fact that automobiles equipped with today's state-of-the-art emission control technology produce close-to-zero emission levels which were not considered achievable when emission control work began.

Although worldwide emission control programs have a common single objective, individual legislative requirements unfortunately developed towards different directions in different countries. By reviewing the corresponding legislative frameworks, emission standards and test methods from their historical evolution to the latest status and future perspectives, the following chapters shall now provide insight in the philosophy and complexity of existing emission control requirements in the key player nations USA, Japan and the European Union.

Keywords. Automobiles, Emissions, Legislation, Standards, Test methods, Driving cycles, California Air Resources Board, LEV II, OBD, Environmental Protection Agency, "Tier 2", Fuel economy, CAFE-Program, Gas guzzler tax, European Union, EU-Directive, ECE-Regulation, Japan MOT, MLIT, Heavy duty vehicles

1 Emission Legislation

1.1 Emission Legislation in California

California deserves the role of being the world's pioneer in the field of emission control legislation for automobiles.

Its roots can be traced back to 1943 when air pollution started to draw public attention in Los Angeles, California. When the problem aggravated, California established so-called "Air Pollution Control Districts" in 1948 which started to attack the problem by reducing emissions from industrial and private combustion processes [1].

Substantial visibility improvement was achieved through the lowering of smoke emissions from these sources by more than 75% down to the level of 1940 when "smog" episodes were still unknown. However, eye-irritation and other health effects remained and even worsened due to high and continuously increasing ozone concentrations in the atmosphere [1].

The fact that industrial processes do not directly emit ozone (at least not in notable amounts) led to the suspicion that some kind of chemical reaction in the atmosphere was the cause of ozone formation [1].

In 1952, A. J. Haagen-Smit at the "California Institute of Technology" could demonstrate ozone formation when a mixture of organic compounds and NO_x was subjected to sun radiation. This famous work was not only a milestone in the understanding of the "Los Angeles Smog" as a phenomenon resulting from precursors like hydrocarbons and nitric oxides which are emitted from the evaporation and combustion of fossil fuels, it drew as well public and political attention to the automobile as the main source of these emissions [2].

From 1950 to 1960 the car population in Los Angeles increased by 70% and the amount of total hydrocarbons emitted by these mobile sources became the major single contribution to the Los Angeles smog and it became clear that legislation was needed to limit and reduce these and other emissions from the automobile [3, 4].

The next logic step towards the establishment of such legislation was the definition of an air quality which could be considered sufficiently safe for public health. With the objective to find limits which would eventually re-establish the air quality of 1940, the "State Department of Public Health" – by assignment of California legislature – worked out the first air quality standards for California. These California "Air Quality Standards", as adopted in 1959, represent as well the first attempt to define a legislation and standards for the limitation of emissions from automobiles [4].

In April 1960 the "California Motor Vehicle Pollution Control Board" (CMVCB) was created and authorized to develop a method for the certification of emission control systems the introduction of which would later be required by law. Such legal requirements should, however, only become mandatory after the CMVCB had certified at least two different executions of such systems [5].

In 1968 the "California Air Resources Board" (CARB) was established combining the responsibility for motor vehicle emission control regulations, air quality standards setting and the coordination of activities by local offices working on emissions from non-mobile sources [6].

Although much of California's basic work and achievements of the early years was taken over by Federal authorities and used as a starting point for the establishment of Federal regulations, the California ARB has maintained its worldwide leading role in the field of mobile source emission control legislation to date. Its work did not only form the basis for legislation dealing with component and new vehicle certification in the US but triggered similar developments in the rest of the world.

California was as well the first state to establish additional requirements for the control of new production vehicles on March 19, 1969. These "conformity of production" (COP) testing requirements became mandatory as of April 1, 1970 and have since then been substantially expanded in scope [7].

Further, on March 3, 1973, California legislature introduced the first "compliance testing" requirements, known as "Title 13"-testing. This testing included a full certification emission test either at CARB's or the manufacturer's premises and was first executed by CARB representatives on new model year 1974 and later vehicles in the US before the cars were delivered to customers [8].

The second part of California's "Title 13"-testing was the so-called "New Dealership Surveillance"-testing mandated as of 1974, allowing CARB to check new vehicles in the dealerships for compliance with the certified status by means of parts checks and an idle emission test [8].

The last link in the chain of California's emission legislation was the "Inspection and Maintenance" (I/M) program, which basically was equivalent to the 3rd stage of its 1973 "Motor Vehicle Inspection" program. The introduction and routine application of an effective I/M-program was a legal (Federal) obligation for all states in the US which did not meet the "National Ambient Air Quality Standards" (NAAQS) determined by the US-EPA.

Although California's auto emission control legislation thus eventually covered the full lifecycle of vehicles from the end of the development status (certification requirements), via series production (COP testing) to field operation (I/M programs), most emphasis was placed for about two decades on its responsibility to regulate and control new vehicle emissions compliance.

In-use testing of vehicles beyond the scope of I/M-programs was only done when the certification test results of a certain vehicle or other indicators suggested to the CARB that the design of subject emission control system may not maintain the emission level in actual field operation as it was demonstrated by the car manufacturer during the official durability test run of the new vehicle certification procedure.

Such in-use testing of a full vehicle according to the certification emission test procedure is a rather time-consuming and costly undertaking for the legislator not only from the point of view of vehicle procurement but also because a statistically valid sample size must be tested if the final test result is intended for use in legal actions against a non-complying car manufacturer. So this kind of controlling and ensuring emission compliance in the field is not very attractive for the legislator.

I/M testing, on the other hand, seemed to be a rather effective and efficient tool to ensure a vehicle's emission potential in actual use. However, with the introduction of increasingly sophisticated emission control technologies in motor vehicles it became more and more difficult for workshop mechanics to understand and correctly service the cars and their emission control equipment. This was especially true when these systems began to include "black box" electronics. When the attempt to keep pace with this development through information and education programs for workshop and garage personnel failed, the ground stone was laid for a legislation requiring from car manufacturers the development of technical means which would allow an automatic self-checking of the emission control system on board the vehicle itself.

Consequently California Legislation defined the world's first "On Board Diagnostic" (OBD I)-requirements [9], an emission-related subject which will later be discussed separately in Sect. 4.2.

However, "the event which changed the world" not only in the field of auto emission control legislation but as well in the corresponding technology section happened in 1990 when the California "Health and Safety Code" was amended by California's famous "Low Emission Vehicle " (LEV)-Program.

The first step of this program, the so-called "LEV I"-regulations [10], defined the following new vehicle classes:

- Transitional Low Emission Vehicles (TLEV),
- Low Emission Vehicles (LEV),
- Ultra Low Emission Vehicles (ULEV),
- Zero-Emission Vehicles (ZEV),

comprising passenger cars (PC) and light duty trucks (LDT) up to 3,750 lbs as well as LDTs and medium duty vehicles from 3,751 to 5,750 lbs.

The LEV-program was oriented at the special air quality conditions in California and its objective is, therefore, the reduction of ozone in the lower atmosphere. It introduced two new elements in its emission control regulations which focus on the determination and consideration of all organic gases (HC-emissions) which are – due to their reaction with NO_x – considered as ozone precursors. These two elements were:

- all measurable HCs not containing O_2 with <12 C-atoms (except methane) and all O_2-containing HCs (ketones, aldehydes, alcohols and ethers) were combined in a so-called "Non-Methane Organic Gas" (NMOG)-standard;
- the NMOG emission test result of vehicles operated on alternative or re-formulated fuels had to be corrected with a so-called "Reactivity Adjustment Factor" (RAF) which reflects the ozone-forming potential of the given exhaust gas.

For the so-corrected NMOG emission value the new California legislation defined an introduction scenario in which the applicable NMOG standard was stepwise reduced from model year 1994 to model year 2004.

Further characteristics of California's new LEV-program are that it did not only – for the first time in the history of auto emission regulations – consider engine and fuel as a unit in which both sides depend on each other with regard to the unit's emission potential and which have, therefore, to be regulated together (which was actually done by requiring from the mineral oil industry improvements of existing fuels and the development of new "clean fuels" for the future). The program specified as well a limit value for CO-emissions at low temperature of 50°F (6.7°C), it established a new formaldehyde standard for the "LEV"-group, methanol vehicles and "clean fuel vehicles" and introduced a more sophisticated test procedure for evaporative emissions, a modified test method for "non-methane organic gases" (NMOG) and substantially more stringent "OBD II" requirements.

But its most surprising element – which came close to a shock for vehicle manufacturers – was the "Zero Emission Vehicle"(ZEV) mandate. This mandate required from large volume manufacturers (car companies producing >35,000 units for sale in California/year like GM, Ford, Chrysler, Toyota, Nissan, Mazda and Honda) to introduce 2% of their California sales program as zero emission vehicles beginning with model year 1998. This requirement should increase to 5% in 2001 and 10% in 2003. Low volume manufacturers got some more time and were allowed to start introducing ZEVs only as of model year 2010, however with the full 10% rate at this point of time.

The technical challenge, to build and introduce in the market a vehicle with zero emissions gave a dramatic push to the development of advanced battery technology since no other solution but an electric vehicle was seen feasible to meet this mandate within the given time frame.

As could be expected, the following years showed that, in spite of intensive efforts, an electric vehicle which would eventually meet customer demands with regard to safety, driving range and cost could not be materialized. This fact was as well acknowledged by CARB which – by amending the first step of its LEV-program ("LEV 1") – set another milestone for California's auto emission control legislation: Following a November 5, 1998 hearing ARB decided about its so-called "LEV II" program which was formally adopted August 5, 1999, filed with the California Secretary of State on 10-28-1999 and became operative on 11-27-1999 [11].

The main new elements of the "LEV II"-program were:

- postponement of the ZEV-mandate to model year 2010 but then requiring from all vehicle manufacturers that 10% of their California sales program must be ZEVs,
- inclusion of sport utility vehicles ("SUVs") and pick-up trucks <8500 lbs in the passenger vehicle regulations,
- extension and further reduction of the NMOG fleet average standards for model years 2004–2010,
- substantial strengthening of the NO_x standards for LEVs and ULEVs (up to –75% vs. the existing standards),
- drastic strengthening of the PM-standard for LEV II Diesel vehicles,
- phase-in of the new standards (2004: 25%; 2005:50%; 2006:75%; 2007: 100%),
- increase of the durability requirement from 100,000 to 120,000 miles for passenger cars (PCs) and light duty trucks (LDTs),
- elimination of less stringent emission standards ("Tier 1"-and TLEV-standards) because CARB expected that these standards would have allowed an increase of PCs, pickups and SUVs with,
  Diesel engines – which it did not want to happen,
- introduction of a "Compliance Assurance Program" (CAP 2000),
- establishment of a further low emission vehicle class, the "Super Ultra Low Emission Vehicle" (SULEV),
- allowance to earn ZEV-credits with vehicles having close-to-zero emissions, so-called "Partial Zero Emission Vehicles" (PZEV).

The latter provision allows manufacturers to substitute a certain percentage of their "base" ZEV-mandate by PZEVs or other substitute vehicle categories as shown in Table 1. This table shows further the extension of the 10% base mandate and its strengthening up to model year 2018 as decided in the Board hearing of January 25, 2001.

During the time after publication of the LEV II-program problems with its materialization became more and more obvious which caused the staff of the ARB to work out further modifications to the existing text of the ZEV-mandate. These modifications – which are summarized in Table 2 – were presented by the ARB staff on 8 December 2000 and adopted (with some further modifications)

Table 1. Partial ZEV-Credits for the Substitution of California "Base"-ZEV-Requirement

Model Years	Minimum total ZEV requirement (of total production)	Minimum share of EV or ERHEV	Minimum share of AT-PZEV (or ZEV or ERHEV)	Maximum share of PZEV (or ZEV or ERHEV or AT-PZEV)
from 2003 to 2008	10%	2%	2%	6%
from 2009 to 2011	11%	2.5%	2.5%	6%
from 2012 to 2014	12%	3%	3%	6%
from 2015 to 2017	14%	4%	4%	6%
from 2018 on	16%	5%	5%	6%

Including modifications from the January 25, 2001 Air Resources Board Hearing.
ZEV = Zero Emission Vehicle.
ERHEV = Extended Range Hybrid Electric Vehicle (HEV with at least 0.2 PZEV) baseline credit and AER ? 20 miles attributable to off-board charging).
AT-PZEV = Advanced Technology Partial Zero Emission Vehicle (with at least 0.4 PZEV credit).
PZEV = Partial Zero Emission Vehicle (with at least 0.2 PZEV baseline credit).

Table 2. Modifications to the California ZEV-Mandate presented in the Board Hearing on January 25, 2001

Technology	Characteristics	PZEV-Credit
Improved SULEV	- Meets SULEV Emission standard - OBD II - Requirements - Meets "Zero-EVAP" Reuqirement - Warranty 150,000 miles/15 years	0.2 (baseline credit)
Emission-Free Mileage	- Has "All Electric Range" (AER) with off-board charging - No credit if AER < 20 miles in city cycle - Max. credit if AER > 100 miles - Additional credit of 0.1 for HEV with AER > 20 miles - Half credit if only one pollutant is zero /Example: Vehicle with on-board methanol reformer and zero NOx	0.3 – 0.6
Advanced ZEV-Technology	- AER < 20 miles (no credit for emission-free mileage) - Vehicle equipped with high performance battery, electric propulsion or fitted with any other progressive ZEV-technology	0.25
Clean Fuel Use	- Fuel cycle related NMOG emissions < 0.01 g/m (depending on % of vehicle miles travelled with clean fuel) - Includes all emissions of fuel production and distribution (Clean Fuel: CNG, LPG, Hydrogen)	max. 0.2

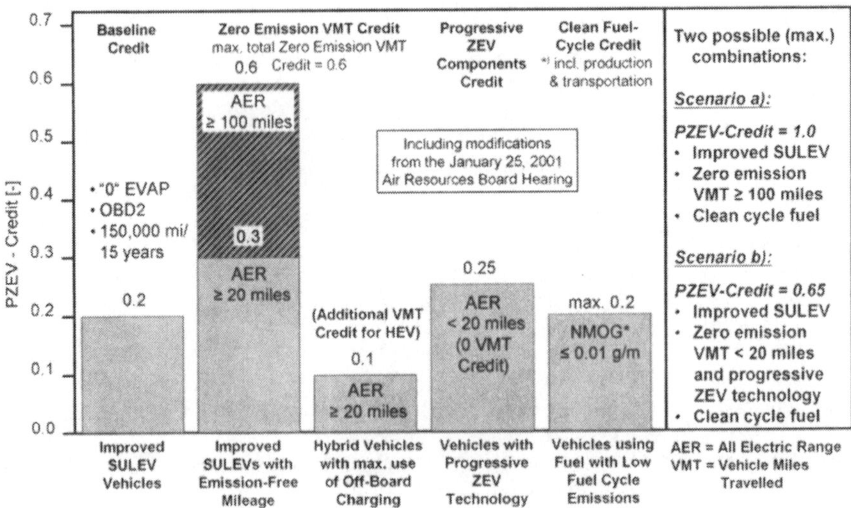

Fig. 1. Final California "Partial Zero Emission Vehicle" (PZEV) – credits

in a Board hearing on January 25, 2001 [12]. The package was again discussed during the Board hearing on June 28, 2001. Figure 1 shows the revised options for PZEV credits which may be earned depending on the applied technology and its zero emission potential.

Opinions about the Board's decision to add several substitution possibilities to its ZEV-program differ. One may consider several relaxations now added to the program as an effort by CARB to save the original mandate vis-à-vis obvious problems connected to an enforced materialization of the ZEV-mandate without losing face.

On the other hand, ARB's allowing additional alternatives to the pure electric vehicle to participate in the program may be seen as a reasonable adaptation to real world conditions without giving up the final objective of introducing extremely clean vehicles throughout the state of California.

The general policy of California's legislature of establishing highly demanding and even "technology-forcing" auto emission standards in connection with schemes allowing their stepwise introduction became the guideline for all other nations when establishing and executing their own emission control legislation for automobiles.

1.2 US-Federal Emission Legislation

The event that triggered Federal efforts towards evaluating the air pollution problem was the Donora incident of 1948 [13]. Following this air quality episode the "Department of Health, Education and Welfare" (DHEW) emphasized that means should be provided to allow for an investigation of such air pollution problems on a nationwide basis. As a consequence, the "Public Health Service Act" and the

"Air Pollution Control Act" were signed in 1955 and can be regarded as the starting point for the Federal auto emission legislation [14].

The "Air Pollution Control Act" authorized the DHEW to suggest research programs and investigations about the effects of air pollution, but still emphasized that " the bill does not propose any exercise of police power by the Federal Government, and no provision in it invades the sovereignty of states, counties or cities" and that "there is no attempt to impose standards of impurity" [14].

Between 1955 and 1963 extensive programs were performed to establish a basis for the determination of possible countermeasures against air pollution and emission sources [15]. Auto emissions came into focus only after 1960 because measurement techniques existing before that time did not allow a meaningful determination of the emissions of a given average vehicle [16].

Eventually, on June 8, 1960 the "Schenck Act", (better known as the "Motor Vehicle Exhaust Act") was signed assigning the "Surgeon General" to establish a study about the composition, effect and tolerable level of pollutant emissions from automobiles and to submit it to Congress within two years [17].

These investigations provided the background for the basic law of the Federal auto emission legislation, namely the "Clean Air Act" signed on December 17, 1963. The Act introduced the principle that the effects of pollutant emissions from automobiles on the environment had to be reviewed on a regular basis and asked oil and auto industry to establish a cooperation for the development of devices and fuels which would lower pollutant emissions from automobiles [18].

Between 1963 and 1967 extensive state and local research programs were performed. The conclusion was drawn that a nationwide program for the control of air pollution caused by automobiles was technically feasible. Accordingly, the "Clean Air Act" of 1963 was amended on October 20, 1965 by the famous "Title II – Control of Air Pollution from Motor Vehicles" [19].

The amendment authorized the DHEW to define the first Federal emission standards for model year 1968 and required emission testing of prototype vehicles before a certificate for sale could be issued. Disablement of parts of the emission control system before the sale of a vehicle was forbidden and production vehicles were considered to meet the applicable emission standards as long as they were technically identical with the certified prototype vehicle [19].

A major change in the philosophy of Federal auto emission legislation showed up in the "Air Quality Act " which was signed on November 21, 1967. The previous provision that the secretary of the DHEW should gather and publish air quality criteria just for information was replaced. From now on he was authorized to inform the states about air quality criteria which he deemed necessary for the protection of public health. He was further entitled to recommend emission control systems which would be needed to meet these air quality objectives [20].

Further, the DHEW was asked to establish national air quality regions for which the individual states then had to define air quality standards and establish a plan how they intended to achieve them.

"Title II" of the "Clean Air Act" was amended by a provision that no individual state (except California) is allowed to introduce special legislation requiring vehicle testing as a prerequisite for sale. However, all states remained free to introduce emission requirements in connection with registration renewal [21].

Oversight hearings, held by the "Subcommittee on Air and Water Pollution" during the time period from 1968 to 1969 revealed that emission control for the conventional engine had not made progress and further legislative steps were necessary to cope with the increasing air pollution problem in US cities. Further hearings in 1970 eventually lead to the "Clean Air Amendments" of December 31, 1970 [22].

Now a major part of the individual state's responsibility was shifted to Federal authorities. The "Environmental Protection Agency" (EPA) was established combining environmental activities which were so far in the responsibility of different departments.

Legislation asked for the development of new fuels which would result in less harmful combustion products, suggested an intensified search for low emission vehicles and put emphasis on short- and long-term effects of air pollution [23]. A detailed discussion of the motor vehicle related provisions in the "Clean Air Act" as amended (1970) is given in [8].

The following time period of the 1970s can be called the "hearing decade". Triggered by the 1970 amendments, legislation proposed stringent – "technology-forcing" – emission standards for the upcoming years for which no emission control technology was available or foreseeable at this point of time. Even if vehicle manufacturers demonstrated "good faith efforts", i.e., if they proved their inability to meet these standards in spite of utmost financial and manpower support, they could only get a temporary relief on the time axis or were allowed to meet a less stringent interim standard. Eventually, the 1970 amendments brought the breakthrough of catalyst technology within the following decade [8].

Further substantial modifications to Federal emission regulations were introduced by the "Clean Air Act Amendments of August 7, 1977 which are as well discussed in detail in [8]. An important provision shall be described here as example for the ever increasing scope and complexity of the Act: It was no longer sufficient for a car manufacturer to develop an emission control system which meets the specified standards. As of model year 1979 the manufacturer was now additionally required to submit sufficient proof that any new emission control technology would not pose, or contribute to, an unreasonable risk to public health and safety.

The practical consequences of this provision were felt soon for Diesel vehicle manufacturers: Results of in-vitro and in-vivo tests with Diesel exhaust gas had to be presented and discussed in the March and June hearings in 1979 which were scheduled to establish the first particulate matter standard for Diesel passenger cars.

After legislation had caused widespread application of the three-way catalyst technology, emission limits of an unprecedented low level could be achieved for CO, HC and NO_x emissions. As a matter of fact, emissions of these regulated pollutants were practically zero for a vehicle with a warmed-up engine. Could emission control legislation go any further than that? It could!

The "Clean Air Act Amendments" of November 15, 1990 introduced modifications to the existing test procedure for evaporative emissions, set even more stringent emission standards, doubled the "useful life" definition from 50,000 to 100,000 miles and required the introduction of an "On-Board Diagnosis" (OBD)-

system. Similar to California's new legislation, the 1990 amendments considered engine and fuel as a unit which had to be regulated together. It consequently established requirements for new clean fuels, introduced a mandate by which fleet owners were required to buy clean fuel vehicles and allowed a favorable fuel economy calculation in case a vehicle was designed as a "dedicated vehicle", i.e., for exclusive use of methanol, or as a "fuel flexible vehicle" allowing a mixed operation with methanol and gasoline [24].

The 1990 amendments did, however, not include a zero emission vehicle mandate and made further legislative steps, envisaged for model year 2003 and beyond, subject to further discussions after the results of ongoing studies about air quality needs had become available.

On December 17, 1999 the so-far latest program for future Federal emission legislation, EPA's "Tier 2" program, was signed [25]. The important changes vs. the former status of Clean Air Act requirements are:

- passenger car emission standards became also valid for mini vans, sport utility vehicles (SUVs) and light duty trucks (LDTs),
- engine and fuel were, for the first time in Federal auto emission control regulations, regarded as a unit and were, therefore, regulated together,
- emission standards were set "fuel neutral", i.e., the same standards applied to all engine/fuel combinations (identical for gasoline and Diesel engines).

In parallel, EPA proposed a substantial streamlining of its certification procedure within the so-called "Compliance Assurance Program" CAP 2000 which it had developed during a 3-year discussion process with manufacturers, the CARB and other interested parties [26]. The objective of this proposal was to shift compliance emphasis from pre-production cars to in-use vehicles. It relieves auto manufacturers to a large degree from administrative burdens like the need for submitting detailed information about certification vehicles to EPA for review and discussion and reduces vehicle emissions and durability testing prior to certification. However, it requires manufacturers to test in-use vehicles after 10,000 and 50,000 miles of field operation to demonstrate that certification standards are still met.

"CAP 2000" was published in the Federal Register May 4, 1999 as Final Rule. It became mandatory as of model year 2001 but manufacturers could "opt-in" to that program already as of model year 2000.

1.3 Emission Legislation in Japan

From 1955 to 1973 the Japanese economy experienced a remarkable boom [28]. When activities towards coordinated traffic and city planning as well as towards safeguarding environmental protection needs could not cope with this development, legislative efforts were undertaken to stop this development. These efforts can be grouped into the following steps:

- establishment of an administration for environmental protection (1955–1963),
- structural improvement of this administration during the late 1960s,

- establishment of an Environment Agency (EA) and strengthening its function during the first half of the 1970s,
- administration of environmental protection policy.

The following section shall describe how Japan managed – within these activities – to get its auto emission related environmental challenges under control by an effective legislation.

When, in the early 1960s, the advantages of a rapidly improving life standard became more and more endangered by an even faster deterioration of environmental conditions, especially in big cities, local authorities were the first to limit smoke and particulate emissions from industrial processes by so-called "prefectural ordinances". The government joined these local efforts by signing the "Smoke and Soot Regulation Law" of 1962 [27].

The next milestone was the establishment of the "Basic Law for Environmental Pollution Control" of 1967 which emphasized that the responsibility to act against the different forms of environmental damage has to be shared among local authorities, the state government and industry. It envisaged as well the establishment of air quality standards which were eventually defined in 1973 [27].

Automobile emissions were first mentioned in the "Air Pollution Control Act" of 1968 which then replaced the "Smoke and Soot Regulation Law" of 1962. The Act required that emission standards for vehicles be determined for CO, HC, lead "and all other substances which might endanger public health" [27].

When the environmental situation continued to deteriorate dramatically with emissions from automobiles playing a major role in this development, a "Pollution Countermeasures Headquarter" was formed in 1970 and led by the Prime Minister. From this point on, legislation changed its philosophy towards a more aggressive approach on future regulatory steps [27].

While the 1967 law expressed its policy by a wording like "an effort shall be made to keep preservation of the living environment in harmony with economical development", this provision was eliminated when the "Basic Law" was revised in 1970 [27].

In order to get the air pollution under control in cities and other areas, the 1970 law included regulations for fuels and the possibility of traffic restrictions. Local authorities retained the right to introduce individual stricter emission control requirements by means of "prefectural ordinances" if such steps were warranted by the existing local situation [8].

Eventually, in July 1971, the "Environment Agency" was established combining environmental responsibilities which were shared among different ministries before. From now on it was the responsibility of this agency to set emission standards for motor vehicles [8].

However, the EA worked closely together with the "Ministry of Transport" (MOT) – later named "Ministry of Land, Infrastructure and Transport" (MLIT) – which was responsible for the enforcement of these standards. MOT's general responsibility for matters concerning road vehicles is based on the "Road Vehicles Act" of July 1, 1951. Details concerning the content and provisions for handling these matters are described in "Ordinances". MOT Ordinance no. 67 contains the "Safety Regulations for Road Vehicles" which, in Article 31, contains the

Japanese emission control regulations for vehicles with gasoline and Diesel engines [8].

How the Japanese EA handles a legislative proposal can be shown by the example of the introduction of the US statutory standards into the Japanese auto emission regulations:

In October 1972 the "Central Council for Environmental Pollution" presented its report "Long-Term Policy for Establishing Permissible Limit on Automotive Exhaust Gas". Based in this report, the EA – in its notification no.1, dated January 21, 1974 – decided to make these standards mandatory in Japan as of April1, 1975 (for HC and CO) and as of April 1, 1976 (for NO_x) [8].

In a following 2-year hearing sequence from June 1974 to August 1975 this decision was extensively discussed between the EA and car manufacturers especially with regard to the feasibility of the NO_x standard. Manufacturers' positions and technical capabilities were evaluated by an "Expert Committee on Automotive Pollution". The committee's findings eventually made the EA revise its previous introduction plan and postpone the mandatory introduction date for the NO_x standard by 2 years (for domestic manufacturers) and by 5 years for (for importers) [8].

1.4 Emission Legislation in the European Union

Before the EU was established through the "Treaty of Maastricht" on February 7, 1992, Common Market Member States were combined as "European Economic Community" (EEC). This body was formed from 6 states on March 25, 1957 through the "Treaties of Rome" which became effective on January 1, 1958 [28]. This treaty formed as well the working basis of the Community insofar that it required the elimination of trade barriers among member states.

The EEC Commission in Brussels with its General Directorate GD III "Internal Market" is responsible for all work directed towards harmonized legislation among member states of the Community. A General Directorate can be compared to a ministry of a member state. Within GD III works the group "Elimination of Technical Barriers to Trade – Motor Vehicles".

Until the "Treaty of Maastricht" the Commission could develop motor vehicle-related "Directives" totally on its own and transmit them to the "Council of Ministers". It was not obliged to take into consideration anybody's comments. For practical reasons, however, experts from governments and industry were invited to provide political, technical or other advice.

After a Directive was developed by a General Directorate, all members of the Commission had to approve its transmission to the Council. After discussion within the "Economic and Social Committee", after hearing of the "European Parliament", and after voting of permanent representatives, the Council made the final decision by unanimous vote.

After publication of the final Directive in the Register of the EEC, the Commission had the right to initiate a proposal to the Council, on what legal basis the Directive should be transformed into national law of the member states. There were two possibilities:

- mandatory harmonization,
- optional harmonization.

If the Council decided to define the Directive's legal status according to the "mandatory harmonization principle", the member state was not allowed to maintain its corresponding national laws/regulations in parallel to the EEC-Directive. If the Directive's legal status was selected by the Commission according to the "optional harmonization principle", the national regulation/law was permitted to continue to co-exist to the Directive. The option to select the Directive or the national law for vehicle certification then was at the discretion of the car manufacturer [29].

After transformation of the Directive into the national legal framework, a member state was obliged to accept, e.g., an operation certificate for a vehicle from another member state and had to issue itself such harmonized certificates upon request.

An important additional provision shall be mentioned in this context: In an agreement of the Council of 1969, member states declared not to modify or strengthen national regulations without consultation. National intentions to introduce, e.g., more stringent emission standards have consequently to be submitted to the EEC Commission in Brussels. The Commission hears all other member states and then evaluates whether the proposal could create barriers to trade within the Community [29].

This legislative procedure underwent a substantial change with the "Treaty of Maastricht" insofar that the European Parliament was no longer only to be heard in the discussion of legislative proposals but became an equal decision power to the Commission. The principle of "co-operation" was replaced by the principle of "co-decision" which is shown in Fig. 2 [29].

The first Directive of the Community dealing with auto emissions was initiated by the Commission with a proposal submitted to the Council on October 22, 1969. The draft was discussed by government representatives of the (then) 6 member states of the Community on December 15/17, 1969 and eventually – on January 23, 1970 – Brussels agreed on its first Emission Directive published under the well-known designation "70/220/EEC" [31].

Starting from this "basic law" further Common Market emission regulations were developed over the following years in a more or less pro-active attitude of the European auto manufacturers. Negotiating about further strengthening of emission standards was – in contrast to similar developments in the US or Japan – a multinational undertaking for the EU: At that point of time Europe comprised 22 independent states working together at the UN in Geneva in all transportation-related matters including emissions. Many of these states differed substantially in historical evolution, economical strength and political intentions or obligations concerning environmental protection and automobile emission control [29].

Ten of these states were combined separately in the Common Market which was based on special agreements among member states, mainly concerned with free trade. Five of these member states had car producing industries with all together 18 auto manufacturing companies serving domestic, European and other markets.

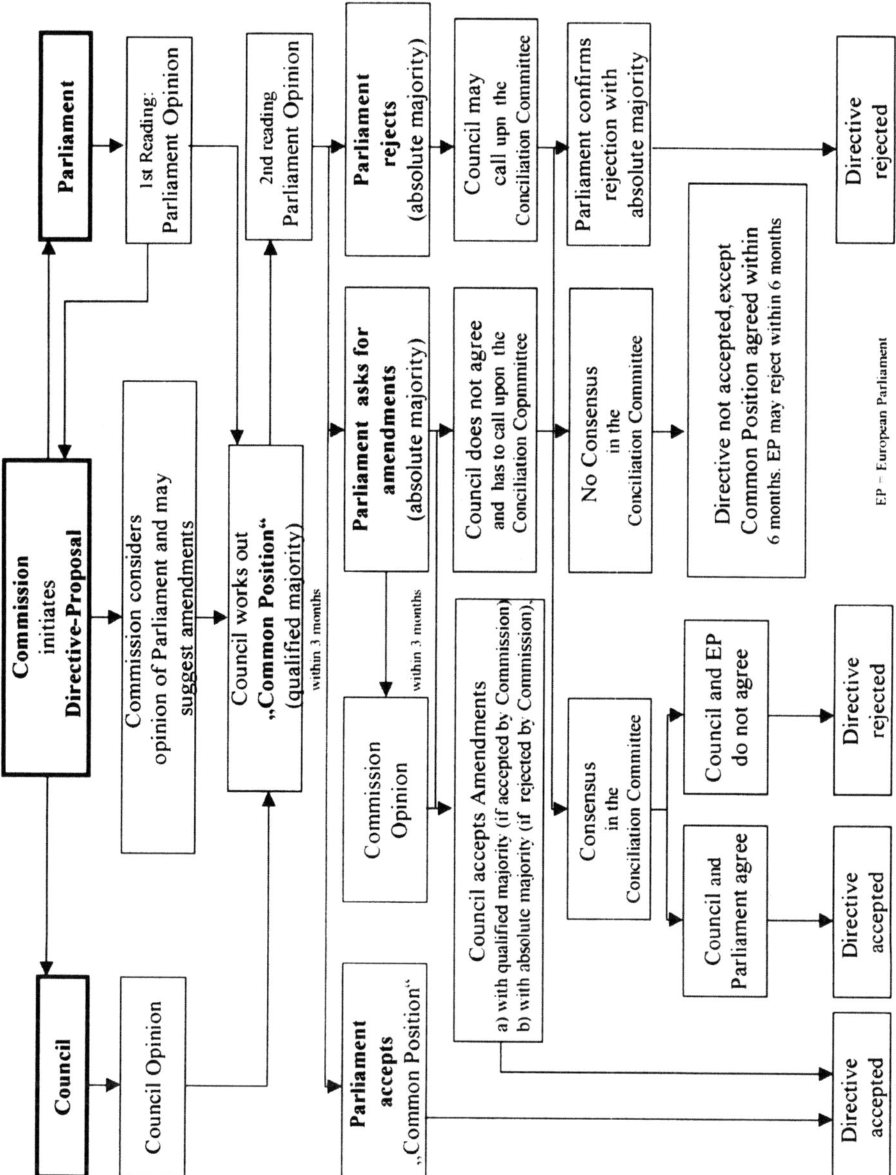

Fig. 2. The co-decision procedure in the European Union (EU) after the "Treaty of Maastricht"

It is not surprising that under these conditions there was almost no chance for an immediate Europe-wide policy change concerning the automobile and its emissions: Existing national legislation or international agreements posed an almost insurmountable hurdle, or made it at least a rather time-consuming event before a consensus among these many independent nations could be reached. This fact must be kept in mind when looking at – and possibly criticizing – the comparatively slow evolution progress of Europe's and the European Union's auto emission control legislation.

Nevertheless, Common Market member states, working together in Brussels towards the development of EEC-Directives, and at the same time working together in Geneva towards the establishment of UN-ECE Regulations, managed to define commonly agreed strengthened auto emission legislation and to strengthen these requirements step-by-step over the years after 1970. These steps are summarized in Table 3 for the time frame from the basic law in1970 to the latest perspective for the year 2000 and beyond [29].

Table 3. Evolution of ECE-Regulations vs. EEC-Directives

EEC-Directive	Introduction of Amendment[a]	ECE-Regulation	Introduction of Amendment[a]	Contents
70/220/EEC	03-20-1970	ECE-R15/00	08-01-1970	Basic Emission Law
74/290/EEC	01-01-1975	ECE-R15/01	12-11-1974	1st standards reduction step
77/102/EEC	04-01-1977	ECE-R15/02	03-01-1977	2nd standards reduction step
78/665/EEC	04-01-1979	ECE-R15/03	03-06-1978	3rd standards reduction step
83/351/EEC	12-091-1983	ECE-R15/04	10-20-1981	4th standards reduction step
88/76/EEC	10-01-1988	ECE-R83/00	01-01-1990	Introduction of "US-Option"
88/436/EEC	10-01-1988	ECE-R83/00	01-01-1990	PM-Directive
89/458/EEC	01-01-1990	ECE-R83/00	01-01-1990	Directive for cars with 1,4 ltr. engines
91/441/EEC	01-01-1992	ECE-R83/01	12-30-1992	EU 1 (1992)
93/59/EC	10-01-1993	ECE-R83/02	07-02-1995	EU1-adaptation for N1-vehicles
94/12/EC	01-01-1994	ECE-R83/03	12.07-1996	EU 2 (1996)
96/69/EC	10-01-1996	ECE-R83/04	11-13-1999	EU2-adaptation for N1-vehicles
98/77/EC[c]	10-22-1998	ECE-R83/03 Amendment 1	05-14-1998	Directive for vehicles with gas engines
98/69/EC	07-13-1999	ECE-R83/05	03-29-2001	EU 3 (2000) und EU 4 (2005)
99/102/EC	12-18-1999	ECE-R83/05 Supplement 1	(09-12-2001)	OBD-Amendments
2001/1/EC	01-23-2001	[b]	[b]	OBD-Requirements for vehicles with gas engines

[a] entering into force of the amendment (specific dates apply to the concerned vehicle categories').
[b] not yet determined.
[c] the Directive 98/77 on gas engines has been handled and published before the Directive 98/69 on EU3 and EU4 although the natural number is higher than that of directive 98/69.

Within this evolution, the Common Market Emission Directive 83/351/EEC represents the endpoint of an era during which legislation for emission control on automobiles could be met by engine modifications but which did not yet require catalytic exhaust gas after-treatment devices.

The next step, however, Emission Directive 91/441/EEC triggered the Europe-wide breakthrough of the catalyst technology. Today the presence of catalysts on EU automobiles is a self-evident fact. However, it had been far from easy to reach a consensus among EU member states about a legislation which would pave the way for this technology.

Looking back to the beginning of emission legislation within the EU one will find a rather long start-up time before member states committed themselves to a common and pro-active approach towards progressive emission reduction on automobiles. But when this initial hurdle was taken, the EU emission legislation caught up in large steps with major elements of today's most advanced emission control requirements in the world established by California. This will be described in detail in Sect. 2.4 for the example of the Community's hard way towards establishing commonly accepted, demanding emission standards.

2 Emission Standards

2.1 California Emission Standards

The first special requirements and standards for the control of emissions from automobiles can be found in the California emission legislation of 1957. This legislation entitled and assigned the "Air Pollution Control Boards" to

- establish "performance standards" for every part, component, system or equipment used for the emission control of an automobile, and to
- authorize the sale of such components or systems [8].

A substantial further step towards the control of vehicle-related emissions was done in 1959 when the California legislature ordered the "Bureau of Air Sanitation" to establish – until February 1, 1960 – maximum allowable limits for pollutant emissions from automobiles [4].

Before such standards could be set, it was necessary to define the existing situation and the envisaged target. Concerning vehicle emissions the average emissions of an "uncontrolled" vehicle had to be determined as a starting point. Concerning air quality, the existing level had to be measured and a target air quality had the to be defined. The difference, expressed in % reduction, would then have to be transformed into a reduction rate for the baseline emissions of the existing vehicle fleet and thereby eventually lead to the target emission standards required by legislation.

The baseline emission levels for HC and CO were determined in a study performed by the "Coordinating Research Council" (CRC) from November 5 to December 7, 1956. From 169 vehicles, which were tested according to a 12-mode test, baseline emission levels were calculated to 1375 ppm HC and 3.8 vol.% CO [5].

On the air quality side, the input data used for the definition of the first emission standards for automobiles were highly uncertain due to the limited measurement capabilities at this point of time. Nevertheless, a serious attempt was made to establish these standards in relation to an envisaged air quality target. The objective was, not to exceed 30 ppm CO as an 8-hour average and 0.15 ppm oxidants as a 1-hour average taking into account all emission sources in the Los Angeles basin. This target should allow re-establishment of California's air quality level of 1940 by the year 1970. In order to achieve this target, an 80% reduction of HC emissions and a 60% reduction of CO emissions from vehicle exhaust gases was needed. Applying this reduction rate to the abovementioned baseline one arrives at the world's first emission standards for automobiles which were 275 ppm HC and 1.5 vol.% CO [4, 32].

These first exhaust emission standards were adopted by the "State Board of Public Health" on December 4, 1959 and became mandatory as of model year 1966. They remained valid for two years and were independent of engine displacement/vehicle size. After these two years, in which California remained the only state in the US with certification standards and a certification procedure, the standards were – in coordination with the then starting Federal standards – adapted to different engine displacements (410/350/275 ppm HC and 2.3/2.0/1.5 vol.% CO for 50–100/100–140/>140 cu. in. engine displacement) [8].

As of model year 1970 unified standards, i.e., standards independent of engine capacity, were re-established and expressed henceforth on a g/mile-basis: 2.2 gHC/m and 23.0 gCO/m).

In 1971 California took the lead again by introducing the first NO_x emission standard (4.0 gNO_x/m) which was added to the 1970 set of HC and CO standards.

A special situation occurred in model year 1972 where two different sets of standards became applicable depending on the test cycle used. If a manufacturer continued to certify according to the "old" 7-mode cycle the standard set was 1.5 gHC/m; 23.0 gCO/m; 3.0 gNO_x/m whereas for the "new" LA4-cycle a standard set of 3.2 gHC/m; 39.0 gCO/m; and 3.2 gNO_x/m applied.

As of model year 1973 the standards were continuously strengthened, while in model years 1975/76 only the HC standard of 0.9 g/m and the NO_x standard of 2.0 g/m were "own" California standards whereas the CO standard of 9.0 g/m was a standard prescribed for California by the Federal Environmental Protection Agency (EPA). Since California legislation did not yet include Diesel regulations at this point of time, the EPA further prescribed for California a separate standard combination for passenger cars with Diesel engines (1.5 gHC/m; 9.0 gCO/m; 3.1 gNO_x/m).

For the model year 1977 the standard set of 0.41 gHC/m; 9.0 gCO/m; 1.5 gNO_x/m California took into account the methane part in the total hydrocarbon (THC) emission test result and allowed the use of a so-called "methane bonus" of 0.89 or applied a less stringent (total)HC standard of 0.46 g/m if hydrocarbons were measured methane-free during the emission test.

Until and including model year 1980 this standard set remained unchanged except for the fact that legislation from now on applied to Diesel vehicles as well. As of model year 1982 California took over EPA's particulate matter (PM) standard of 0.6 g/m without developing such a standard separately in its own legislation.

For model years 1980 to 1982 California introduced a "100,000 miles option" for the durability run which normally required a running distance of 50,000 miles. The new option was combined with a relaxed NO_x standard (1.5 g/m instead of 1.0 g/m) until and including model year 1983. In model year 1984 the NO_x standard was tightened to 1.0 gNO_x/m and was maintained at this level until and including model year 1994. The option was highly welcome for Diesel vehicles which could offer pronounced long-time emission durability but have, due to the high combustion temperatures of their fuel-efficient combustion process, severe difficulties in meeting stringent NO_x standards.

As of model year 1981 manufacturers had the choice between two sets of emission standards: Option A (0.41 gHC/m; 3.4 gCO/m; **1.0** gNO_x/m) and an Option B (0.41 gHC/m; 7.0 gCO/m; **0.7** gNO_x/m). The important difference was the relaxed NO_x standard of Option A.

If a manufacturer selected the less stringent NO_x standard of Option A in model year 1981 he automatically obliged himself to meet a more stringent NO_x standard of Option AA (0.41 gHC/m; 7.0 gCO/m; **0.4** gNO_x/m) in model year 1982. If he selected the more stringent Option B (0.7 gNO_x/m) in model year 1981 he was allowed to apply the same set of standards (i.e., a less stringent NO_x standard than it would otherwise have applied in model year 1982).

If the manufacturer decided to extend his emission warranty period from 50.000 miles/5 years to 75.000 miles/7 years he could also continue to certify according to the 0.7 gNO_x/m standard until and including model year 1992. Otherwise he had to meet the stringent 0.4 gNO_x/m standard as of model year 1983.

The previously mentioned Federal PM standard introduced in California in model year 1982 became an "own" California standard of 0.4 gPM/m in model year 1985 and was further tightened to 0.2 gPM/m in model year 1986 and to 0.08 gPM/m in model year 1989.

As of model year 1993 a so-called "Tier 1" standard set with a drastically strengthened HC-standard (0.25 gHC/m; 3.4 gCO/m; 0.4 gNO_x/m) became applicable.

The most important milestone in the development of California's emission standards was, however, the date of August 13, 1990 when the "California Air Resources Board" (CARB) published its "Low Emission Vehicle" (LEV)-program.

This program did not only replace the "Tier1"-standards as of model year 1994 but introduced a newly defined "Low Emission Vehicle" group (TLEV, LEV, ULEV and Zero Emission Vehicles.

For these vehicles the following standards applied:

- Transitional Low Emission Vehicles (TLEV): 0.125 gHC/m; 3.4 gCO/m; 0.4 gNO_x/m;
- Low Emission Vehicles (LEV): 0.075 gHC/m; 3.4 gCO/m; 0.2 gNO_x/m;
- Ultra Low Emission Vehicles (ULEV): 0.040 gHC/m; 1.7 gCO/m; 0.4 gNO_x/m.

California's "LEV"-program focuses on the reduction of HC emissions which are considered as the most critical exhaust gas constituents causing California's air quality problem. So the program included the additional requirement that manufacturers had to meet a "Non-Methane Organic Gases" (NMOG) standard as an average for their vehicle fleet of a given model year.

The NMOG average standard started in model year 1994 at 0.250 g/m and was then stepwise strengthened to 0.062 g/m in model year 2003. The manufacturer could compose its yearly fleet as it fits best as long as its fleet average NMOG value met these limits.

As of model year 2004 the provisions of California's "LEV II"-regulations apply including another drastic strengthening of emission standards while at the same time providing means by which vehicle manufacturers may get some limited relief from the ZEV mandate.

Within the LEV II-program, LEV I-standards were substantially strengthened and a new category, the "super ultra low emission vehicles" (SULEVs), was added resulting in the following standard set:

- Low Emission Vehicles (LEV): 0.075 gHC/m; 3.4 gCO/m; **0.05** gNO_x/m;
- Ultra Low Emission Vehicles (ULEV): 0.040 gHC/m; 1.7 gCO/m; **0.05** gNO_x/m;
- Super Ultra Low Emission Vehicles (SULEV): 0.010 gHC/m; 1.0 gCO/m; 0.02 gNO_x/m;

This standard set includes a PM standard for vehicles with Diesel engines of **0.01** gPM/m.

The NMOG fleet average standard was as well strengthened from 0.053 g/m in model year 2004 to 0.035 g/m in model year 2010. The complete NMOG standards reduction program can be seen in **Fig. 3.**

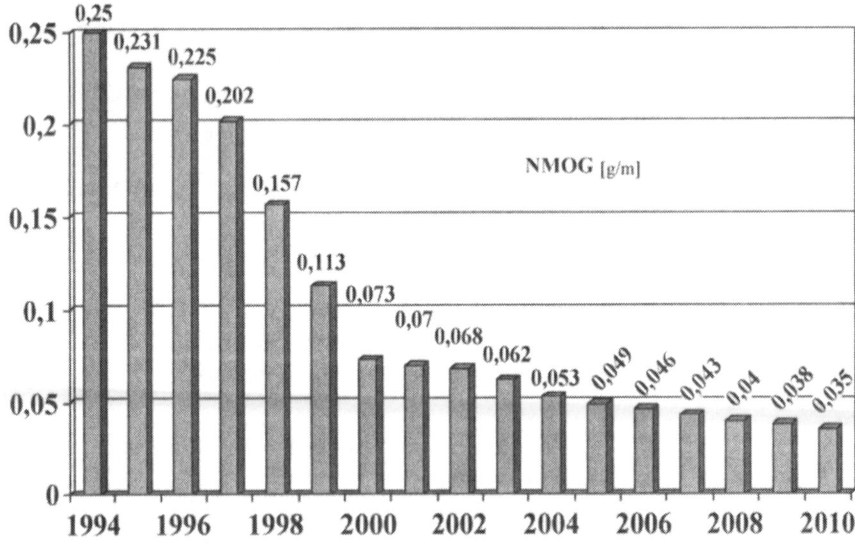

Fig. 3. Introduction scheme of California's NMOG Fleet Average Standard

2.2 US-Federal Emission Standards

Federal exhaust emission regulation for passenger cars was first published in the Federal Register in March 31, 1966. The emission standards mentioned therein were, however, not deduced from defined air quality targets but were simply taken over from California relying on the work done there before [33].

The first "own" Federal exhaust emission standards for passenger cars were published on June 4, 1968 for CO and HC and became mandatory as of model year 1970. These standards were also not derived from air quality targets but were calculated from auto emission baseline levels as it was done before for the first California standards [8].

The corresponding baseline study, which was performed between 1968 and 1969, comprised 1,500 "uncontrolled" vehicles and its evaluation resulted in average emissions for a 1968 vehicle of 979 ppm HC and 3.4 vol.% CO. The reduction rates were set close to those which had been used by California some years before namely 82% for HC and 71% for CO. When these reduction targets are applied to the above mentioned baseline level one finds the emission standards envisaged for 1970 to be 180 ppm HC and 1.0 vol.% CO [34].

However, since on June 4, 1968 the transition from concentration-based emission measurement to mass-based measurement was decided to start with model year 1970, these values had to be adapted accordingly. The final, mass-based Federal emission standards for model year 1970 are, therefore, 2.2 g/m HC and 23.0 g/m CO.

After air quality targets for photochemical oxidants, CO, HC and NO_x became available in 1970, it was possible to calculate the rate of further emission reduction, as it was initially done by California, but which could now be done on the basis of an improved "rollback" method.

The calculation which is described in detail in [8] showed that reductions of 92.5% for CO, 99,0% for HC and 93.6% for NO_x would be needed if the envisaged air quality was to be achieved in the 1980 target year [35].

It shall be pointed out here that these reduction rates were the orientation point for the statutory emission standards of the 1970 "Clean Air Act Amendments". These so-called "statutory standards" were calculated from the above-mentioned % reduction rates to the following mass emission-based figures: 0.41 gHC/m; 3.4 gCO/m and 0.4 gNO_x/m.

Due to still deteriorating ambient air conditions, the previously envisaged target date for the introduction of these standards was advanced from 1980 to 1975 for HC and CO and to 1976 for NO_x [36]. The introduction schedule could, however, not be met due to the non-availability of suitable emission control technology and the eventually applied standards became the result of hearings in which interim standards were approved and/or postponement of standards was decided.

These hearings took about the full decade from 1970 to 1980 and covered both the feasibility/postponement of the "statutory standards" for vehicles with gasoline engines as well as the feasibility/postponement of the 1.0 g/m NO_x standard for vehicles with Diesel engines which was initially envisaged for introduction in the Federal legislation as of model year 1982.

The possibility to meet the initial statutory standards came in model year 1977 when Volvo was the first car manufacturer in the world to certify a vehicle with a three-way catalyst, a technology which became possible after the introduction of the newly developed Bosch O_2 sensor.

After this breakthrough there was no urgent need to define additional reduction steps for the existing emission standards since three-way catalyst equipped vehicles met these standards "automatically" with a notable safety margin.

A substantial further reduction of emission standards came nevertheless with the "Clean Air Act Amendments of 1990" which were signed on November 15, 1990. The amendments reduced the HC standard from 0.4 to 0.25 g/m, the NO_x standard from 1.0 to 0.4 g/m (but left the CO standard unchanged at 3.4 g/m) for vehicles with gasoline engines. For vehicles with Diesel engines the NO_x standard remained unchanged at 1.0 g/m but the particulate matter standard was drastically strengthened from 0.2 g/m to 0.08 g/m.

The latter PM standard was seen very critical by manufacturers because it would lead to an elimination of the Diesel engine as a power plant for passenger

Table 4. "Phase-In" Scenario of the EPA-"Tier 2" Average-NO_x-Standards

[%]	2001	2002	2003	2004	2005	2006	2007	2008	2009 and later	NO_x-Std. [g/m]
LDV/LLDT (Interim)	NLEV	NLEV	NLEV	75 max.	50 max.	25 max.				0.30[f] average
LDV/LLDT (Tier 2 + Evap)	early banking [b]	[b]	[b]	25	50	75	100	100	100	0.07 average
HLDT/ MDPV (Tier 2 + Evap)	early banking [b]	[b]		[b]	[b]	[b]	[b]	50	100	0.07[d] average
HLDT (Interim)	Tier 1 [b]	Tier 1 [b]	Tier 1 [b]	25	50	75	100	50 max.		0.20[a, d] average
MDPV (Interim)	HDE	HDE	HDE	[c,e]	[e]	[e]	[e]			

[a] 0.60 g/m NO_x-cap applies to balance of LDT3s/LDT4s, respectively, during the 2004–2006 phase-in years.

[b] Alternative phase-in provisions permit manufacturers to deviate from the 25/50/75% 2004–2006 and 50% 2008 phase-in requirements and provide credits for phasing-in some vehicles during one or more of these model years.

[c] Required only for manufacturers electing to use optional NMOG values for LDT2s or LT4s and MDPV flexibilities during the applicable interim program and for vehicles whose model year commences on or after the fourth anniversary date of the signature of this rule.

[d] MDPVs and HLDTs must be averaged together.

[e] Diesels may be engine-certified through the 2007 model year.

[f] Beginning with the 2004 model year, all new LDVs, LDT1s and LDT2s not incorporated under the Tier2 phase-in will be subject to an interim corporate average NO_x standard of 0.30 g/m.

cars on the US market unless new technology (like, e.g., a particle filter) could be developed in time for model year 1994 when this standard was scheduled to become applicable.

The latest modification of Federal auto emission standards was announced on December 12, 1999 when EPA's so-called "Tier 2" program was signed and published.

In contrast to California's focus on HC reduction, the underlying objective of the Federal program with regard to emission standards is to achieve an average NO_x level of 0.07 g/m on 100% of a manufacturer's fleet in model year 2007. The standard will be phased-in via 25% steps starting in model year 2004 as shown in Table 4.

The NO_x standard is not a fixed limit like it has been in the past. The manufacturer has the choice to select among several sets of NMOG, CO, HCHO (formaldehyde), NO_x and PM levels, the so-called "bins", those which fit best to the potential of its individual vehicles as long as it achieves 0.07 g/m NO_x in model year 2007 as an average for its full fleet. Table 5 summarizes the "bins" which are available for a vehicle's full useful life certification.

It is important to note that the new standards have to be met over 120,000 miles.

While in the 1990 "Clean Air Act Amendments" the PM standard of 0.08 g/m appeared to be the decisive criteria for the Diesel, the "Tier 2" regulations of 1999 made the NO_x standard an even more critical hurdle. This NO_x standard practically wiped out the light duty vehicle with Diesel engine from the US market and a re-introduction can only be envisaged if exhaust gas after-treatment systems become

Table 5. EPA "Tier2" Bin Groups for Full Useful Life Certification

Final "Tier"-Stndards – Combination ("Bins")						Remarks
Bin-Nr.	NMOG	CO	HCHO	NO_x	PM	
10A	0.230	6.4	0.027	0.6	0.08	The higher temporary NMOG, CO and HCHO value(s) of Bin 10A, 9A and 8A apply only to HLDTs and expire after 2008. For vehicles certified to these "Final Tier 2"-Bins (as of MY 2004) the "full useful life" for LDVs/LDTs is 10 years or 120,000 miles. The "full useful life" for non-Tier 2 LDVs/LDTs is 10 years/1000,00 miles. "Interim-Standards" (Bin-groups) apply for vehicles which are not yet certified to the "Final Tier2"-Standards. In this case, NO_x-fleet average values of 0.30 g/m (for LDVs & LDTs from 2004 to 2007) resp. 0.2 g/m (for HLDTs & MDPVs from 2004 to 2008) apply.
10	0.156	4.2	0.018	0.6	0.08	
9A	0.180	4.2	0.018	0.3	0.06	
9	0.090	4.2	0.018	0.3	0.06	
8A	0.156	4.2	0.018	0.20	0.02	
8	0.125	4.2	0.018	0.20	0.02	
7	0.090	4.2	0.018	0.15	0.02	
6	0.090	2.1	0.018	0.10	0.01	
5	0.090	2.1	0.018	0.07	0.01	
4	0.070	2.1	0.011	0.04	0.01	
3	0.055	2.1	0.011	0.03	0.01	
2	0.010	2.1	0.004	0.02	0.01	
1	0.000	0.0	0.000	0.00	0.00	

Bin 8A and 10A may be used for vehicles running on alcohol or gas.

available which bring a technical breakthrough in NO_x reduction allowing the Diesel to achieve the same low NO_x emission level like its gasoline counterpart.

2.3 Emission Standards in Japan

Before going into details of the evolution of Japanese emission standards, a unique characteristic of the Japanese emission control legislation shall be mentioned. In the course of emission standards setting, Japanese authorities (Ministry of Transport, Ministry of International Trade and Industry and Environment Agency) have to a large extent allowed different mandatory introduction dates for these standards for domestic manufacturers and importers. The technical possibilities of importing companies have been carefully evaluated through individual contacts and a series of hearings with the affected car manufacturers and the findings of theses hearings were actually taken into account during the standard setting process in order to allow for necessary development lead time and undisturbed international trade without giving up the ultimate objective of introducing very strict environmental norms.

So from 1975 onwards importers got one to three years, in the case of the 1976 NO_x standard even 5 years, additional lead time. Further, the standards are generally set in two groups, in so-called "mean" and "max" standards. The more stringent "mean" standards have to be met during type approval testing and as average values for serial production vehicles for models which are sold at >1,200 units/year. The "max" standards are valid for the certification of models with sales <1,200 units/year and as individual limit for serial production cars.

Historically seen, Japan started to set its first emission standard for passenger cars with **gasoline** engines parallel to California, i.e., in 1966. Its legislation started with a 3.0 vol.% CO limit for the "designated vehicle" group which covered all domestic car manufacturers. This value had to be determined by means of a special "4-mode exhaust emission test" (as described later in Sect. 3.2) and became mandatory as of September 1 of that year.

The standard was lowered as of September 1, 1969 to 2.5 vol.% and as of September 1, 1971 to 1.5 vol.%. As of April 1, 1973 – when the 4-mode test was replaced by the new 10-mode hot start test – the limits for CO, HC and NO_x were again reduced but expressed as mass emissions from now on. The first Japanese mass-based standards were set at (mean/max values): 2.94/3.80 gHC/km; 18.4/26.0 gCO/km and 2.18/3.0 gNO_x/km.

These "1973 Standards" included also the requirement that all vehicles already in use had to be retrofitted with ignition timing adjustment systems which would lower HC emissions by 8% and NO_x emissions by 20%. The systems should be installed via a phase-in program beginning on May 1, 1973 and ending March 31, 1975.

On January 1, 1974, the Japanese Environment Agency (EA) had initially decided to introduce the US "Muskie-Standards" for CO and HC as of April 1, 1975 (the so-called "1975 Standards") and for NO_x as of April 1, 1976 (the so-called "1976 Standards") as they were originally proposed in the US but expressed for Japan in [g/km]. While the Agency maintained the introduction date of the

"1975 (HC and CO) Standard" it decided, however, on February 24, 1975, to postpone the "1976 (NO_x) Standard" introduction date for domestic manufacturers to April 1, 1978 (the standard set was now called "1978 Standards") since adequate NO_x emission control technology could not be developed in time. Based on individual hearings the Japanese Cabinet decided on January 1, 1977 to postpone the mandatory introduction date of this standard for importers for another 3 years, i.e., until 1981. Importers thus had received in total a 5-year extension against the originally planned introduction date for the "1976/1978 (NO_x) Standard" [37–40].

As in the US, this stringent NO_x standard eventually led to the introduction of the three-way-catalyst in Japan as well. The remarkable progress in emission reduction achieved by this technology can be seen when the former "1978 Standards" are compared with the emission standards defined by the Japanese so-called "2000 Standards" shown in Table 6.

As already announced during the adoption of the "2000 Standards", Japan has in the meantime defined the expected further tightening of this legislation which will take place in two steps both for gasoline and Diesel engines.

Comparing the figures with other legislations is not possible due to different test procedures. However, a stringency comparison can be made on the basis of the technology which is needed to meet these standards. For EU vehicles with gasoline engines such comparison reveals that "Step1" of the Japanese "beyond-2000"-standards" can be met by Euro 3 technology whereas "Step 2" obviously needs Euro 4 technology. For passenger cars with Diesel engine today's Japanese NO_x standards are already more stringent than Euro 3 requirements while the Japanese PM standard is less stringent. Further tightening is as well foreseen in two steps as "short-term targets" and as "long-term targets". It is expected that the "long-term targets" will be decided before the end of 2001.

Table 6. Emission Standard Reduction in Japan since 1978 (Introduction of Three-Way Catalyst Technology) up to recent "2000 Standards"

Standards for Passenger Cars with Gasoline Engines					
Test-Procedure	Emission	1978 Standards		2000 Standards	
		mean	max.	mean	max.
10.15 – Mode Hot Start Test [g/km]	HC	0.25	0.39	0.08	0.17
	CO	2,1	2,7	0,67	1,27
	NO_x	0,25	0,48	0,08	0,17
11-Mode Cold Start Test [g/Test]	HC	7,0	9,5	2.20	4.42
	CO	60,0	85,0	19,0	31,1
	NO_x	4,4	6,0	1,40	2,50

Mean-Value: To be met during vehicle type approval and as average standard in series production.

Max.-Value: type approval standard if sales of the vehicle model is < 2,000 units/year and as individual standard for series production.

From the technological point of view it seems at present that the standards which were initially foreseen for introduction only in 2007 but which shall now be advanced to 2005 cannot be met in time. Auto manufacturers are in the process of explaining their technical development status and negotiate an adequate introduction date during hearings in mid 2001 before the Japanese authorities.

With regard to passenger cars with **Diesel** engines, the first emission limit was established as a filter blackness standard of 50% which was valid for the Japanese 3-mode full load test from July 1, 1972 to March 31, 1994. As of April 1, 1994 the standard was reduced to 40% and lowered again to 25% within the so-called "1997 Standard" which became applicable for importers as of April 1, 2000.

Following the limitation of smoke emissions, gaseous emissions for passenger cars with Diesel engines were first regulated as of September 1, 1974 on the basis of a (stationary) "6-mode test". The standards (mean/max) were: 510/670 ppm HC; 790/980 ppm CO and 450/590 ppm NO_x for pre-chamber engines or 770/1000 ppm NO_x for direct injection engines.

As of August 1, 1977 the NO_x standards were lowered to 380/500 ppm NO_x (pre-chamber engines) or 650/850 ppm NO_x (direct injection) and again within the "1979 Standards" which were postponed for importers until April 1, 1981 to 340/450 ppm NO_x (pre-chamber engines) or 540/700 ppm NO_x (direct injection engines). As of April 1, 1984 pre-chamber engines had to meet 290/390 ppm NO_x while the next reduction step of the direct injection engines standard became applicable one year later as of April 1, 1985 with 470/610 ppm $NO_{x.}$

The next important step was the establishment of the so-called "1986 Standards" which were based on the new "10-mode (hot start) test" and which became mandatory as of October 1, 1986/87 for domestically produced vehicles and as of April 1, 1988/89 for imported vehicles (the two different years refer to vehicles with mechanical/automatic transmission). The standards (mean/max) were: 0.4/0.62 g HC/km; 2.1/2.7 g CO/km; 0.7/0.98 gNO_x/km (vehicles with reference mass <1265 kg) and 0.9/1.26 gNO_x/km (vehicles with reference mass >1265 kg). When the new "10.15-mode (hot start) test" became applicable as of April 1, 1993 (for importers), the standards remained unchanged.

The next reduction steps affected the NO_x standard, lowering the mean/max limits for importers to 0.5/0.72 (0.6/0.84) for vehicles <1265 kg (>1265 kg) as of April 1, 1996. Eventually, as of April 1, 2000 all standards became independent of vehicle reference mass at the level of 0.4/0.62 g HC/km: 2.1/2.7 g CO/km and a further reduced NO_x limit 0.4/0.55 gNO_x/km.

Particulate matter (PM) emissions were first regulated (independent of vehicle reference mass) to 0.2 g/km (mean value) and 0.34 g/km (max. value) within the so-called "1994 Standards" which became valid for importers as of April 1, 1996. The applicable test procedure for the determination of theses particulate matter emissions was the 10.15 mode test which replaced the former 6-mode test. The standard was lowered for importers to 0.08 g NO_x/km as of September 1, 2000 and will further be reduced to 0.056 as of April 1, 2004. A long-term target around 0.028 is presently under discussion but the exact value, date and test procedure were not yet fixed when this chapter was written.

2.4 Emission Standards in the European Union

The evolution of emission standards in the European Union (Brussels) has to be seen in close connection to the corresponding developments at the UN-ECE (Geneva) where all transportation-related issues for Europe, including automobile emission control, are discussed by competent experts on a broader international platform. However, it was the basic work of some individual states like France, Sweden and the Federal Republic of Germany which eventually triggered activities concerning auto emissions in Geneva and Brussels. This shall now be shown on the example of the Federal Republic of Germany which in fact took the leading role in moving the EU towards ever strengthened emission standards for motor vehicles.

A first step on the political level in Europe towards determination and limitation of emissions from automobiles was done in Germany in January 1956 when the "Association of German Engineers" (VDI) was assigned by an "Interparliamentary Committee" to develop directives and recommendations for lowering air pollution. As a consequence the VDI established its commission "Purification of the Air" ("Reinhaltung der Luft") [30].

In order to provide a competent discussion forum for auto experts which could cooperate with the VDI commission the German "Auto Manufacturers Association" ("Verband der Automobilindustrie", VDA) established the "Subcommittee Exhaust Gases from Gasoline Engines" ("Unterausschuss Abgase von Otto-Motoren"). The assignments of this subcommittee were to establish allowable limits for pollutants in a gasoline engine's exhaust gas which do not yet cause damage or nuisance to the public, to evaluate possibilities for pollutant reductions and to develop necessary measurement techniques.

In November 1962 the VDI issued its draft for the first VDI-Directive (No. 2282) about gasoline engine emissions entitled "Limitation of CO-Emissions for Motor Vehicles with Gasoline Engines" and France announced as well its own regulation about the limitation of CO emissions.

On December 15, 1964 the VDI informed the VDA that the German Government had asked about possible introduction dates for emission standards. The VDA outlined a skeleton schedule according to which the auto industry offered to meet the first German emission standards 3 years after the definition of a test cycle.

The more progress was made with the development of this test cycle the easier it became for politicians to materialize their national plans for establishing emission standards. This was exactly what happened in Europe during these days. Triggered by trends in Germany and France towards individual national emission control legislation the "Hallstein Commission" in May 1967 submitted to the EEC Council a proposal for an EEC Directive about control of air pollution in order to harmonize corresponding regulations already in effect or in preparation in most of its member states.

Now an intensive discussion developed within the member states about the question which basis should be used for the envisaged emission on standards to be determined. Germany strongly favored the vehicle's fuel consumption as reference value whereas France preferred vehicle mass. While the two countries still negotiated to find a common proposal the Federal Ministry of Health ("Bundesgesundheitsministerium", BGM) in September 1967 – finding itself under strong public pressure – announced that it would have to apply an unilateral German regulation if the delays on a European level continued.

On December 19, 1967 for the first time the draft of such a separate German solution was presented and discussed within the special committee "Air Pollution by Exhaust Gases from Motor Vehicle Engines" together with the Ministry of Transport and the Ministry of Health and when public pressure had further increased, Germany issued its comprehensive proposal announcing the introduction of the following attachments to its Road Vehicle Law:

- Attachment XI: Idle CO, affective as of July 1, 1969,
- Attachment XII: Closed Crankcase Ventilation, effective as of January 1, 1969,
- Attachment XIII: Emissions in a driving cycle, effective as of July 1, 1970.

The first quarter of 1968 passed by with continuing discussions about the definition of emission standards but the experts could not find a common proposal.

However, the German activities caused severe concern in other EEC member states and eventually the group "Elimination of Barriers to Trade – Motor Vehicles" met in Brussels on July 11/12, 1968 and heavily opposed the planned German regulations which it considered a severe interference into the development of motor vehicles and as a barrier to trade. Germany pointed to similar developments in France but indicated that it did not intend to fight but was prepared to compromise. An agreement was reached to exchange opinions among Common Market member states before national regulations were promulgated. It was further agreed that ECE Regulations should serve as a basis for a common EEC Directive.

Since the German plan for a national solution could hence be stopped if an ECE Regulation was published on time, all parties opposing such unilateral German action accelerated the European decision-making process: On August 22, 1968 representatives from the German car manufacturers met at the VDA in Frankfurt and declared acceptance of the emission standards discussed during the 4th GRPA ("Group de Rapporteurs sur la Pollution de l'Air") session in Geneva but which were not agreed upon during that meeting. Eventually during the 26th meeting of WP 29 (UN "Working Party 29" in Geneva) on September 2/6, 1968 all parties agreed to these emission standards (except Germany and Sweden) and the chairman of the GRPA was requested to formulate a draft for an ECE Regulation including these standards.

Eventually the document was accepted unanimously and submitted by France and Spain to the UN on April 11, 1969. It became the famous basic law "ECE Regulation No.15", commonly known as "ECE-R15" which was published on August 1, 1970.

At the same time, actually even before the document submitted to the UN was published as ECE-R15, the EEC Commission had become active. On October 22, 1969, a draft for an EEC Directive was submitted from the Commission to the Council in order to prevent national solutions like the German "Attachment XIII".

The draft was subject to a discussion among Government representatives from the 6 member States of the Community on December 15/17, 1969, and eventually on January 23, 1970, Brussels agreed on its Emission Directive which was published on March 20, 1970 under the well-known designation "70/220/EEC".

According to the introduction schedule connected to EEC Directives, Germany now had to incorporate this Directive until June 30, 1970 into its national legislation and in fact did so on April 4, 1970 in Form of "Attachment XIV" to §47, StVZO (Road Vehicle Approval Regulation). It could, however, according to the EEC schedule not enforce this regulation before October 1, 1970.

After the risk of an isolated national solution and, thereby, the risk of an internal trade conflict in the Community did no longer exist, the EU could begin to build future emission control requirements on a solid basis.

Besides some non-EEC member states it was mainly the Federal Republic of Germany which continued to ask for further and substantial progress in the field of auto emission control as well during negotiations in Brussels and in Geneva where all EEC member states took part in the discussion and formulation of ECE regulations. However, progress with regard to the definition of emission standards did only take place at a slow pace and Brussels did not move faster than Geneva. In fact, Brussels just transformed the content of established ECE Regulations for about 10 years into EEC Directives.

In its "Government Declaration" of October 28, 1969, Germany had committed itself to the political target of improving air quality. Accordingly it formulated a comprehensive "Environment Program" in 1971 in which it defined as a long-term target to lower emission levels until 1980 by 90% compared to the uncontrolled status of 1969. This program was made known to other European countries and in its preamble Chancellor Brandt said:

"Environment problems do not stop at borders. Cooperation with neighboring countries is necessary", and "Cooperation in a European frame is highly important. The Environment Program of the Federal Government is an offer to all states to try to find a common solution to avoid the danger of an environmental crisis."

In order to evaluate questions arising for the auto industry from the "Environment Program" an expert group was set up which included representatives from auto and oil industry and government. In a 1-week session from August 10 to 15 1972 in Baiersbronn the group evaluated a basis on which the German auto industry could develop its opinion towards the 1980 goal of the Federal Government. The meeting result became known as the "Baiersbronner Program" ("Baiersbronn Program") and said that catalyst technology should be feasible and asked for unleaded gasoline and for a materialization of the 1980 goal at least on an EEC-wide basis. Emission standards were suggested reflecting the envisaged 90% reduction vs. the uncontrolled level.

On July 24, 1974 the German Ministry of the Interior (BMI) established the "German Environment Agency" (Umweltbundesamt, UBA) and assigned it to evaluate and submit material on which detailed decisions could be based. The UBA defined a set of standards, including a combined ($HC+NO_x$) standard which again reflected the 90% reduction goal but was oriented at internationally valid air quality standards and focused on that portion of air pollution which was really caused by motor vehicles in Europe.

The UBA-report 7/76 and the corresponding proposal for modification of ECE-R15 (equivalent to 70/220/EEC) were submitted by the German delegation to the ECE in Geneva during the 19th GRPA session on September 18/21, 1978.

The German proposal was discussed together with a similar Swiss proposal which as well asked for a substantial emission reduction beyond ECE-R15/03 but did not find support. Eventually, during the 1st GRPE session on February 11/15, 1980 an agreement could be reached for a further reduction step, but to a rather modest one of –15% for CO and –15–25% for ($HC+NO_x$). The new standards were published as ECE-R 15/04 on October 20, 1981 and later incorporated into EEC Directive 83/351/EEC of June 16, 1983.

However, the original German proposal was re-submitted in form of an EEC memorandum of the Federal Government to the EEC in June 1981 which showed that Germany accepted the above compromise according to its EEC obligations but considered the standards only as an intermediate step on its way to meet the objectives of the 1971 "Environment Program".

A consideration of further drastic emission reductions beyond ECE 15/04 (or 83/351/EEC) by the member states would most probably not have started before the 1990s if the Federal Republic of Germany had not taken the lead and submitted again its above proposal to Brussels on November 4, 1983. The discussion triggered by the German initiative revealed that – at that point of time – no EEC member considered it necessary to discuss new emission legislation which would entail the catalyst technology.

The major obstacle why available US-emission control technology could not just be taken over to Europe was the non-availability of unleaded fuel. So the discussion in Brussels first focused on the "Lead Directive" (78/611/EEC). After long and controversial discussions the presidents of the German auto manufacturers met in Paris and agreed on March 7, 1984 that the auto industry would no longer object to the introduction of unleaded gasoline if this event did not entail a short-term catalyst introduction on more than a few large vehicles only.

The Commission was informed accordingly and the way was cleared for a revision of the "Lead Directive". On May 16, 1984 the Commission proposed that unleaded gasoline may be made available by member states as of January 1986 and that this fuel must be made available at the gas stations as of 1989. The way was then free for the introduction of the catalyst technology and a new chapter of outstanding emission reduction potential was opened in the history of Europe's emission legislation for automobiles. Vehicles with gasoline engine could lower the emission of regulated pollutants in their exhaust gas under normal operating conditions down to levels close to zero and could thus no longer be regarded as a threat to the environment. This breakthrough was reached with the introduction of Directive 91/441/EEC.

It did, however, not introduce a consolidation phase, e.g., by a moratorium concerning a furtherance of the corresponding legislation, but was immediately used by legislation as a basis for a new chapter in the historical development of auto emission standards in the European Union which required passenger car engines to achieve unprecedented low emission levels. This chance was taken up by legislating authorities which met with auto industry and other stakeholders on September 21/22, 1992 in Brussels to agree on a strategy for the next millennium [41].

This conference "Auto Emissions 2000" did, however, not only highlight the low emission potential of future (three-way catalyst equipped) vehicles with gasoline engines, but brought as well into focus that its Diesel counterpart would not be able to follow with regard to NO_x and PM emissions.

Especially the Diesel engine's PM emissions became a matter of controversial discussions. Many assumptions and often contradictory theories were discussed about possible health effects from Diesel particle emissions and it has been far from easy for the legislator to define a PM standard which would sufficiently address the health concern. Therefore, this standard became a "moving target" for auto manufacturers over the years as shown in Table 7 [42].

From the intensive scientific and political discussions about the effect of particles in Diesel exhaust gas it can be anticipated that the Diesel will not come out of critique unless control technology becomes available which allows one to practically eliminate particulate matter emissions. Similarly, legislation will continue to push for more stringent NO_x standards for the Diesel engine until it eventually meets the same levels as its gasoline counterpart.

The development of EU emission standards for automobiles from the baseline setting in 1970 to the latest so-called "EU 2000" legislation is shown in Fig. 4 [43].

As can be seen from this figure, the first reduction steps of the "Basic Emission Directive" 70/220/EEC of March 1970 and of its successor 74/290/EEC of May 28, 1974 focused on HC and CO emissions only. Since the corresponding emission control measures mainly consisted of adjusting the engines to a lean mixture ratio, NO_x emissions increased. This deficiency was corrected as of October 1977, when additional NO_x standards became applicable and the Diesel engine was included in this legislation. Particulate matter emission became regulated as of 1988 and standards requiring catalysts on all vehicle classes were introduced with the 1992 Directive 91/441/EEC (Euro 1).

More detailed information about the emission standards of this Directive and their further reduction steps is given in Fig. 5 for vehicles with gasoline engines and in Fig. 6 for vehicles with Diesel engines.

These figures show as well the so-called "EU 2000" legislation specified in the two steps of emission Directive 98/69/EC which includes tax incentive possibil-

Table 7. Political Focus during Standard Setting for the Reduction of Diesel particulate Emissions: A Moving Target

Period	Political Focus	Technical Measure	Effect
before 1987	Smoke	Combustion Optimization	Reduction of Particle Mass
as of 1987	Mutagenic Effect	Oxidation Catalyst	Combustion of Particle-bound Organics
as of 1992	Epigenetic Effect	Combustion Optimization	Reduction of Particle Mass
as of mid 90s	Fine and Ultra Fine Particles	Combustion Optimization; Particulate Filter?	Reduction of Particle Mass and Particle Number

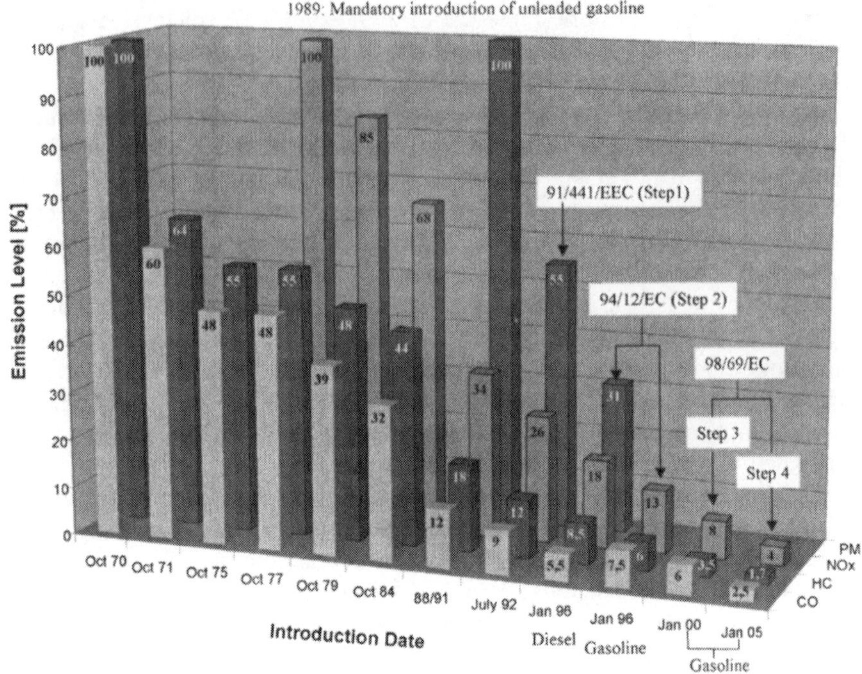

Fig. 4. Development of EU Emission Standards

ities for countries which want to advance the introduction date of cleaner vehicles.

The upcoming "Step 4" (EU4) of Directive 98/69/EC may be the final legislative requirement towards lowering exhaust emissions from vehicles with gasoline engines. It seems, however, very likely that a "EU5"-Directive will follow for vehicles with Diesel engines with PM and NO_x standards drastically reduced vs. EU4 requirements.

Latest emission control technology has proven that close-to-zero emission levels can be achieved on development and new production vehicles. In the future legislation will focus on measures which ensure that this potential is maintained in actual vehicle life, i.e., on the road under the many varying driving habits in different climatic and geographic environments.

3
Emission Test Methods

3.1
Emission Test Methods in the USA

California was not only the first state in the world to introduce emission standards for automobiles, it took the leading role as well in establishing the first test

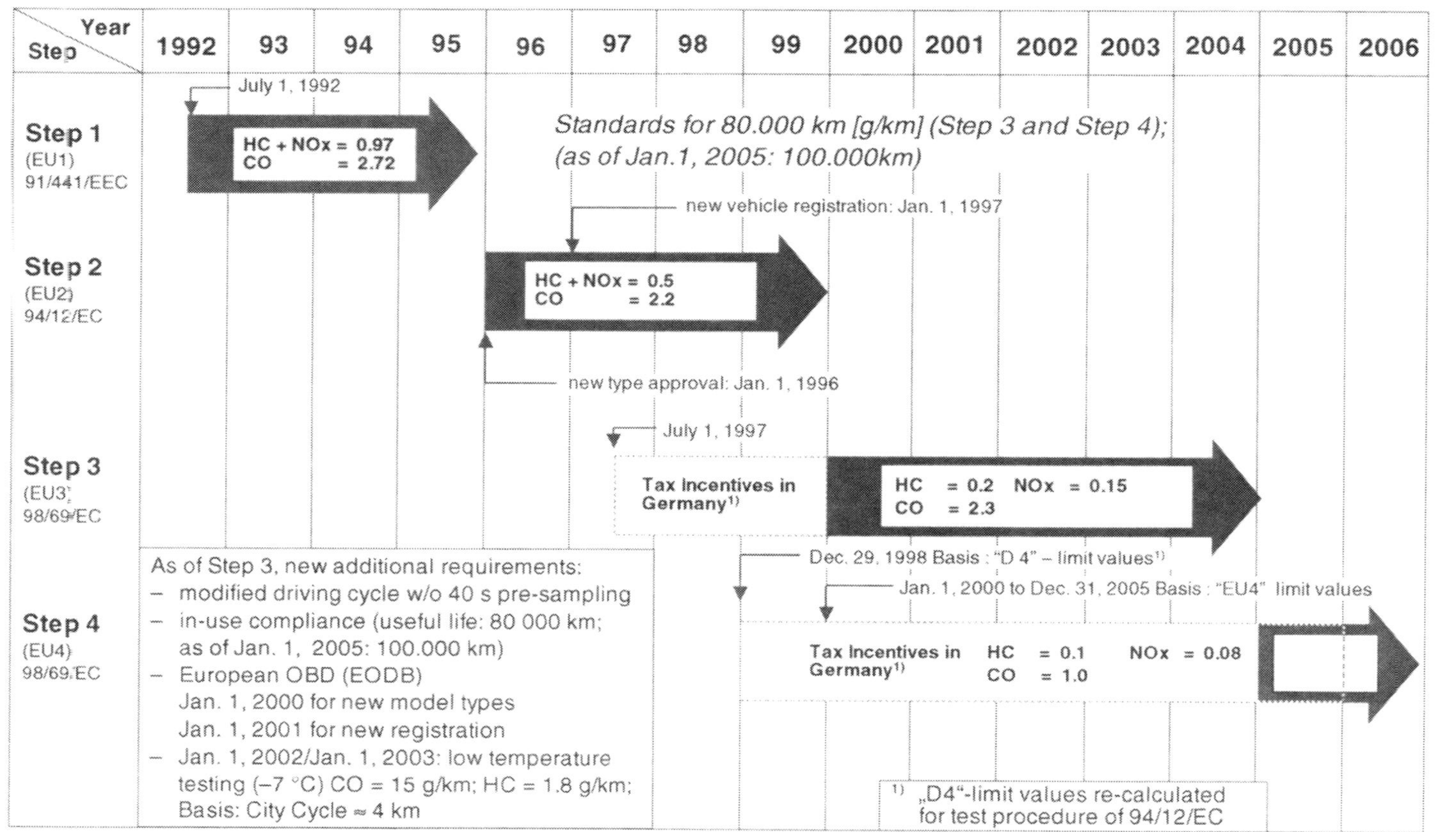

Fig. 5. Introduction scheme of EU Emission Standards for passenger cars with gasoline engines

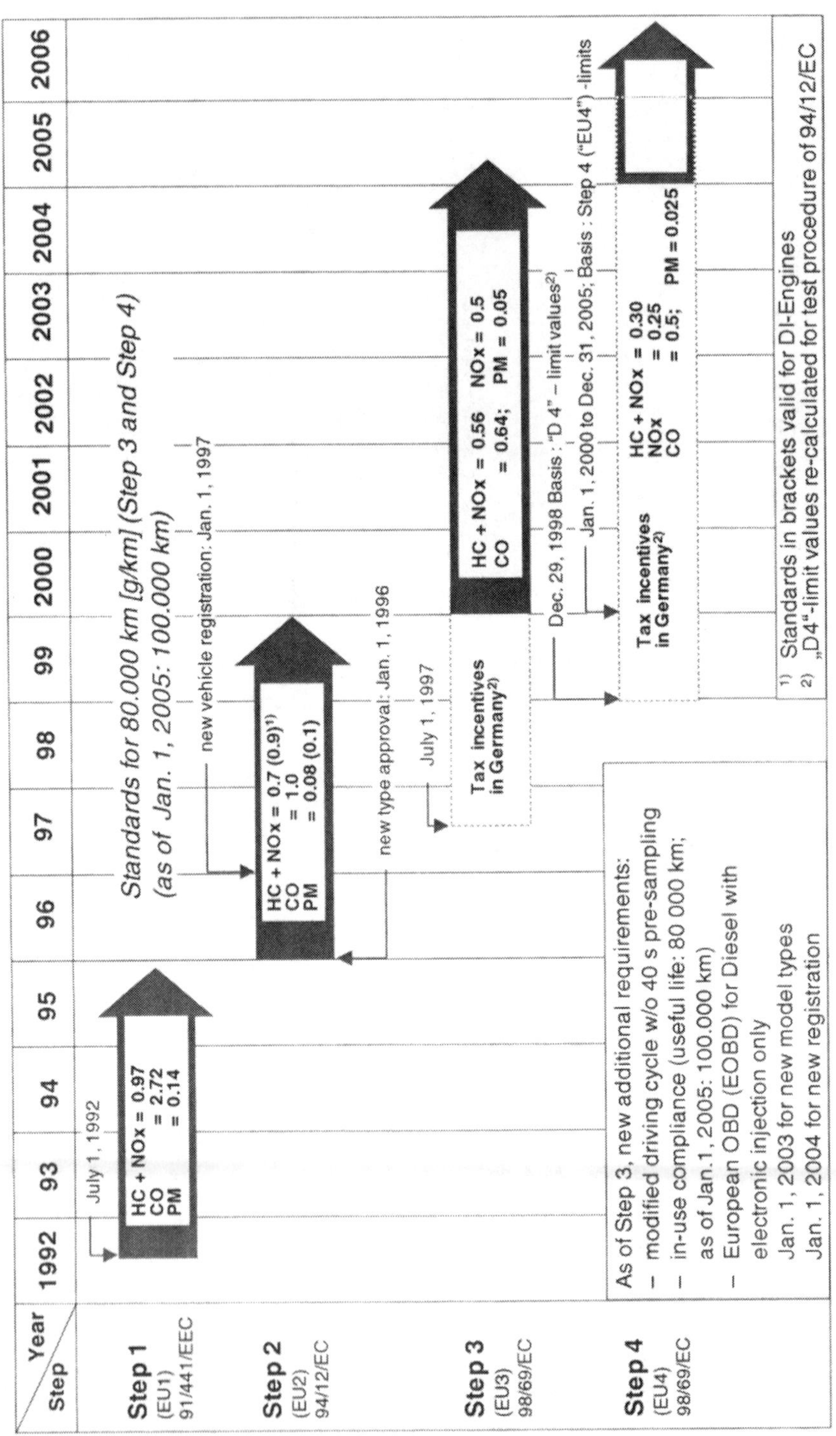

Fig. 6. Introduction scheme of EU emission standards for passenger cars with Diesel engines

procedure for the determination of vehicle emissions during a driving cycle on a chassis dynamometer together with the necessary measurement technique.

Detailed requirements for emissions testing of passenger cars were first published by the "California Motor Vehicle Pollution Control Board" (CMVPCB) on May 19, 1961.

In the beginning of emissions determination, whether this was done during driving modes on the road or on a chassis dynamometer, concentrations of HC and CO emissions were determined in the undiluted exhaust gas by means of "Non-Dispersive Infrared"(NDIR) instruments with hydrocarbon emissions measured as *n*-hexane.

For the calculation of the test result a weighing factor was applied to the concentration values of HC and CO after these values had been corrected to the stoichiometric air/fuel ratio before according to the formulae $15/(CO+CO_2)$ (for all driving modes except decelerations) and $15/(6HC+CO+CO_2)$ for deceleration phases. The weighing factor corresponds to the %-share of each of the driving modes of the 7-mode test to the total exhaust gas volume which was statistically determined.

Because this "California Test" represents a historic mile stone it is shown here in Fig. 7.

The corresponding measurement technique which was used for emissions determination during first vehicle certification testing of passenger cars in California in model years 1966/67 is shown in Fig. 8.

Since, however, the actual exhaust gas volume of a vehicle only unsatisfactorily correlates to the statistical %-share, the "California Test" could not reflect the real mass emissions of HC and CO.

Federal legislation applied this test method only for model years 1968/69. However, since engine displacement as well as performance and fuel consumption varies largely, the exhaust volume depends as well on the vehicle model tested. Federal test procedures as adopted on March 30, 1966 for model years

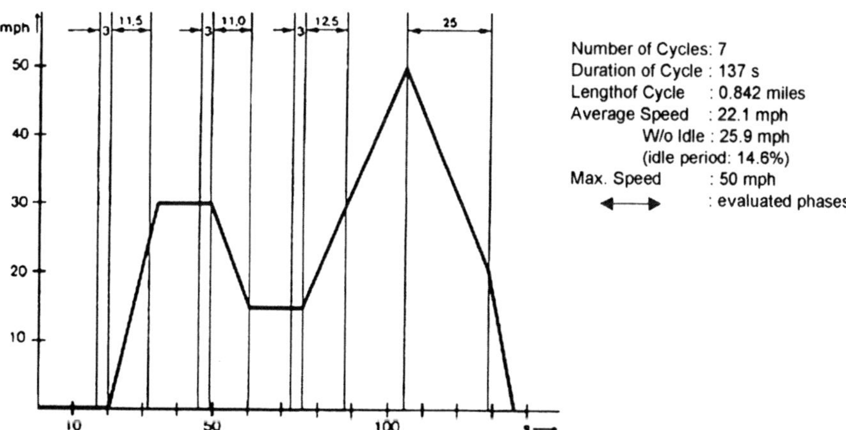

Fig. 7. The first driving cycle for emission certification testing of vehicles on a chassis dynamometer: the "California 7-Mode Cycle"

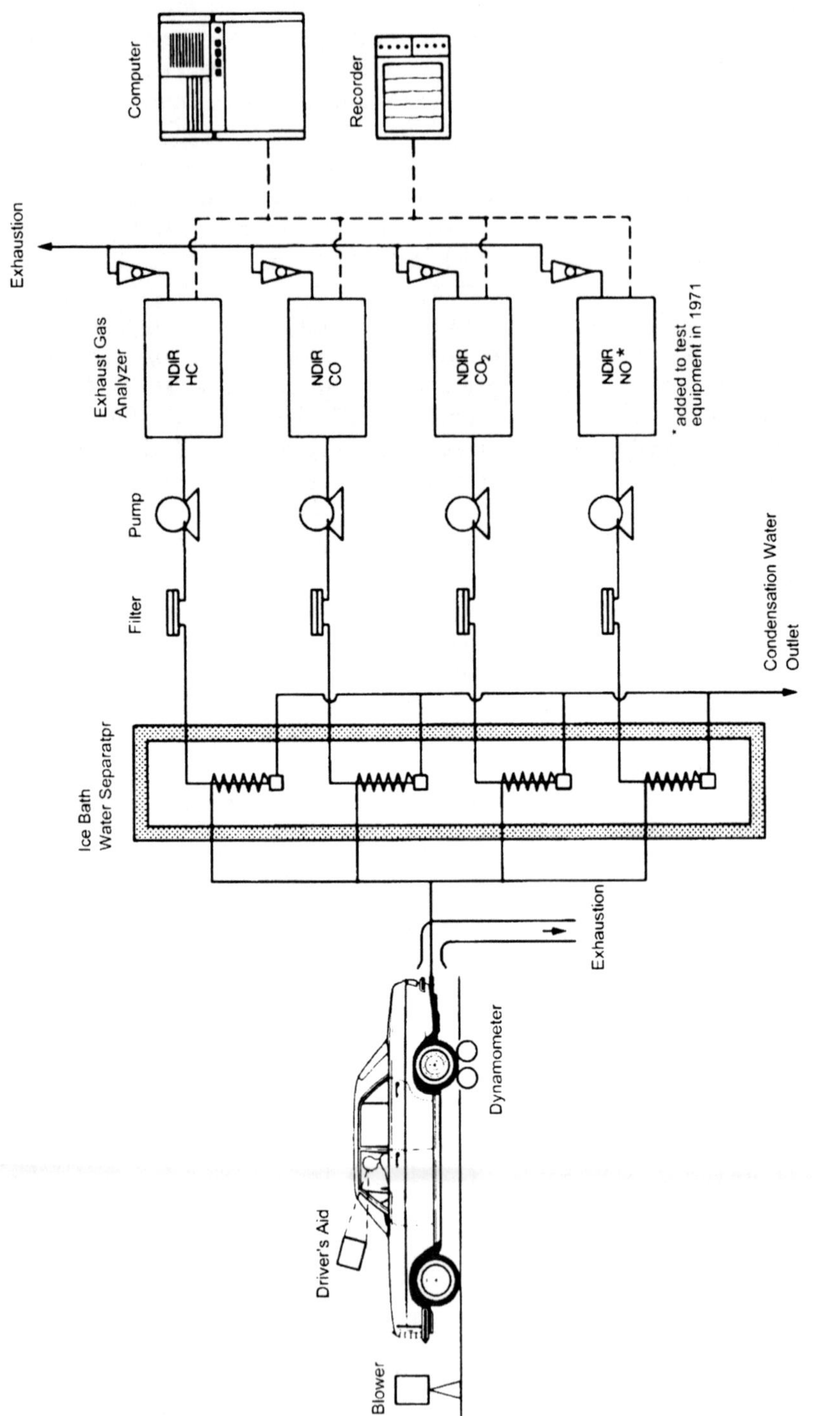

Fig. 8. First emission measurement technique for vehicle testing on a chassis dynamometer (California 1966)

1968/69 tried to correct this deficiency by determining the emission standards in relation to engine displacement.

On June 4, 1968 45 Federal requirements specified in CFR Part 85 were again amended to revise the test procedure for model years 1970/71. Although the new procedure was still based on the California 7-mode test, it allowed, however, to convert concentration measurements to mass emission values. This was achieved by transforming the mass emission values with an empirically determined formula which assumed the exhaust gas volume as a function of vehicle mass and transmission type for an average vehicle of the 4,000 lbs inertia weight class. This method allowed one to set again unified emission standards, however mass-based now.

An additional modification to the test procedure was the transition from NDIR to "Flame Ionization Detection" (FID) measurement for the determination of HC emissions. The problem with NDIR was that it could only determine the concentration of the paraffin *n*-hexane while olefins and aromatics, which play an important role in smog formation, could not be detected.

Since the abovementioned mass-based emission standards were established in relation to an average exhaust gas volume, they could not represent an identical requirement for all vehicle sizes. So auto manufacturers were informed on April 3 and June 25, 1969 that as of model year 1972 a true mass emission test method would be established. The new procedure applied a "Constant Volume Sampler" (CVS) for the determination of the exhaust gas volume with proportional sampling and FID technique for the measurement of total HC.

Introduction of the CVS method allowed henceforth the application of a "non-repetitive" and "closed" driving cycle as it had been developed in the meantime and which was officially proposed in the Federal Register on July 15, 1970 as the so-called "LA4-cycle".

As of model year 1973 the test method was expanded to include chemiluminescence (CL) measurement technology for the determination of NO_x emissions.

During work performed within the APRAC ("Air Pollution Research Advisory Committee") project "CAPE-10" of the "Coordinating Research Council" (CRC) it was found that vehicles in Los Angeles city were used on average for 4.7 trips per day. One of these trips started after the vehicle had stood over night, the other 3.7 trips included cold and hot starts. It was determined that one of the 3.7 trips with cold engine whereas the rest of 2.7 trips were mere hot starts. The hot/cold weighing factor thus resulted to 2.0:4.7=0.43 for the cold start portion and to 2.7:4.7=0.53 for the hot start portion of the test cycle. These factors became part of the CVS-CH (Cold/Hot) test procedure used as of model year 1975.

Together with this new cold/hot weighing method of the test result the test procedure was also expanded, from a two-bag analytical principle to the 6-bag system, and the "Critical Flow Venturi" (CFV) developed by Philco-Ford was introduced as an alternative to the existing "Positive Displacement Pump" (PDP) dilution principle. Both test equipment set ups are shown in Fig. 9 [8].

With the inclusion of Diesel vehicles in the Federal emission legislation as of model year 1975 the sampling and analytical system of the test procedure had to be modified again. In order to avoid condensation of unburned hydrocarbons in the relatively cool Diesel exhaust gas, the sampling pipe had to be heated up to

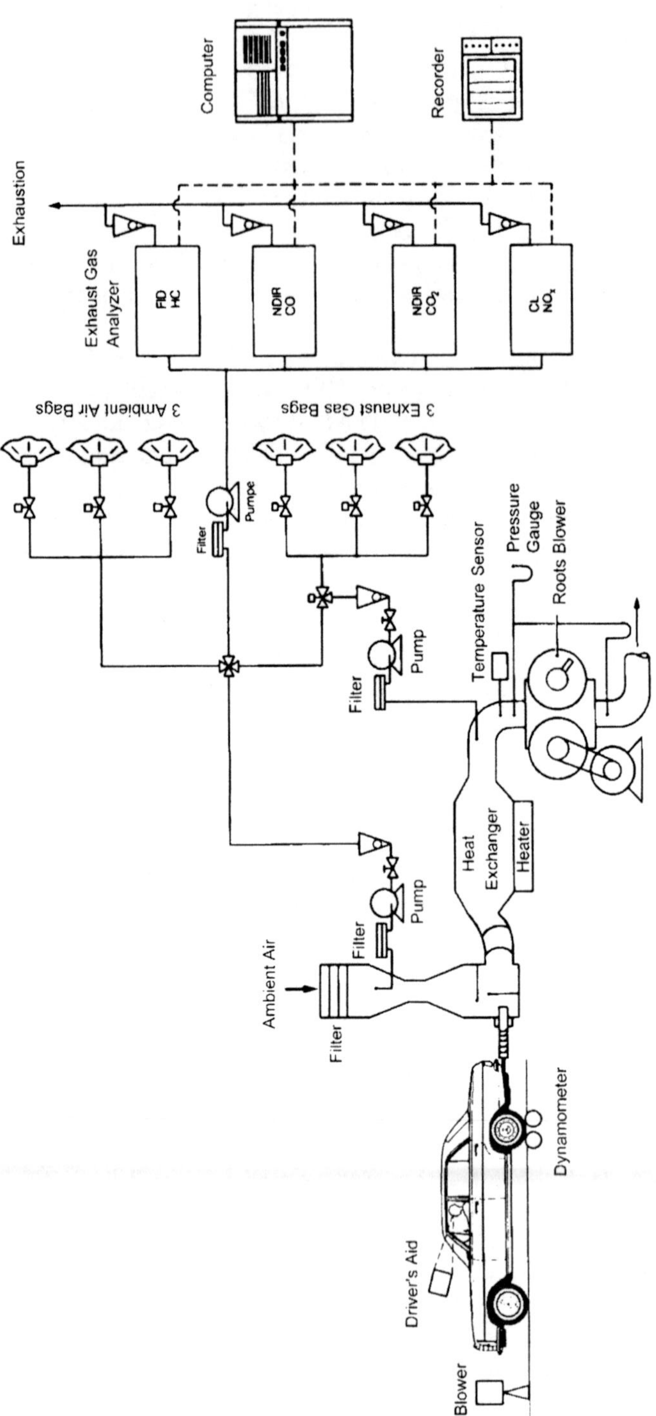

Fig. 9A. Emission measurement technique with positive displacement pump (PDP) or Critical Flow Venturi (CFV) as used in the Federal Test Procedure 1975 for vehicles with gasoline engines

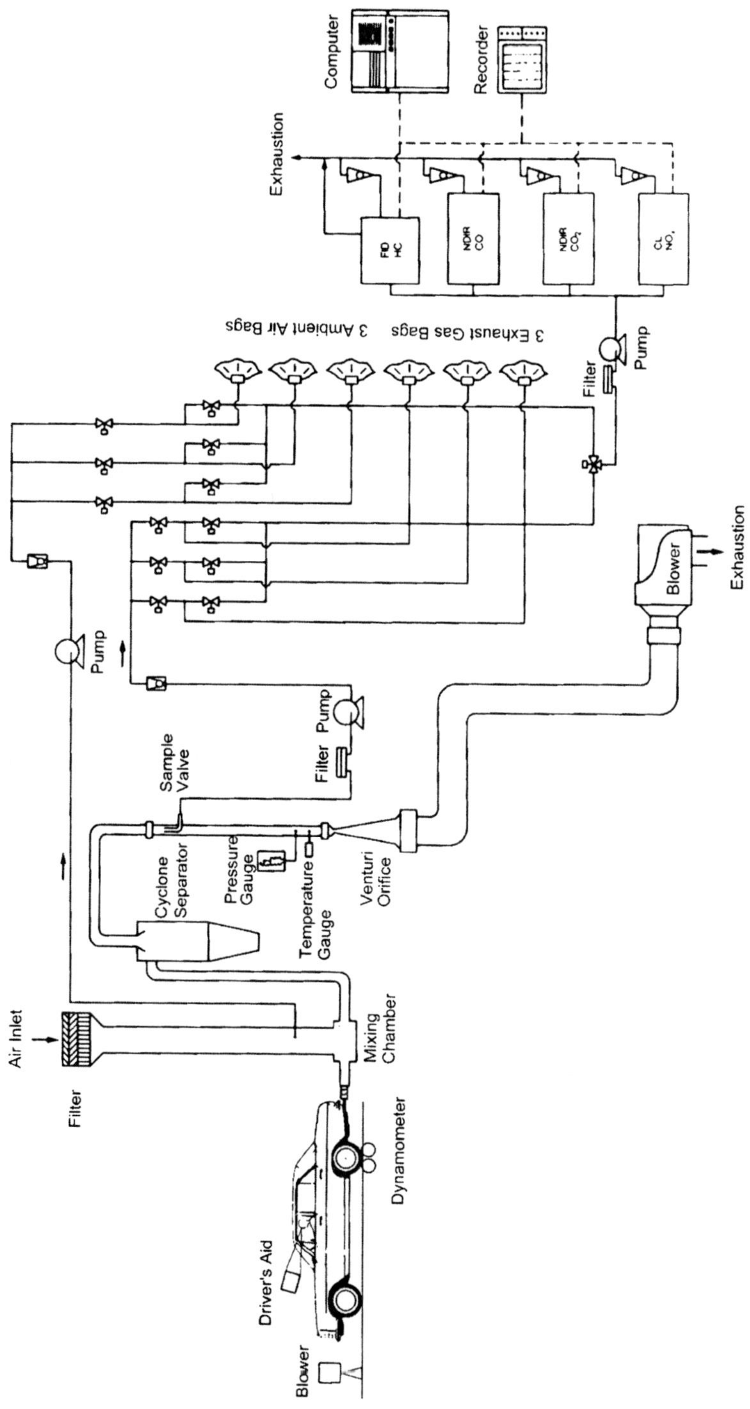

Fig. 9B (continued)

the also heated FID. As of model year 1980, Diesel vehicles were also included in California's emission legislation and consequently this test procedure became applicable here as well.

The last modification of the CVS-CH test procedure was introduced nationwide as of model year 1982 when Diesel vehicles had to additionally meet particulate matter (PM) standards. In order to simulate realistic conditions during the measurement of particulates suspended in the exhaust gas stream, a dilution tunnel with high inner turbulence (Reynolds number >40,000) and the necessary particulate filtering devices had to be integrated in the measuring system which is shown in Fig. 10.

The here described test methods have eventually become the general standard for modern emission test facilities worldwide. Their practical application takes place for vehicle certification testing, which also includes the determination of the vehicle's evaporative emissions in a sealed enclosure.

A variation to the measurement technique for Diesel exhaust gas may, however, be introduced in the future: After emphasis had shifted for some time from particle mass to particle-adsorbed hydrocarbons, it was recently put on fine and ultra-fine particles. Hence a discussion has started whether today's measurement technique for Diesel PM should be changed against a method which evaluates particle size and numbers in order to better address possible health effects.

3.2 Emission Test Methods in Japan

The first Japanese exhaust emission standards for passenger cars with gasoline engines were based on a concentration measurement of CO during the unique Japanese "4-mode test". This test – which is shown here in Fig. 11 for historical purposes – was applied from 1966 to 1972 and was replaced as of 1973 when Japan changed its test method from concentration measurement in the undiluted exhaust gas to the before described CVS-measurement technique together with the introduction of the new "10-mode test".

When Japan started to test passenger cars with Diesel-engines in 1972 its test method only consisted of a special "3-mode test" in which smoke emissions were determined as filter blackness during full load at three different engine speeds. From 1974 onwards, the test method was modified by including a unique "6-mode test" during which gaseous emissions (HC, CO, NO_x) were determined on a concentration basis (using NDIR instrumentation for CO and NO and HFID technique for HC). As of 1975 the "smoke test" part of the test method was amended by an additional "Free Acceleration Test". When the "10-mode test" became applicable as well to vehicles with Diesel engines as of October 1, 1986/87 for domestic manufacturers and vehicles with mechanical/automatic transmission and as of April 1, 1988/89 for importers and vehicles with mechanical/automatic transmission, the "6-mode test" became obsolete [8].

The Japanese emission test methods differed from the rest of the world right from the beginning of emission testing on automobiles – and still differ today – especially with regard to the applied driving cycles. This may result from the fact that Japan has put high efforts in the development of driving cycles and the

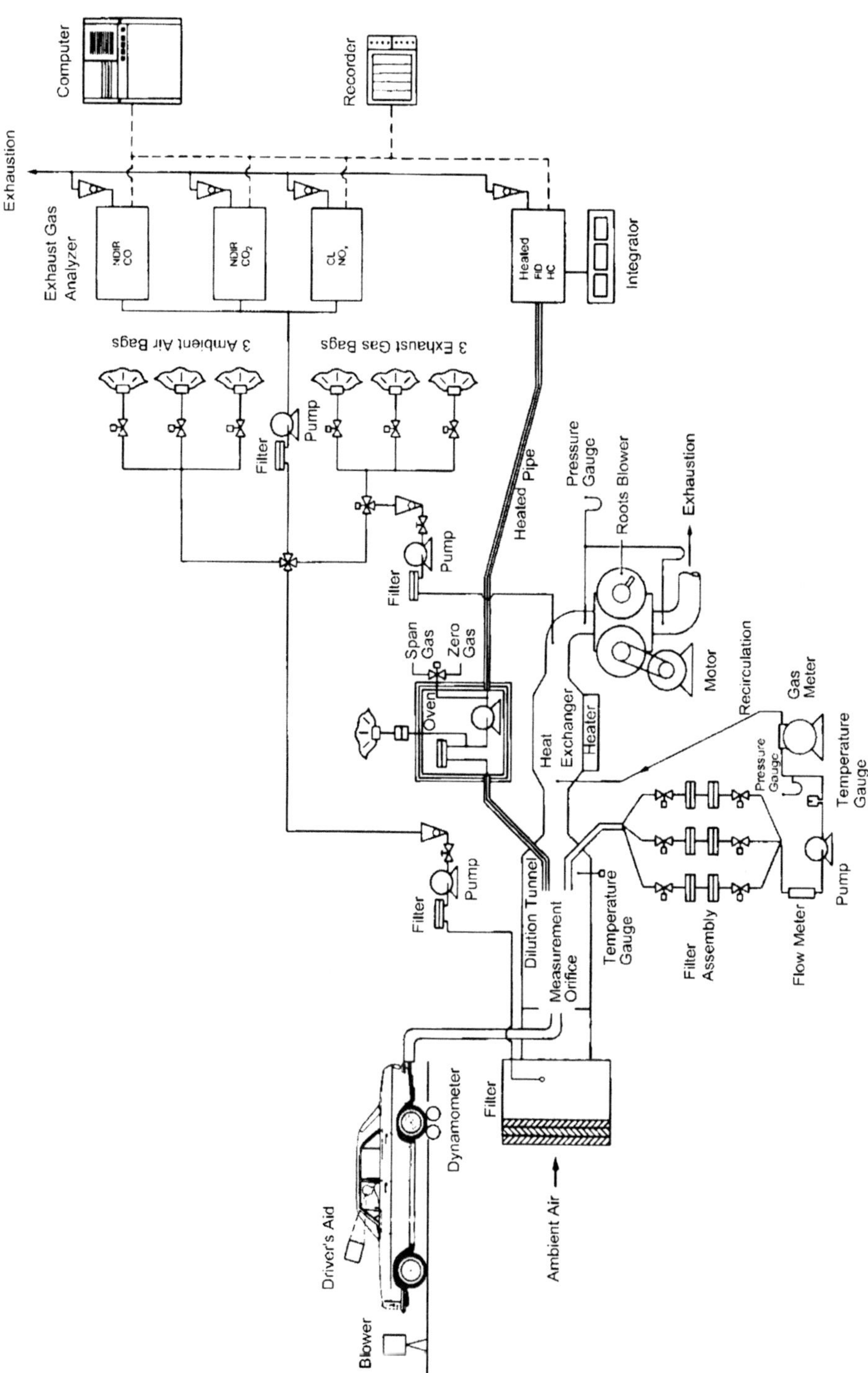

Fig. 10. Emission measurement technique with dilution tunnel for vehicles with Diesel engines used in the Federal Test Procedure as of model year 1982

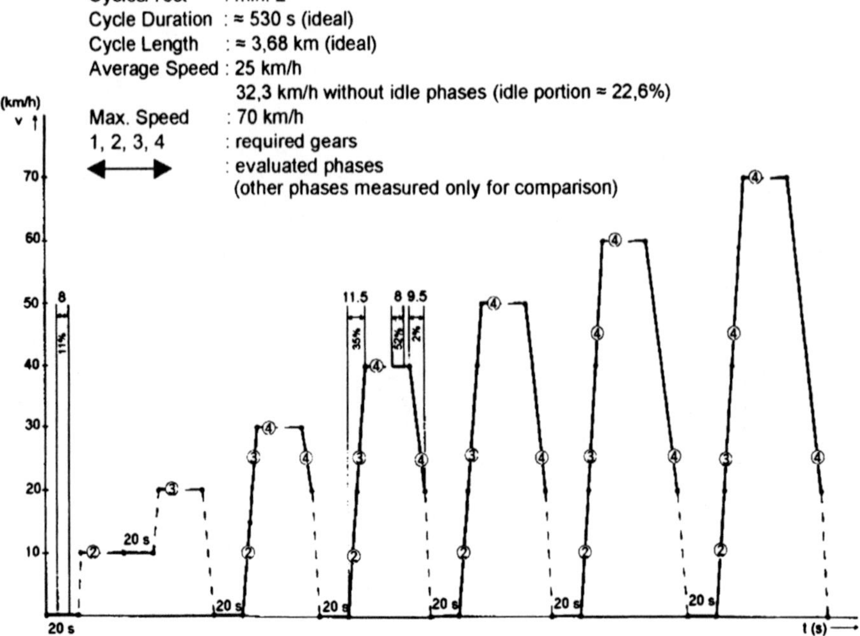

Fig. 11. The first Japanese driving cycle for emission testing of vehicles on a chassis dynamometer: the "4-Mode Test" as of 1966

determination of a driving sequence for the emission test which represents its special congested traffic conditions best. A brief look on the historical evolution of the driving cycle used in today's emission test method shall therefore be a added here.

From 1963 to 1965 the "Senpaaku Gijutsu Kenkyujo" (SENKEN) – today's "Traffic Safety and Nuisance Research Institute of Automobile Type Approval Standards" (TRIAS) – was the first to evaluate driving habits in the city of Tokyo. Four driving modes were evaluated, idle, acceleration, constant speed and deceleration. On July 14, 1966 the result of this investigation was adopted as the standard driving sequence of the first Japanese auto emission test method which was in effect from September 1, 1966 through April 1, 1973 [44].

Since the first Japanese emission test method was limited to the determination of CO, a comparison of the new test with the California "7-mode test" was only possible for this component. The "Shigen Gijutsu Laboratory" (today's "Environmental Nature Resources Laboratory" of MITI) which performed such comparison work established, e.g., the ratio of CO [vol.%] in the 4-mode test to CO [vol.%] in the 7-mode test at 1.25 which means that the CO result from the Japanese test was 25% higher than the result from a California test [45].

Further work was done during a test series performed in June/July 1967 by the city government of Osaka where an even higher idle portion of 40% was found compared to the 35.6% established for Tokyo's congested traffic. Consequently Osaka suggested to use the so-called "Osaka Cycle" derived from this test for its individual emission test method but eventually used the cycle only for research purposes [45].

Since the official "4-mode test"-based emission test method did not take into account that exhaust gas volume depends on vehicle mass and engine displacement, the SENKEN laboratory in 1967 started to investigate again driving behavior of vehicles in actual traffic but this time with the objective to establish a test method for the determination of mass emissions instead of average concentration levels [39]. The program went on in 1968 and resulted in a so-called "8-mode cycle" which represented typical traffic in the city and outskirts of Tokyo.

Although this cycle and two other 8-mode cycles did not find a direct inclusion in the test procedure, they formed the basis for the Japanese "10-mode test" which was eventually adopted for the official emission test method as of April 1, 1975 [45].

As of April 1, 1975 the official test procedure was further amended through addition of an "11-mode test" to simulate vehicle operation including an engine cold start. Therefore, investigations about driving habits performed in 1973 included early morning rush-hour traffic with cold-started vehicles coming in to Tokyo from outside the city area.

Parallel to corresponding developments in Europe – where the European driving cycle was modified to include an extra urban higher speed section and was eventually introduced as the "New European Driving Cycle" (NEDC) together with "Euro 1" (91/441/EEC) – Japan added a high speed sequence to its 10-mode hot start tests as well. The resulting "10.15-mode test" became mandatory as of November 1, 1991 for domestic manufacturers and as of April 1, 1993 for importers.

The last attempt to establish a driving cycle which would reflect real world conditions was made in 1976 by the "Tokyo Metropolitan Research Institute for Environmental Protection" (TMRIEP). All Japanese test cycles evaluated before were "repetitive" cycles and consisted of linear and stationary driving sequences. In contrast, the new "Tokyo Metropolitan Cycle" developed by TMRIEP was a non-repetitive cycle composed from 760 trip sections derived from Tokyo's ring artery road Meiji Dori [46]. It took into account the frequency distribution of average vehicle speeds in relation to road characteristic and day time. Its speed vs. time trace and duration (1466 s) resembled very much the US LA4-Cycle (1372 s), however the average speed was only 22.5 km/h compared to 31.67 km/h and its maximum speed was only 57.8 km/h compared to 91.2 km/h of the LA4-cycle. The cycle was not introduced in Japan's official emission test method but was used by TMRIEP for the evaluation of effects of emission control measures.

With the introduction of modern three-way catalyst technology the question about the most representative driving cycle has – at least for today's passenger cars with gasoline engines – practically lost its importance from the test result point of view since only a few seconds after engine start decide about passing or failing applicable emission standards.

3.3 Emission Test Methods in the European Union

Work about emission test methods for automobiles started in Europe in the mid 1950s when, e.g., in Germany the VDA Subcommittee "Exhaust Gases from Gasoline Engines (Unterausschuss "Abgase von Otto-Motoren") was assigned to establish emission standards, to evaluate possibilities for pollutant reduction and to develop necessary measurement techniques. Until October 1958 the German Ministry of Traffic ("Bundesministerium für Verkehr", BMV) had distributed research assignments in the field of automobile emissions (e.g., testing towards the development of lead-tolerant catalysts, pollutant reduction in the exhaust gas of gasoline engines, air quality measurements in German cities) to different parties but it soon became clear that a coordination of these activities – which went on in similar fields at the same time in France and Sweden – had to be coordinated to proceed in an effective manner.

So on February 25, 1959 Prof. Luther (University of Clausthal) called together representatives from the VDA ("Verband der deutschen Automobilindustrie"), from the VDMA ("Verein Deutscher Maschinenbauanstalten"), from the FVV ("Forschungsvereinigung Verbrennungskraftmaschinen") and from research institutes and oil industry in an effort to achieve this work coordination [46].

A further milestone is April 7, 1959 when the before described VDA subcommittee met to discuss the need and possibilities for chassis dynamometer testing. It is important to note that the German development work towards emission measurement under actual driving conditions right from the beginning concentrated on chassis dynamometer testing which in later years became the acknowledged and applied principle throughout Europe.

On December 18, 1963 the German VDI ("Verein Deutscher Ingenieure") started to evaluate work under progress on the development of an appropriate driving cycle. In Germany such work was performed by Prof. Luther who evaluated typical driving conditions in six German cities using five vehicles with different engine displacements. Similar investigations were in progress in Great Britain, Sweden and France [30]. On April 14, 1964 Prof. Luther defined the main criteria he felt necessary to be applied for such driving cycle. First of all the cycle should be performed on a chassis dynamometer. The French driving sequence developed by the "Union Techniques de l'Automobile, du Motorcycle et du Cycle" (UTAC) – which was already established when the work on the German driving cycle started – was only applicable to road testing.

Although Prof. Luther, the German TÜV ("Technischer Überwachungsverein") and the German auto manufacturers would have preferred if the German driving cycle had turned out to be similar to the California cycle so that the California cycle could just be taken over, the findings of Prof. Luther, presented on December 1, 1965, showed that vehicles were idling in German city traffic with 45% whereas the corresponding portion in the California cycle was only 14% (later evaluations showed 35% which came pretty close to the UTAC road driving sequence which had used a value of 31%).

The driving cycle discussion had been monitored by WP29 which in its 20th session on December 20, 1965 assigned the BPICA ("Bureau Permanent International des Constructeurs d'Automobile") to propose a unified European driving cycle. The first draft of this cycle was presented by BPICA during the 1st session of the GRPA in Paris on July 6/8, 1966. After some modifications (e.g., a reduction of the average speed from 21.2 to18.9 km/h which was requested by Great Britain and after an evaluation in the London laboratories of the BPICA)

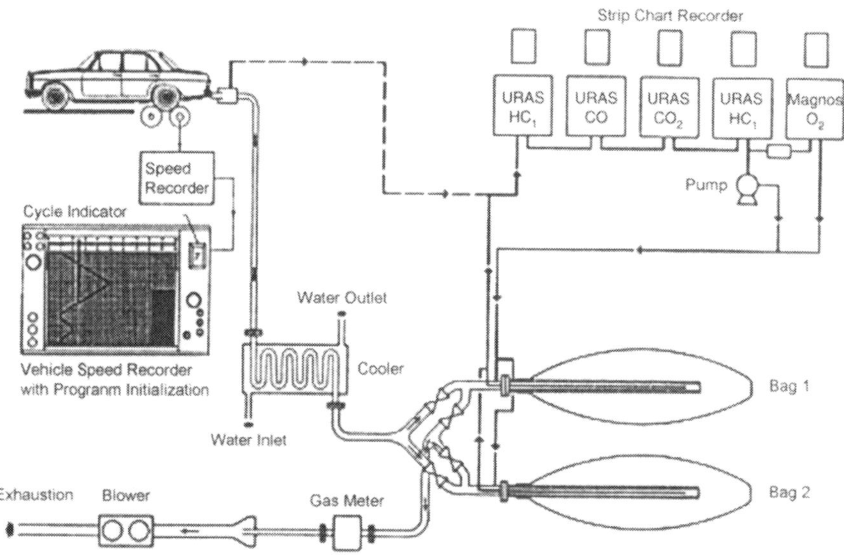

Fig. 12. The first ECE/EEC emission measurement technique for passenger cars with gasoline engines: the "Big Bag"-method as of 1970

the cycle was eventually accepted during GRPA's 2nd session on January 9/11, 1967. It became known as "ECE cycle" (which can be seen as "Part1" of the test shown in Fig. 13) and formed the basis of vehicle type approval emission testing within the first ECE Emission Regulation ECE-R15 and the first EEC Emission Directive 70/220/EEC described already in Sect. 1.4.

These regulations required the determination of CO and HC emissions in the vehicle's exhaust gas by means of a so-called "big bag" test [47]. This test method is shown for historical purposes here in Fig. 12.

In the course of further strengthening of emission standards this test method was, however, substantially changed both with regard to the measurement technique as with regard to the driving cycle: As of October 1, 1982 the "big bag" method was replaced by the CVS measurement technique and the driving sequence was amended by introducing the so-called "New European Driving Cycle" (NEDC) – as shown in Fig. 13 – in conjunction with Emission Directive 91/441/EEC which became mandatory as of July 1, 1992 ("Euro 1"). The test procedure was again slightly modified as of January 1, 2000 in conjunction with Emission Directive 98/69/EC ("Euro 3") insofar that the first 40 seconds after engine start were henceforth included in the emission measurement. Previously exhaust emissions were not sampled during the first 40 seconds after engine start.

As minor as this modification may seem to be, it represented nevertheless another substantial strengthening element in the transition from Directive 94/12/EC ("Euro 2") to Directive 98/69/EC ("Euro 3") since the emission test result of modern gasoline engines fully depends on the catalyst performance during these first seconds after engine start.

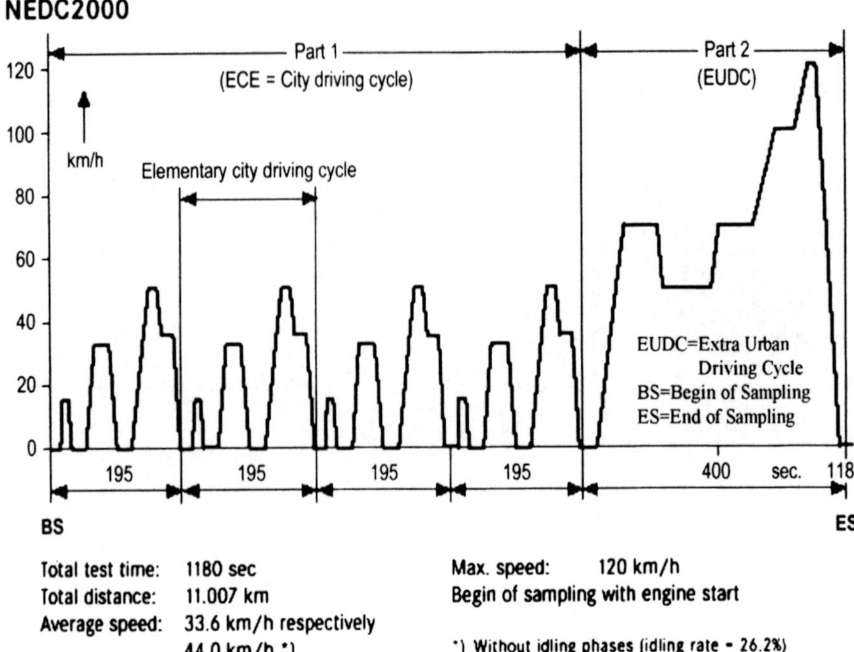

Fig. 13. The "New European Driving Cycle" (NEDC) as modified with step "EU 3" of directive 98/69/EC

The NEDC in combination with the modified test procedure form the basis of emission testing within the European Type Approval Procedure. While in the US an "Executive Order" (US-California) or a "Certificate of Conformity" (US-Federal) require just one test during which the vehicle meets applicable emission standards (independent of the safety margin between actual test result and standard), granting of an EU "Type Approval" may need up to three emission test runs depending on the quality of the first test. The corresponding test sequence and emission test evaluation procedure are shown in Fig. 14.

4 Additional Requirements for Passenger Cars

The extremely stringent emission requirements for new motor vehicles and the increasing importance attributed to the stability of a vehicle's emission characteristic over its lifetime already represent a highly demanding task for the automobile industry. However, three additional areas have to be mentioned in this context since they are directly related to – and substantially impact on – overall exhaust emission control efforts:

- efforts towards the control of evaporative emissions,
- efforts towards the reduction of CO_2 emissions, or – in other words – for the reduction of fuel consumption, and
- additional efforts resulting from "On-Board Diagnostics" (OBD)-requirements.

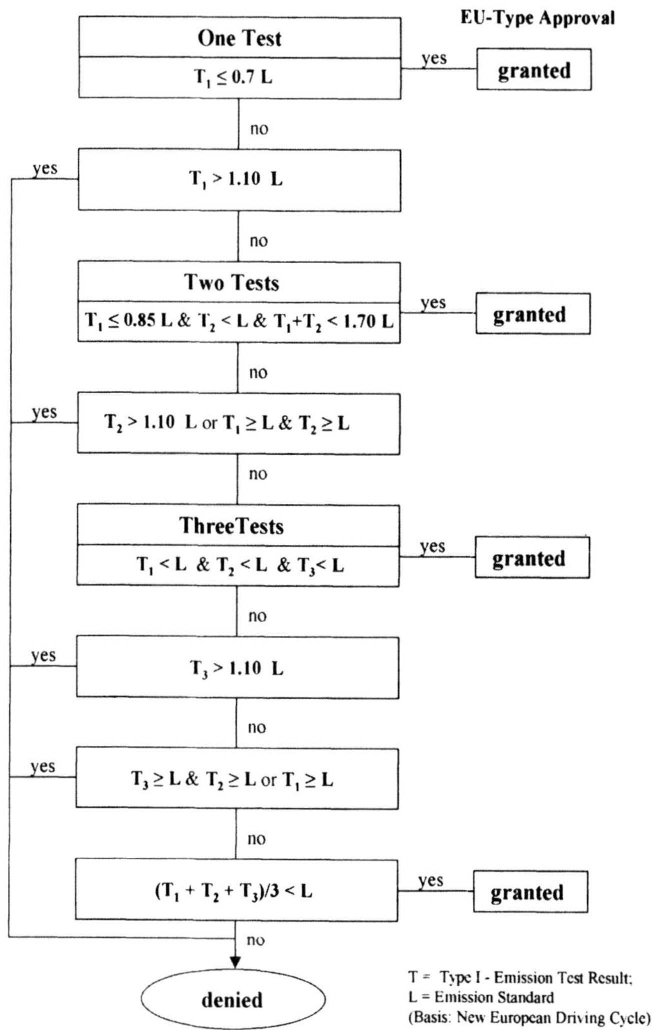

Fig. 14. Evaluation procedure for type approval emission testing in the European Union

The combined effect of future stringent exhaust and evaporative emission standards, CO_2/fuel consumption reduction obligations or commitments and OBD legislation represents a major technical challenge for the concerned industry.

4.1 Control of Evaporative Emissions

One of the first publications about the measurement of evaporative emissions dates back to the summer of 1937 [48]. Already at that early point of time an attempt was made to determine the amount of fuel gases escaping from the carburetors of three test vehicles. However, it took about another two decades before evaporative emissions once again became the subject of detailed investigations [49].

When the deadline for legally required installation of evaporative emissions control systems was approaching, the concerned industry was still confronted with substantial problems with the measurement of these emissions since a suitable measurement technique was not at hand. Substitute methods like, e.g., density measurement, were used to estimate the amount of evaporated fuel mass. While this method could be applied to carburetors, it did not work for the determination of evaporative emissions from the fuel tank on certain vehicle designs, e.g., on vehicles with flow back lines from the fuel pump to the tank [50].

Another method was the "cold trapping" method [49] by which the evaporative emissions from the carburetor were gathered in a plastic bag which was then cooled down in order to achieve gas condensation and thereafter allow a mass measurement of the evaporated fuel.

In 1959 the AMA ("Automobile Manufacturer Association") initiated a program for the determination of evaporative losses from carburetors and fuel tanks, the results of which were published in 1961 [51]. In 1965 the CRC ("Coordinating Research Council") started a follow-up program which focused on the reliable determination of evaporative emissions from carburetors and was meant to expand the knowledge obtained from the AMA investigation. The results of this program reflect the problems which still existed with the measurement techniques available at that point of time:

- "cold trapping" was found to be rather time consuming and was, therefore, not considered a viable technique for field testing of vehicles, further this method did not allow the true evaporative emissions on certain vehicle designs;
- indirect methods (like, e.g., density measurement or determination of fuel/vapor ratio) are faster, less expensive and more precise in their results.

The report concludes with the recommendation that the CRC should develop as soon as possible an intensive program to further develop methods for the determination of evaporative emissions [52]. This program was initiated in 1966 and when its results were published in January 1967 [53] the authors still concluded that "the evaluated methods had shown faults and imprecision which were substantially higher than expected".

In February 1967 the automotive industry heard about a proposal of the DHEW ("Department of Health, Education and Welfare") which for the first time mentioned that the evaporative emissions of a vehicle should be measured in a SHED ("Sealed Housing for Evaporative Emissions Determination") and General Motors could prove the functionality and superiority of the SHED versus all other existing test methods [54].

Although the SHED technique was practically available already at this point of time it took another decade before it became – triggered by the US-EPA ("Environmental Protection Agency") – a legally required test method for vehicle certification testing as of model year 1978.

The first evaporative emissions standard of 6 g/test became applicable for model year 1970 in California and for model year 1971 in Federal regulations. It was reduced to 2 g/test within both legislations as of model year 1972. Until the end of model year 1977 emissions were determined by the so-called "carbon trap", small active carbon-filled canisters, which were attached to all accessible openings of the vehicle's fuel system (like, e.g., carburetor, air filter inlet, tank filler cap, tank ventilation valves).

With the nationwide introduction of the SHED test procedure as of model year 1978 the standard was adapted to the new method which resulted in a figure of 6.0 g/test. The standard was lowered to 2.0 g/test for model year 1980 (California) and 1981 (Federal).

The SHED test in this initial form – which is shown in Fig. 15 – was applied until the end of model year 1994 (California) or until the end of model year 1995 (Federal). The test sequence of this figure shows the steps encountered as the test vehicle undergoes the procedure of a diurnal heat build and a hot soak test [55].

The diurnal evaporative emission (heat build) test was introduced to quantify evaporative emission losses occurring during a simulated 1-hour diurnal temperature rise. The measurements were added to HC emission losses from the hot soak test to obtain a measurement of total HC vapor losses occurring during motor vehicle operation.

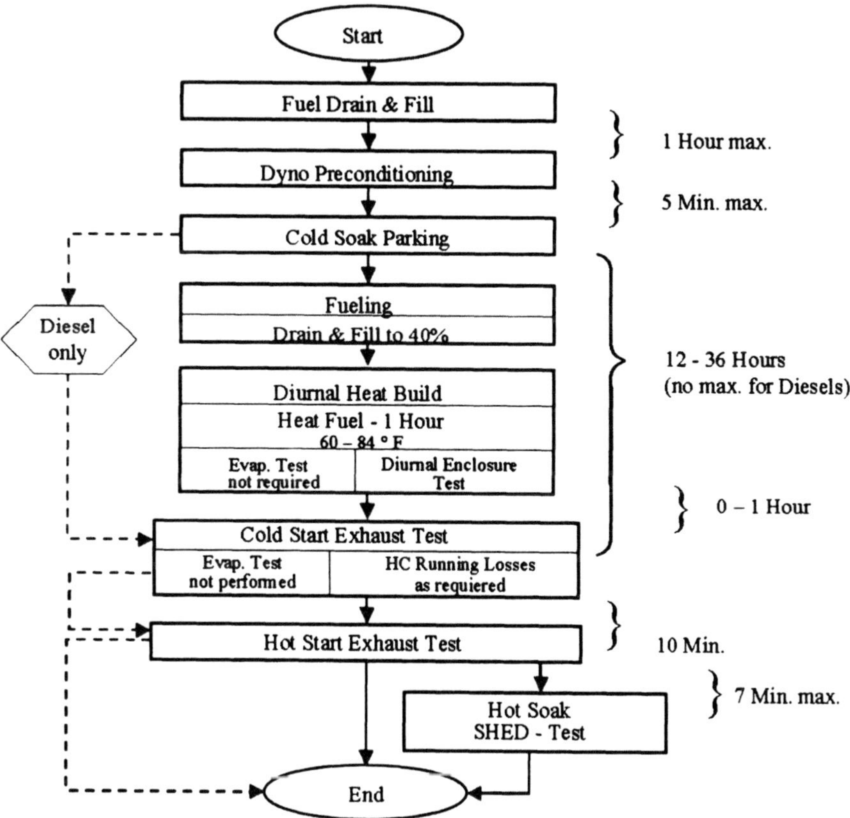

Fig. 15. SHED test initial form as of US-model year 1978

The purpose of the hot soak test was to quantify hydrocarbon evaporative emission losses which occur when a vehicle is parked and the hot engine is off. These measurements are added to those obtained by the diurnal test.

An emission standard of 2 gHC/test was set for the sum of emissions from the diurnal test and the hot soak test.

When the California ARB became more and more concerned about evaporative emissions from various sources in the state, it started to reconsider its existing test procedures. With regard to automobiles it introduced an enhanced SHED test procedure by combining the existing diurnal and hot soak portions of the test sequence with a new running loss test and by setting new emission standards [56]. The new test sequence of the accordingly modified Federal Test Procedure is shown in Fig. 16.

The purpose of the enhanced diurnal evaporative emission test was to quantify evaporative emission losses occurring during 48-hour, 72-hour, and various other simulated diurnal temperature cycles. The test procedure was designed to measure diurnal emissions resulting from daily temperature changes (as well as relatively constant resting losses), measured by the enclosure technique. The enclosure itself was adapted to the new test and was modified into a variable temperature housing (VTSHED).

Emissions were measured according to a temperature vs. time sequence for each 24-hour cycle, with the highest emission level added to the hydrocarbon amount measured during the 1-hour hot soak test.

This substantially revised SHED procedure was first introduced with model year 1995 in California by means of a phase-in via the following steps: 10% for model year 1995, 30% for model year 1996, 50% for model year 1997and 100% for model year 1998. Under Federal regulations the introduction started one year later via the following steps: 20% for model year 1996, 40% for 1997, 90% for 1998 and 100% for model year 1999.

Applicable emission standards for the different test portions together with further strengthening steps within EPA's "Tier2" and California's "LEV2" legislation are summarized in Table 8.

The latest enhancement with regard to evaporative emission control regulations occurred in model year 1998 when the requirement for "On-Board Refueling Vapor Recovery" became applicable. The new regulation was also introduced – on a nationwide basis – via a phase-in starting with 40% of the manufacturer's fleet in model year 1998, 80% in model year 1999 and 100% as of model year 2000 [57]. The applicable standard was 0.20 gHC per gallon (0.053 gHC per liter) of fuel dispensed and has to be met over 10 years or 100,000 miles.

Due to the high emphasis put on the reduction of HC emissions by the California ARB the control of evaporative emissions became recently an important aspect of the zero emission vehicle credit system within California's "LEV2" regulations: A SULEV (Super Ultra Low Emission Vehicle) meeting the extremely stringent exhaust emission standards applicable to this vehicle category of 0.010 gNMOG/m, 1.0 gCO/m, 0.02 gNO_x/m, 0.01 gPM/m and 0.004 gHCHO/m, meeting also the "OBD2" requirements and not emitting any evaporative emissions over 150,000 miles may get a 0.2 ZEV credit. This means that a manufacturer may replace one zero emission vehicle within his

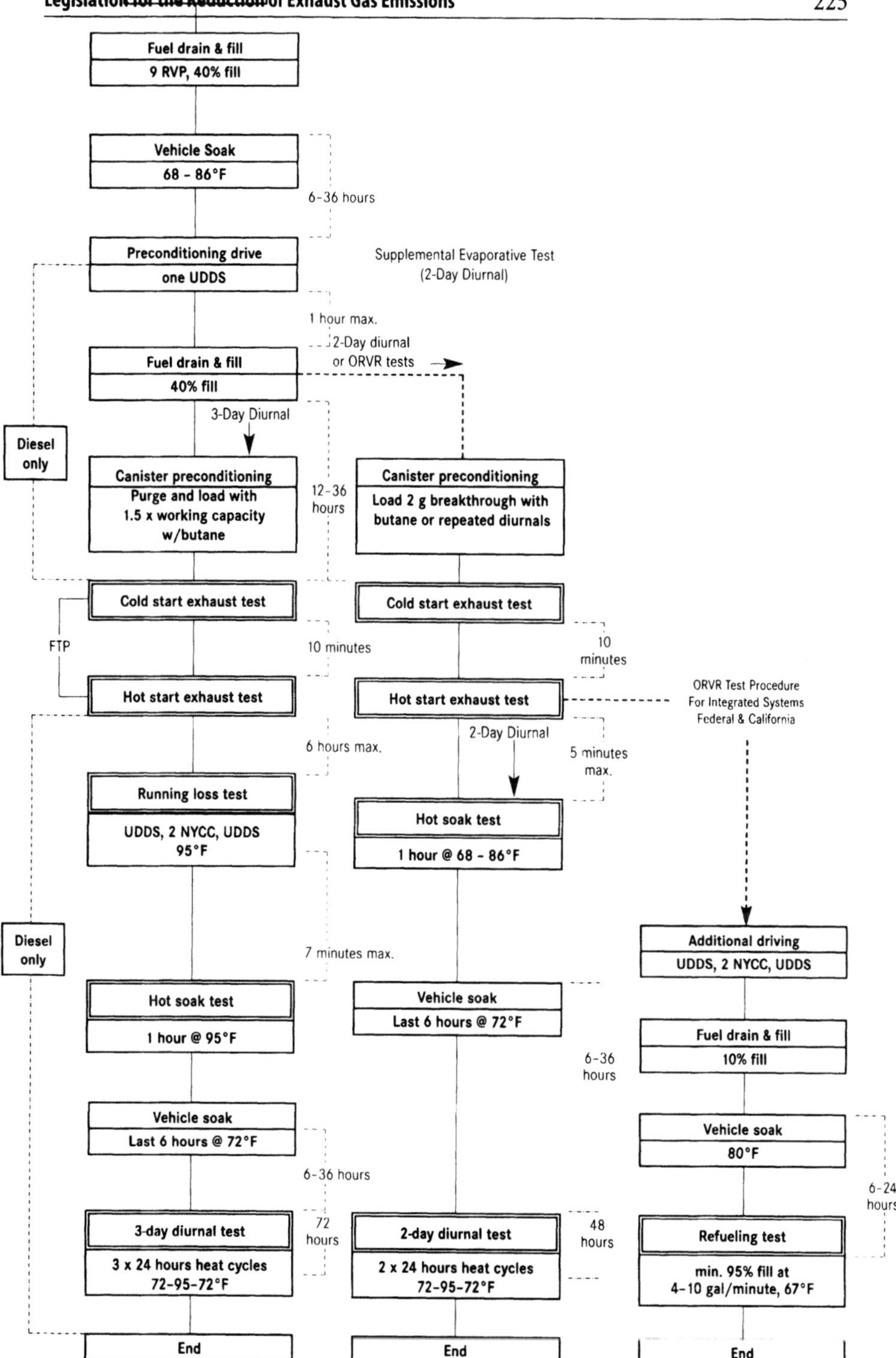

Fig. 16. Exhaust and evaporative emissions plus on-board vapor recovery testing within the US Federal Test Procedure for light duty vehicles and light duty trucks

Table 8. Development of Evaporative Emission Control Requirements in the USA

USA – Federal (EPA)	USA – California (CARB)
Step 1 and **2** First limitation as a model year 1971: **6.0** g/test. ("Cold Trapping"-method); Standard valid for 50.000 miles. As a model year 1972: **2.0** g/test	**Step 1** and **2** First limitation as 1 model year 1971: **6.0** g/test. ("Cold Trapping"-method); Standard valid for 50.000 miles. As a model year 1972: **2.0** g/test
Step 3 and **4** As a model year 1978: Transition to SHED-method. Limit for basic SHED Test (consisting of "Diumal Test" and "Hot Soak Test"): **6.0** g/test. As of model year 1981: **2.0** g/test.	**Step 3** and **4** As a model year 1978: Transition to SHED-method. Limit for basic SHED Test (consisting of "Diumal Test" and "Hot Soak Test"): **6.0** g/test. As of model year 1981: **2.0** g/test.
Step 5: "Enhanced" Evaporative Emissions Control **3-Day Diurnal Test** plus **Hot Soak Test**: **2.0** gHC/test **2-Day Diurnal Test** plus **Hot Soak Test**: **2.5** gHC/test **Running Loss Test**: **0.05** gHC/mile These standards are valid for 10 years/100.000 miles. They were introduced from model year 1996 to 1999 in steps of 20/40/90 and 100%. Test fuel: Indolene: Test temperature: 22.2–35.6 °C	**Step 5: "Enhanced" Evaporative Emissions Control** Standards, tests and durability requirements are the same as Federal but are introduced already from model year 1995 to 1998 in steps of 10/30/50 and 100%. Test fuel: California "Phase 2" – Reformulated Fuel; Terst temperature: 18.3–40.6 °C
Step 6: "Tier 2" – Requirements **3-Day Diurnal Test** plus **Hot Soak Test**: **0.95** gHC/test **2-Day Diurnal Test** plus **Hot Soak Test**: **1.2** gHC/test **Running Loss Test**: **0.05** gHC/mile The standards are valid for 10 years/120.000 miles. They will be introduced from model year 2004 to 2007 in steps of 25/50/75 and 100%.	**Step 6a: "LEV 2" – Requirements** ("Near Zero" – Standards) **3-Day Diurnal Test** plus **Hot Soak Test**: **0.50** gHC/test **2-Day Diurnal Test** plus **Hot Soak Test**: **0.65** gHC/test **Running Loss Test**: **0.05** gHC/mile **Step 6b: "LEV 2" – Requirements** ("PZEV" – Standards) Emissions from the fuel system: **0.0** gHC/test. Permitted vehicle-related ("background"-)emissions for the Diurnal plus Hot Soak Tests: **0.35** gHC/test. These standards are valid for 15 years7150.000 miles and will be introduced from model year 2004 to 2006 in steps of 40/80 and 100%.

Limitation of Evaporative Emissions from the Refueling Process

"Fuel Spit-Back" – Test[a]	EPA: **1.0 gHC**/test; CARB: **1.0 ml fuel** dispensed at fill rate of 10 gal./minute
ORVR – Test[b]	0.20 **gHC**/gal. (0.053 g/ltr) at fill rate of 4–10 gal/minute. Introduction from model year 1998 to 2000 in steps of 40/80/100%. Not valid for gas-fueled vehicles. Diesel vehicles may be exempt if requirements are met without onboard storage system

[a] up to 1998; [b] Onboard Refueling Vapor Recovery.

ZEV introduction obligation by 5 of such zero evaporative-emissions SULEVs [58].

As has already been shown in Fig. 16 in connection with the description of evaporative emission testing in the US, complete vehicle emission certification testing includes both an exhaust emission test and a test for the determination of the vehicle's evaporative losses. A similar flow chart describing the corresponding EU test procedure shall be added here for comparison purposes in Fig. 17.

4.2 Reduction of CO_2 Emissions and Fuel Consumption

CO_2 emissions are a normal constituent of the atmosphere since approx. 96.5% of all global yearly CO_2 emissions stem from natural sources [59]. It is assumed that these emissions stay in an equilibrium between emission sources and emission sinks and that the remaining 3.5% CO_2 emissions from anthropogenic sources disturb this equilibrium and contribute to the global warming effect (see details in chapter "Means of transportation and their effect on the environment").

Concerning motor vehicles, it is estimated that road transport accounts for about 11% of these 3.5% global anthropogenic CO_2 emissions [59]. Taking further into account that only a certain portion of these 11% can be avoided through a reduction of fuel consumption, the limited role the automobile can play for the global situation becomes obvious.

On the other hand, a reduction of CO_2 emissions means a reduction of fuel consumption and will, therefore, directly support the indisputable objective of saving resources.

Insofar, related governmental programs towards reducing CO_2 emission automatically cover steps for fuel economy improvement, i.e., saving resources, as well as measures against the anticipated global warming effect.

Some major aspects and consequences of these programs shall now be described on the examples of the European Union, Japan and the United States.

4.2.1 *Focus on CO_2 Reduction in the EU*

Requirements for the reduction of CO_2 emissions and fuel consumption established in the EU result from the global climate conference of Kyoto where the participating nations have agreed about political targets for the reduction of their national CO_2 output: It was agreed to reduce greenhouse gases for 2008/2012 compared to 1990/1995 levels by 5% for all industrialized countries with the European Union accepting a reduction target of 8% which has to be shared among member states. These political commitments were transformed into national targets which shall now be described on the example of the Federal Republic of Germany.

In order that the German Government can achieve the national CO_2 reduction target of 21% resulting from the Kyoto commitment [60] all involved CO_2 emission sources have to contribute. The German automotive industry has agreed to

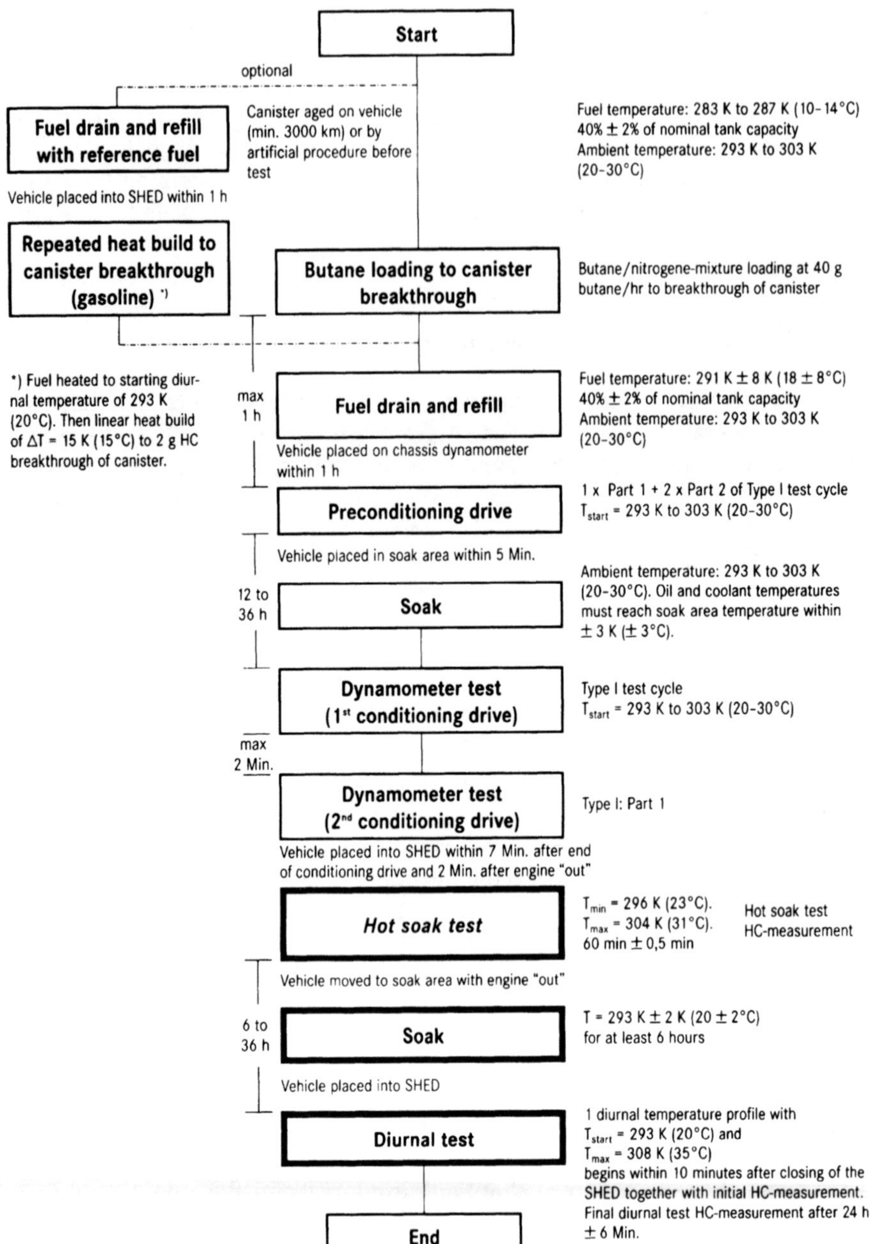

Fig. 17. Exhaust and evaporative emission testing within the type approval system of the European Union

On February 5, 1999 EU Commissioner Bjerregaard signed the following Commission Recommendation to the ACEA on the reduction of CO_2-emissions from passenger cars (doc. 1999/125/EC):

- The members of the European Automobile Manufacturers Association ACEA should, mainly by technological developments and market changes linked to these developments, collectively achieve a CO_2 emission target of 140 g/km, as measured according to Directive 93/116/EEC, for the average of their new cars sold in the Community (category M1 as defined in Annex I to Council Directive 70/156/EEC) by 2008.
- The ACEA should evaluate in 2003 the potential for additional fuel-efficiency improvements with a view to moving further towards the objective of 120 gCO_2/km by 2012
- Individual members of the ACEA should place on the market in the Community models emitting 120 gCO_2/km or less, as measured according to Directive 93/116/EEC, by the year 2000
- The members of the ACEA should make every effort to achieve collectively an intermediate CO_2 emission target in the range of 165 – 170 g/km, as measured according to Directive 93/116/EEC, by 2003
- The ACEA should cooperate with the Commission in the Monitoring of its Commitment

Fig. 18. CO_2 reduction in the EU: recommendations by the EU Commission

take its share in form of a self-commitment announced by its association "Verband der Automobilindustrie" (VDA). This so-called "VDA-Commitment" obliges the German car manufacturers to reduce the fuel consumption of their new vehicle fleets in 2005 by 25% compared to the year 1990.

Responding to the recommendation from the EU Commission shown in Fig. 18, the association of the EU car manufacturers ("Association des Constructeurs Européens d'Automobile", ACEA) issued a similar self-commitment within the EU frame. This so-called "ACEA Commitment" determines that the EU car manufacturers intend to reduce the average CO_2 emissions of their 2008 new vehicle fleet to 140 g/km. This corresponds to a 25% CO_2 reduction in 2008 compared to 1995. In 2003 an intermediate level of 165–170 g/km is envisaged as an indicator whether progress occurs as planned.

In spite of differences in setting the objective, both commitments can be seen as equivalent in the final result. The involved car manufacturers try to achieve this goal through a market-conform share ratio of gasoline and Diesel vehicles which are both facing – and have to overcome – the inherent technical conflict between NO_x emission reduction and simultaneous engine efficiency improvement shown in Fig. 19 [61].

4.2.2
Fuel Economy Programs in the US

The US have not signed the Koyto protocol because the government did not want to impose CO_2 reduction requirements on the US economy the justification of which – namely the global-warming effect of CO_2 emissions – it obviously doubts. However, past energy crisis and the dependency of the US from oil im-

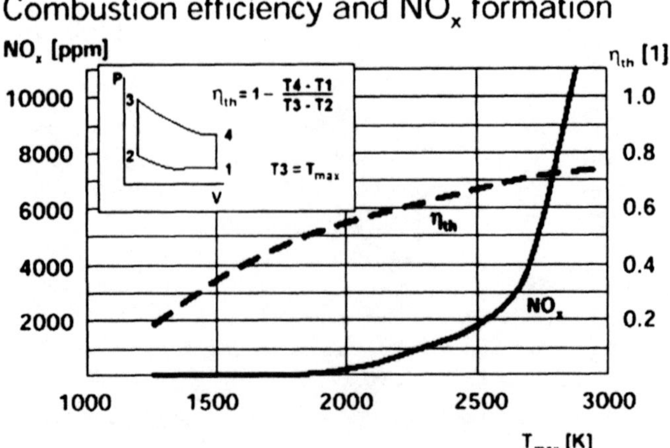

Fig. 19. Basic "Conflict of Interest" between NO_x reduction and fuel economy improvement

ports (and from the unforeseeable behavior of some oil-producing countries) eventually became the driving forces behind the government's efforts to reduce the nation's energy consumption.

Concerning motor vehicles, fuel consumption has for a long time only played a minor role for the buying behavior of US car drivers. On the other hand, the fact that fuel economy values determined by US car makers on the basis of in-house test procedures which differed from manufacturer to manufacturer were not comparable and were, therefore, only of limited help for the consumer, eventually became a matter of increasing concern.

As a consequence, the "Society of Automotive Engineers" (SAE) in 1974 published its recommendation SAE J 1082 "Fuel Economy Measurement – Road Test procedure" with the objective to establish uniform test conditions for the determination of road fuel consumption of passenger cars. The "Federal Trade Commission" (FTC) announced that it would, in the future, only allow fuel economy values to be published for advertising if they were determined according to this new procedure in order to ensure comparability [62].

Even before, a parallel development of a uniform method for the determination of fuel economy values had started at the "Environmental Protection Agency" (EPA). In fall 1972 (begin of model year 1973) the EPA issued its "Gas Mileage Guide for New Car Buyers" which it had established together with the "Federal Energy Administration". For model years 1973 and 1974 this brochure contained the fuel economy as calculated from the exhaust emission values determined during the city driving cycle of the official vehicle certification chassis dynamometer test.

When, as of model year 1975, an additional highway cycle was added to the certification test, fuel economy values from both cycles were also mentioned in the guide. As of model year 1976 a combined value was calculated from these two figures for use in official fuel economy labeling and taxation programs according to the following formula:

Equation 1: Fuel economy calculation from exhaust emission test results in US:

$$FE = \frac{1}{0.55/CFR+0.45/HFE} \text{ [mgp]} \quad (1)$$

FE: fuel economy,
CFE: FE calculated from emissions during city cycle of Federal Test Procedure (FTP-75),
HFE: FE calculated from emissions during Highway Cycle of Federal Test Procedure (FTP-75),
mpg: miles per gallon.

Up to model year 1977 the participation of car manufacturers in the so-called "Fuel Economy Labeling Program", i.e., their consent to the publication of their fuel economy values in the above-mentioned guide, was voluntary. However, participation became mandatory with the introduction of the first legal fuel economy standards in model year 1978.

Since it was not expected that competition alone would yield substantial fuel economy improvements in the upcoming years, such mandatory fuel economy targets were set up by legislature: On December 22, 1975, Subchapter V of the "Motor Vehicle Information and Cost Savings Act" was replaced by the "Energy Policy and Conservation Act" which made the fuel economy values shown in Table 9 mandatory for a manufacturers new passenger car fleet as of model year 1978.

A civil penalty of $5.50 is imposed for each tenth of a mpg by which a manufacturer's CAFE level falls short of the standard, multiplied by the total number of passenger automobiles or light trucks produced by the manufacturer in that model year. Credits earned for exceeding the standard in any of the three model years immediately prior to or subsequent to the model years in question can be used to offset the penalty.

Table 9. US-"Corporate Average Fuel Economy" (CAFE) – Standards of the "Energy Conservation Act" – 1975

Model	Fleet Fuel Economy [mpg]	Fuel Consumption Equivalent [1/100 km]
1978	18	13.07
1979	19	12.38
1980	20	11.76
1981	22	10.69
1982	24	9.80
1983	26	9.05
1984	27	8.71
1985	27.5	8.55
1986	26	9.05
1987	26	9.05
1988	26	9.05
1989	26.5	8.55
1990 and later	27.5	8.55

Due to its averaging character, the CAFE program did not, however, have a notable effect on the improvement of individual vehicle's fuel economy of passenger cars in the US. Since Congress had blocked the Department of Transportation (DOT) from considering a revision, the program has not been furthered since middle of the 1990s. This restriction has been removed in the meantime and the DOT is presently (end of 2001) in the process of revising the program. As a result it is likely that both light duty vehicles (LDVs) and light duty trucks (LDTs) CAFE will be strengthened [63].

In 1978, the "Energy Tax Act" established an additional tax, the so-called "Gas Guzzler Tax" on the sale of new model year vehicles whose individual fuel economy failed to meet the statutory level of 22.5 mpg. This tax is shown in Table 10 and only applies to cars (the exemption of the light truck category from this tax has raised critique because the sport utility vehicles – SUVs – which are sold in increasing numbers and driven like passenger cars, also fall into this group and so do not have to pay such tax in spite of their comparably poor fuel economy). The tax, although being collected by the Internal Revenue Service (IRS) from car manufacturers, is practically paid by the vehicle buyer.

By showing the amount to be paid on the window sticker of new cars the tax shall achieve its purpose, namely to discourage the production and purchase of fuel-inefficient vehicles.

While neither the CAFE program nor the gas guzzler tax triggered substantially improved, i.e., more fuel efficient technology, serious efforts to develop such technology and alternatives to existing propulsion systems for passenger cars were undertaken within the Department of Energy's (DOE) "Partnership of a New Generation of Vehicle" (PNGV) program.

The PNGV program was initiated by President Clinton on September 29, 1993 as a cooperative research development program between the Federal Government and the "United States Council for Automotive Research" (USCAR), whose members are DaimlerChrysler Corporation, Ford Motor Company and General Motors Corporation. When established, the program created a unique industry and government partnership with the broad objective of strengthening U.S. com-

Table 10. US – "Gas Guzzler Tax" ("Energy Tax Act" – 1978, US Code Title 26, Sec. 4064)

Combined Fuel Economy [mpg]	Gas Guzzler Tax [US$]
22.5 or higher	0
22.4–21.5	1,000
21.4–20.5	1,300
20.4–19.5	1,700
19.4–18.5	2,100
18.4–17.5	2,600
17.4–16.5	3,000
16.4–15.5	3,700
15.4-14.5	4,500
14.4–13.5	5,400
13.4–12.5	6,400
12.5 or lower	7,700

Goals of the 1993 PNGV-Program

- **Goal 1:** Significantly improve national competitiveness in manufacturing for future generations of vehicles
  Improve the productivity of the US manufacturing base by significantly upgrading US manufacturing technology, including the adoption of agile and flexible manufacturing and reduction of cost and lead times, while reducing the environmental impact and improving quality.
- **Goal 2:** Implement commercially viable innovations from ongoing research on conventional vehicles
  Pursue technology advances that can lead to improvements in fuel efficiency and maintain safety performance. Research will focus on technologies that reduce the demand for energy from the engine and drive train. Throughout the research program the industry has pledged to apply those commercially viable technologies resulting from this research that would be expected to increase significantly vehicle fuel efficiency and improve emissions
- **Goal 3:** Develop vehicles to achieve up to three times the fuel efficiency of comparable 1994 sedans.
  Increase vehicle fuel efficiency up to three times that of the average 1994 Concorde/Taurus/Lumina automobiles with equivalent cost of ownership adjusted for economics („80 mpg target")

Modified Focus after the 2001 Program Revision

In its 7th report the National Research Council suggested to shift the focus of the PNGV-Program from the original goals towards progress in the development of:

- Advanced Combustion Engines - Light Weight Materials - Electric Motors
- High Power Energy Storage - Power Electronics - Fuel Cells with Infrastructure

Fig. 20. US department of energy (DOE)-program "Partnership for a New Generation of Vehicles" (PNGV)

petitiveness in manufacturing and in the implementation of energy-saving innovations in passenger vehicles. In addition, it was intended to develop a new generation of vehicles by setting a stretch goal to achieve fuel economy up to three times (80 mpg gasoline equivalent) that of comparable 1994 family sedans without sacrificing size or utility or increasing the cost of ownership. The contents and objectives of the program are summarized in Fig. 20.

Although the 80 mpg goal of the program could not be materialized within the target time frame of 10 years, the "National Research Council" (NRC) – which was assigned with a periodic review of the program – concluded in its 7th report [64]:

"The issues addressed by the program are still relevant. The need to reduce the fuel consumption and CO_2 emissions of the U.S. automotive fleet is more urgent than ever. Since 1993 there has been a 20% increase in the petroleum use in highway transportation which consumes 27% of the total energy in the U.S., the transportation sector is 95% dependent on petroleum and consumes 67% of all the petroleum in the U.S. [65]. The percentage of U.S. petroleum use derived from imports has risen to 52%, and in many parts of the world concerns about the potential for climate change associated with greenhouse gases are even more acute".

However, the surrounding conditions have changed since the program started: EPA "Tier2" emission standards have rendered it substantially more difficult to achieve the program because their stringent PM and NO_x standards practically eliminate the use of the Diesel engine – on which the program had strongly counted – and public emphasis in the buying habits has shifted from sedans – the program's target vehicle class – to sport utility vehicles, vans and pick-up trucks. Consequently the NRDC recommended in its 7th report that the goals of the program should be re-examined.

While the original program was focused on hybrid-electric vehicles a strategy is now being considered for the recently revised PNGV program which eventually leads to the use of fuel cells powered by domestically derived hydrogen [65] as also mentioned in Fig. 20.

On the heavy-duty vehicle side a similar effort towards reduced fuel consumption is undertaken with the "21st Century Truck Program" which has been set up in the meantime as a partnership between government and leading heavy vehicle and engine manufacturers. Its goal is to develop advanced technologies to double the fuel economy of long-haul trucks and triple the fuel economy of busses and other vehicles while also reducing emissions and improving safety [64].

Another indication for increasing concern about the subject "fuel consumption" in the U.S. can be seen in the "National Energy Policy" (NEP) published in May, 2001 which includes recommendations to:

- Increase funding for renewable energy and energy efficiency research and development programs that are performance-based and cost-shared, and
- Create an income tax credit for the purchase of hybrid and fuel cell vehicles to promote fuel-efficient vehicles [64].

4.2.3
CO_2 Commitment and Fuel Economy Targets for Cars in Japan

First requirements for the limitation of vehicle fuel consumption were established in 1979 and contain target values which should be met for individual vehicle classes as of 1985.

Since an increasing demand for larger vehicles, automatic transmissions, air condition and power steering led to an increase in energy consumption in the road transport sector, the MOT (which changed its name in 2000 to "Ministry of Land Infrastructure and Transport", MLIT) and the MITI ("Ministry of International Trade and Industry") established a common study group assigned with the evaluation of possibilities to reduce fuel consumption from automobiles.

On June 4, 1992 this group published recommendations from which target values were derived for the time period from 1992 to 2000. If a manufacturer does not achieve these target values, MLIT could require the manufacturer to present his plans how he will achieve the targets in the future.

However, the underlying "Law for the Efficient Use of Energy" was revised in June 1998 after Japan had obliged itself during the December 1997 Kyoto Conference ("COP3") to reduce the nation's output of emissions of so-called "climate gases" (CO_2, CH_4, N_2O and CFCs) from 2008 to 2012 by 6% compared to 1998 in order to specify more stringent requirements for automobiles as well.

The so-called "Action Plan for the Prevention of Global Warming" eventually set the target to reduce CO_2 emissions in 2010 by 6% compared to 1990 and

Table 11. Japanese Fuel Economy Target Values for Passenger Cars with Gasoline- and Diesel-Engines

Fuel Economy Target Values for Passenger Cars with Gasoline Engines
(to be achieved by manufacturers sales-weighted in each vehicle weight class in 2010)

EIW/[kg]	750	875	1000	1350[a]	1500[a]	1750[a]	2000	2250	2500
km/l	21,2	18,8	17,9	16,0[a]	13,0[a]	10,5[a]	8,9	7,8	6,4
l/100 km	4,7	5,3	5,6	6,2[a]	7,7[a]	9,5[a]	11,2	12,8	15,6
CO_2 [g/km]	110	125	132	148[a]	181[a]	224[a]	264	301	367

Fuel Economy Target Values for Passenger Cars with Diesel Engines
(to be achieved by manufacturers sales-weighted in each vehicle weight class in 2005)

EIW/[kg]	1000	1350	1500	1750	2000	2250	2500
km/l	18,9	16,2	13,2	11,9	10,8	9,8	8,7
l/100 km	5,3	6,2	7,6	8,4	9,3	10,2	11,5
CO_2 [g/km]	141	164	202	223	246	271	306

[a] 88% Imports from Europe (25,6% Improvement); 90% Imports from the USA (25,1% Improvement); 59% Japanese Manufacturers (21,2% Improvement).

MLIT and MITI started to discuss with the concerned automobile manufacturers the technological feasibility of the resulting fuel economy target values which were now clearly set in relation to Japan's CO_2 reduction objectives. These target values were established by means of the so-called "top runner" method, which used the fuel economy value, as measured in the 10.15-mode test, of the best Japanese vehicle in each of the specified weight classes as the "master" level which was then reduced by a certain percentage to arrive at the final target value [66].

These fuel economy targets are shown in Table 11. They represent – similar to the CO_2 reduction commitments made by auto manufacturers in the EU – an additional highly demanding challenge within the overall emission legislation scenario.

Table 12. Expected Fuel Consumption Reduction of Passenger Cars through Achievement of the Japanese Fuel Economy Target Values

PC with Gasoline Engines Fuel Consumption determined in 10.15-Mode Test.	1995 [km/l]	[l/100 km]	2010 [km/l]	[l/100 km]	Fuel Economy Improvement [%]
	12,3	8,13	15,1	6,62	22.8

PC with Diesel Engines Fuel Consumption determined in 10.15-Mode Test.	1995 [km/l]	[l/100 km]	2005 [km/l]	[l/100 km]	Fuel Economy Improvement [%]
	10,1	9,9	11,6	8,6	14,9

Under the assumptions that the model mix of the 1995 vehicle fleet will not change until the target year 2010 and the abovementioned fuel economy target values are achieved, the MLIT expects an average fuel consumption saving of 22.8% for the fleet of passenger cars with gasoline engines and 14.9% saving for the fleet of passenger car vehicles with Diesel engines as shown in Table 12.

4.3 "On-Board Diagnostics" Requirements

A completely new and highly challenging chapter in the development history of automobile emission control technology was opened with the requirements for self-diagnosis of irregularities and defects in the vehicle's emission control systems. These "OBD requirements" were initiated by California and became mandatory for the first time as the so-called "OBD I" within the certification of model year 1988 vehicles.

The new requirement asking for a special system which automatically performs a self-checking of emission-related components and control mechanisms was triggered when California's ARB realized that the capability of workshops and mechanics could no longer keep pace with the development of modern emission control systems which frequently incorporated "black box" electronics.

The OBD legislation can be considered as one of the most important and consequential steps in the evolution of auto emission regulations. The contents of California's OBD II requirements shown in Table 13 which were eventually required from model year 1994 onwards [67], are extremely demanding.

Although they entail high efforts by the vehicle/engine manufacturer and its supplier of electronics on the one side, a proper functioning OBD system, on the other side, gives important benefits.

Continuous self-monitoring of the vehicle's emission level at any time, under all driving conditions and over the full useful life of the vehicle is not only the best possible means to meet legislators' expectations from the compliance point of view, it provides as well the basis for effective and efficient maintenance in the workshops which may result in substantial time savings and cost benefit for the vehicle owner.

Within the EU similar OBD requirements were first introduced as of emission regulation 98/69/EC – Step 2 (the so-called "EU 3" legislation) as shown in Sect. 2.4. These European (EOBD) provisions require a self-checking of the parameters summarized in Fig. 21.

The storage of any irregularity detected on these items is required. During vehicle certification the following defects must be simulated to demonstrate proper failure indication by a light-up of the "malfunction indicator" (MI):

- change of the catalyst against an aged or defective one or electronic simulation of this failure,
- inflammation failure within a defined engine operation range,
- change of the O_2 sensor against an aged or defective one or electronic simulation of this failure,
- electrical disconnection of the circuit of any emission-related component which is connected with a power train computer,

Table 13. On-Board Diagnostics Requirements in California (OBD II) for Low Emission Vehicles with Gasoline Engines

Item to be monitored	Malfunction Criterion
Catalyst System	emissions increase > 1.75 × emission standard[a]
Catalyst Heating System	catalyst does not reach designed temperature within necessary time which would ensure that emissions do not exceed 1.5 × applicable FTP-standard
Oxygen Sensor(s)	voltage, response rate or other criteria which ensure that emissions do not exceed 1.5 × applicable FTP-standard are not met
Engine Misfire (including detection of misfiring cylinder)	misfire rate at defined rpm & load exceeds manufacturer's specified rate to avoid catalyst damage and misfire rate at defined rpm steps exceeds rate keeping emissions from a durability vehicle test < 1.5 × emission standard
Secondary Air System	flow rate < flow rate ensuring that emissions stay < 1.5 × FTP-standard
Exhaust Gas Recirculation (EGR)	a) any system component out of specs and/or flow rate > specified high or low rate such that emissions do not exceed 1.5 × applicable FTP-standard
Fuel System	delivery rate insufficient to ensure that emissions stay < 1.5 × FTP-standard
Evaporative Emissions Control[b]	a) no air flow, b) leak as if there was an orifice of 0.020 in. (0,5 mm) diameter
Positive Crankcase Ventilation (PCV)[c]	no connection to crankcase or intake manifold sensed
Thermostat[d]	a) coolant temp. does not reach mex. temp. required to enable diagnostics or b) does not reach warm-up temp. within 20 °F of designed regulating temp.
Comprehensive Component Monitoring	Any electronic power train part which provides or receives a command from the on-board computer. Input command: lack of circuit continuity or out of manufacturer specifications; Input command: Proper functional response to computer command does not occur
Air Condition Refrigerant System	any loss of refrigerant

[a] phase-in: MY 98: 20%, MY 99: 40%, MY 00: 60%, MY 01: 80%, MY 02: 100%.
[b] phase-in: MY 00: 50%, MY 01: 75%, MY 02: 100%.
[c] phase-in: MY 02: 30%, MY 30: 60%, MY 04: 100%.
[d] phase-in: MY 00: 30%, MY 01: 60%, MY 03: 100%.

- electrical disconnection of the purge system of the evaporative emissions control system.

In spite of obvious benefits of a properly designed and reliably functioning OBD system, legal OBD requirements should not become over-sophisticated. Unrealistic requirements would not only entail excessive development costs but could severely undermine the confidence of vehicle owners in the system – and thus

- **At a minimum the following items have to be diagnosed and indicated to the driver in case of failure by illumination of the malfunction indicator (MI):**
  1. Emissions during certification testing exceed the following threshold levels:
     3,2 gCO/km – 0,4 gHC/km – 0,6 gNO_x/km
  2. Engine (cylinder) misfiring within a specified engine map range
  3. Deterioration of the O_2-sensor
- **The following systems have to be diagnosed for correct functioning of the electric circuit:**
  a. Systems or components of the emission control system which are connected to an on-board computer and would - in case of failure - cause the vehicle to exceed the emission thresholds
  b. Other systems or components of the emission control system which are connected to an on-board computer
  c. Systems or components of the power train system which are connected to an on-board computer and would - in case of failure - cause the vehicle to exceed the emission thresholds
  b. Other systems or components of the power train system which are connected to an on-board computer
  c. Electronic fuel injection system
  d. The purge system of the evaporative emissions control system has at least to be checked for proper functioning of the electric circuit

Fig. 21. Contents of the European OBD Requirements (EOBD)

eventually endanger the potential benefit of the whole technology – if they led to misleading or unnecessary malfunction indications.

OBD regulations, although having already achieved an extremely high level of complexity (about 80% of the on-board computer power is used for OBD purposes) are subject to continued review by legislators in the here discussed frameworks of the European Union, Japan and the USA. Within this scenario, the California Air Resources Board remains the driving force towards the invention and implementation of modifications to existing or establishment of new requirements. Japan is basically accepting OBDII or EOBD systems within its vehicle certification procedure. The EU strives at limiting the complexity of the systems for the sake of reliability and has materialized this strategy successfully with regard to its OBD requirements for the vehicle's evaporative emissions control system.

5 Emission Control Legislation for Heavy Duty Vehicles

An overview of the historical development of measurement techniques and driving cycles for the determination of exhaust gas emissions from the legislative point of view was given in the previous sections using the example of passenger cars. Since the corresponding regulatory work also provided the basis for relevant developments on the heavy duty vehicle side, the next sections will mainly deal, with the evolution of emission standards for the latter.

Heavy duty vehicles for which engine dynamometer certification testing applies are equipped in practical terms only with Diesel engines – at least in Europe and in Japan. For this engine type, the critical emission components are NO_x- and PM independent of its application in either passenger cars or heavy duty trucks. Therefore, the following discussion focuses on these two exhaust gas components.

5.1 Requirements in the US

In the US, the first standards for gaseous and PM-emissions for heavy duty Diesel engines became applicable in 1987 in California and as in 1988 in Federal legislation. Emissions were measured on the basis of the US-Transient Test which consisted of 4 driving sequences simulating New York city driving, Los Angeles city driving, Los Angeles Highway driving and again New York city driving. A summary of these standards is given in Table 14.

As can be seen from this table, California's legislation differentiates its emission limits for hydrocarbons between non-methane hydrocarbons (NMHC) and total hydrocarbons (THC) while otherwise being identical in stringency with Federal requirements for HC and CO. For NO_x and PM, however, California introduced the standard combination of 6.0 g/bhp/h NO_x and 0.6 g/bhp/h 3 years before it became applicable also under Federal regulations. These standards have been eventually reduced in both regulations to the same level of 5.0 g/bhp/h NO_x and 0.10 g/bhp/h PM as of model year 1994. While EPA reduced the NO_x-standard one more step to 4.0 g/bhp/h NO_x as of 1998 for heavy trucks, California applied this standard to urban buses only but here as early as 1996.

The most consistent regulatory change happened when emission standards were established on October 6, 2000 for the Step 2004 and on January 18, 2001 for the Step 2007 respectively. Though the 1994 PM standard did not change for 2004 it has to be seen in combination with the NO_x-standard which was reduced for 2004 from the previously valid Federal level of 4 g/bhp/h to a combined (NO_x + NMHC) standard of 2.4 g/bhp/h. This level of stringency may already require for the first time a significant exhaust gas after-treatment. The 2004 regulation is already fixed but the 2007 standards (0.2 g/bhp/h for NO_x and 0.01 g/bhp/h for PM) are subject to a review process. This review will happen in 3 stages so that legislature can continuously keep track of technological progress and – maybe – consider an adaptation of its legal requirements.

Emission control technologies which will meet this regulatory scheme are still under development and their introduction also depends on the availability of fuel with very low sulfur content. However, since both passenger cars and heavy duty Diesel vehicles today use direct injection engines, the corresponding possibilities and means of reducing NO_x- and PM-emissions are very similar. This is true as well for engine-internal emission reduction means as for exhaust gas after-treatment techniques. Since Diesel engine application in heavy duty vehicles has the main objectives of low fuel consumption and durability, it is a major development target to maintain these characteristics as far as possible when emission control systems have to be applied.

Engine modification measures for lowering NO_x-emissions, like e.g. retarded ignition, or certain exhaust gas re-circulation (EGR) applications may increase

Table 14. Federal and California emission standards for heavy duty Diesel vehicles

Model Year	NMHC		THC		CO		NO_x		PM	
	Federal	Calif.	Federal	Calif.	Federal	Calif.	Federal	Calif.	Federal	Calif.
1987	–	–	–	1.3	–	15.5	–	6.0	–	0.60
1988	–	–	1.3	1.3	15.5	15.5	10.7	6.0	0.60	0.60
1990	–	–	1.3	1.3	15.5	15.5	6.0	6.0	0.60	0.60
1991	–	1.2	1.3	1.3	15.5	15.5	5.0	5.0	0.25	0.25
1994	–	1.2	1.3	1.3	15.5	15.5	5.0	5.0	0.10	0.10/0.07 [a]
1996	–	1.2	1.3	1.3	15.5	15.5	5.0	4.0	0.10	0.10
1998	–	1.2	1.3	1.3	15.5	15.5	4.0	4.0	0.10	0.10
Common standards valid for Federal & California as of 2004 (California differs in OBD requirements)										
2004	$NMHC+NO_x$		NMHC		15.5		n/a		0.10	
	Option 1	Option 2	Option 1	Option 2						
	2.4	2.5	n/a	0.5						
2007	n/a	n/a	0.14		15.5		0.20		0.01	

[a] Emission averaging may be used to meet the PM standard.

fuel consumption. So in order not to reduce the engine's efficiency while at the same time meeting stringent NO_x-standards, manufacturers may prefer to apply after-engine NO_x-control technologies (e.g. in form of selective catalytic reduction catalysts).

When discussing steps for reducing PM-emissions one has to bear in mind that any external add-on device which is based on filtering tends to increase the backpressure in the exhaust system and may thus impact on the engine's fuel consumption.

At this point, an attempt will be made to describe the most probable scenario of how heavy duty Diesel manufacturers may approach upcoming 2004 and later emission standards in the US (and – due to comparable stringency of standards – also in the EU). Table 15 gives a summary of possible technologies and puts them into the perspective of time and stringency.

Since both the NO_x after-treatment and the PM-filtering devices need low sulfur fuel in order to maintain their designed emission durability, the US regulations require that as of June 1, 2006 refiners must start to produce Diesel fuel with <15 ppm sulfur content and make it available in the market as of September 1, 2006.

While emission standards for the certification of new heavy duty Diesel vehicles have been strengthened to a level which should no longer represent a matter of environmental concern about this vehicle category, older vehicles with higher emissions will continue in use for some time. Here, smoke emissions are especially a nuisance and a smoking truck will always trigger associations with the discussion about thr health effects of particulate matter emissions.

So the individual states in the US have introduced additional in-use smoke test programs for heavy duty Diesel vehicles. A survey about the corresponding standards, test methods and results is given for all US-states in a study which was made for the "Association des Constructeurs Européens d'Automobiles" (ACEA) in 1998 [68] with the objective of arriving at recommendations for legislation concerning future policies concerning emission performance of this vehicle group in the field. As for passenger cars, the results of such in-use testing have been generally positive even when applied methods consisted only of a visual inspection of parts and a simple free acceleration test.

5.2
Requirements in Japan

Concerning heavy duty Diesel vehicles, Japan – like other countries – first started to regulate smoke emissions as of September 1972 on the basis of a 3-mode test and as of September 1974 additionally on the basis of a free acceleration test. For domestic manufacturers, the standard of 50% filter blackness was reduced for the 3-mode test to 40% as of October 1994, and to 25% as of October 1997/98/99 (vehicles <3.5 t/ <3.5 t to 12 t/ >12 t). For the free acceleration test this standard was lowered to 40% as of October 1994 and to 25% as of October 1997/98/99 (vehicles <3.5 t/ <3.5 t to 12 t/ >12 t).

The first standards for gaseous emissions became applicable on the basis of a 6-mode test in October 1988 when NO_x-emissions were first limited to

Table. 15. Heavy duty vehicle emission control for future legislation (most probable scenario)

US-2004 1.1.2004		Euro 4 1.10.2005			US-2007 1.1.2007		Euro 5 1.10.2008		
NOx+NMHC	PM	[g/kW h]	NOx	PM	NOx	PM	[g/kWh]	NOx	PM
2.4 [g/bhp/h]	0.10	ESC&ELR ETC	3.5 3.5	0.02 0.03	0.2 [g/bhp/h]	0.01	ESC&ELR ETC	2.0 2.0	0.02 0.03
Technologies					Technologies				
Heavy EGR for NO_x-control High Pressure Charging (as a prerequisite for heavy EGR application) Oxidation Catalyst Particulate Filters are under development but may not become necessary Development for US-2004 provides the basis for Euro 4 Both regulations are fixed					Selective Catalytic Reduction Catalysts (SCR) Particulate Filters NO_x-Storage Catalysts US-EPA tends to NO_x-Storage Catalysts for chassis dyno and transient test but technology may not be available Fuel with <10 ppm S necessary for SCR Both regulations are still under review				

Table 16. Japanese emission standards for heavy duty vehicles >3.5 t

Year	GVW [kg]	Test	Dim.	CO		HC		NOx		PM	
				Mean	Max	Mean	Max	Mean	Max	Mean	Max
1988 to 03-31-1996	<2.500	6-mode	ppm	790	980	510	670	400 (260)	520 (350)	–	–
04-01-1996	>2.500 <3.500	13-mode	g/kWh	7.40	9.20	2.90	3.80	6.00 (5.00)	7.80 (6.80)	0.70	0.96
10-01-1997	>2.500 <3.500			7.40	9.20	2.90	3.80	4.50	5.80	0.25	0.49
10-01-1998	>3.500 <12.000										
10-01-1999	>12.000										
2004 [a]	all > 3.5 t			2.22	2.76	0.87	1,14	3.38	4.35	0.18	0.35
2005 [b]				2.22	?	0.17	?	2.0	?	0.027	?

Values in brackets: Indirect Injection Engines.
[a] New Short Term Targets (=70% reduction vs. current HC- and CO-values; 25%/28% reduction for NOx/PM resp.
[b] New Long Term Targets, advanced to 2005 from 2008 (HC-Standard as NMHC); Max-Standards not yet defined.
[c] New Japanese Transient Test under discussion.

400/520 ppm for direct injection and 260/350 ppm for indirect injection engines. The first simultaneous NO_x and PM limitation to started in October 1994 when the former 6-mode test was replaced by the new 13-mode test. The standards were further lowered as of October 1997/98/99 as shown in Table 16.

In the 4th revision step they were further reduced to 0.35/0.18 g/kW h (max/mean) as of October 2003 for vehicles <12 t and as of October 2004 for vehicles >12 t. The 5th revision defines a standard of 0.027 g/kW h which is to be met by all weight classes >3.5 t as of October 2005.

The latest decisions about future heavy duty Diesel vehicle standards were based on the "Fifth Report of the Central Environment Council" [69]. This report refers to the suggestions already made in the 4th report – which shifted the emphasis in Japan's legislation from NO_x to PM emission control – and recommended producing legislation to bring forward the so-called "long-term targets" from 2007 to 2005, and to subject the vehicles concerned to a new Japanese transient test. Its introduction date is, however, still under discussion. These recommendations were made under the assumption that Diesel fuel with a sulfur content as needed by the expected emission control technology will be available. The introduction of the new Japanese transient test is still under discussion.

A special situation exists in Japan with regard to the promotion of "low emission vehicles". There are presently 3 different promotion programs in effect: a local system in 7 prefectures as of April 1, 1999, a national system as of April 1, 2000 and a special "ultra low PM-emission " system for Diesel vehicles >3.5 t valid as

Table 17. Japanese Promotion Program for "Low Emission Heavy Duty Vehicles" (Vehicles >3.5 t)

Standards (Local Program) for Acknowledgement as "Low Emission Vehicle" (7 prefectures/cities, as of April 1, 1999)

Test	NOx			HC			PM		
13-mode [g/kW H]	TLEV	LEV	ULEV	TLEV	LEV	ULEV	TLEV	LEV	ULEV
	2.54	1.69	0.85	0.65	0.44	0.22	0.14	0.09	0.05

Standards (National Program) for Acknowledgement as "Low Emission Vehicle" (as of April 1, 2000)

Reduction vs. Year 2000	Test Mode	CO	HC	NMHC	NOx	PM
25%	13-mode [g/kW H]	16	0.65	0.52	2.54	0.14
50%	13-mode [g/kW H]	16	0.44	0.35	1.69	0.09
75%	13-mode [g/kW H]	16	0.22	0.18	0.85	0.05

Standards (National Program) for Acknowledgement as "Ultra-Low PM-Emission Diesel Vehicle" (as of September 1, 2002)

The year 2003 is the basis for the PM-reduction figures [%]		CO	HC	NOx	PM	
	[g/kW H]	2.22	0.87	3.38	75%	85%
	PM-standard for 2003: 0.18 g/kW H				0.05	0.027
	(Present PM-standard : 0.25 g/kW H					

of September 1, 2002. The manufacturer whose vehicles comply with these standards has an image gain for his product, the vehicle owner, however, may receive a tax benefit. An overview about the different promotion programs is given in Table 17.

In addition to these local and national promotion programs, the Tokyo Metropolitan Government (TMG) has also established restrictions for the operation of heavy duty Diesel vehicles. Trucks which do not meet special in-use emission limits for PM (in 2003: 0.25 g/kW h; in 2005: 0.18 g/kW h) are not allowed to enter Tokyo metropolitan area. Trucks equipped with TMG-certified particulate filter systems are considered to meet these requirements.

5.3 Requirements in Europe

Within the European scenario, legislative developments have to be considered for both the ECE- and the EU-regime. While the ECE-Regulations represent a regulatory framework which may be adopted by any member state of the UN and used instead of or as an integral part of its national legislation, EEC-Directives are binding for all member states of the Community and replace corresponding national legislation at prescribed dates.

On the ECE-side, emission control requirements for heavy duty vehicles - above 3500 kg maximum permissible mass - started on April 15, 1982 when Regulation ECE-R49 came into force defining standards for gaseous emissions (HC, CO, NO_x). These standards had to be applied to new engines as of March 15, 1982. The first PM-requirements appeared in version ECE-R49/02 which established the so-called "Euro 1"- and "Euro 2"-standards. The latest version came into force as ECE-R49/03 on December 27, 2001 and defined the so-called "Euro3"- and "Euro4"- standards for the years 2000 and 2005 respectively. Up to "Euro2" heavy duty vehicle emissions were measured on the basis of the 13-mode test which is shown in Fig. 22.

On the EU-side, the corresponding EU-Directives were established. The basis EU-emission law appeared parallel to ECE-R49/01 in form of Directive 88/77/EEC on February 9, 1988 defining the so-called "Euro 0" standards for application as of October 1, 1990. The next version was Directive 91/542/EEC which described – as its ECE-counterpart R49/02 – Step A ("Euro1") which became applicable as of July 1, 1992 and Step B ("Euro2") which became applicable as of October 1, 1995. A summary of ECE-and EU-legislation is given in Table 18.

As Table 18 shows, the base regulations (ECE-R49/00 and 88/77/EEC) were amended several times order to take account of technical progress of HDV engine technology. The last significant amendment of the EU-Directive entered into force on December 16, 2000 and included substantial modification of the emission limits and of the certification procedure by introducing two new test cycles for emissions: The "European Stationary Cycle" (ESC) which was applied within Euro 3 (for 2000) to conventional Diesel vehicles (with oxidation-catalyst and/or EGR) and the "European Transient Cycle" (ETC) for Diesel engines equipped with advanced catalysts or particulate filter systems. For gas engines the gaseous emissions is to be determined on the ETC test.

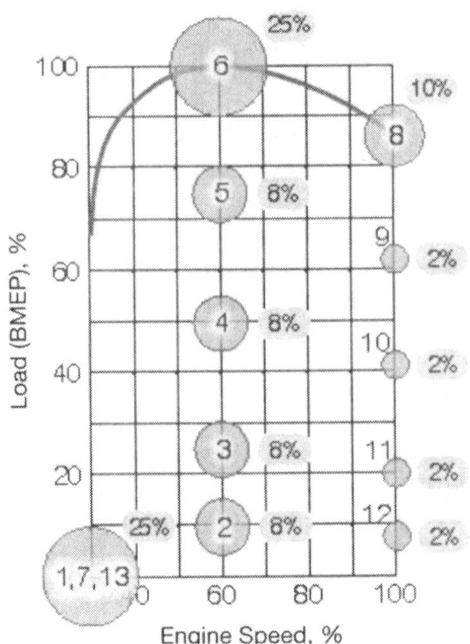

Fig. 22. The 13-mode test for heavy duty vehicle engines of ECE-R49 valid from 1982 to 1989

In addition, so-called "mystery points" were added for the ESC which are to be selected by the testing authority to check the homogeneity of the engine map design – valid for engines w/o external anti pollution technology. Finally, the "European Load Response Test " (ELR) was introduced to allow a better monitoring of the engine's soot emission. The rationale for these additional tests was to make sure that engine maps are set up in a manner which reflects throughout the complete operation range real world driving conditions. It was applied within Euro 3 in combination with the ESC. Within Euro 4 (for 2005), all Diesel are to be tested with the ETC and the ESC/ELR tests. The new cycles are shown in Fig. 23.

In a further step which came into force in April 21, 2001 existing provisions to prevent the use of defeat devices and/or an irrational emission control strategy were clarified. As compared to the original wording the new requirement now defines the tool to evaluate and thus prevent using defeat devices and/or irrational emission strategies. The text of the ECE-R 49 is currently being harmonized with the last relevant EU amendment.

As for passenger cars, emission control regulations for heavy duty vehicles also include provisions about on-board diagnostics. In this field, OBD-technology for heavy duty vehicles will certainly benefit from extensive experience gained on the passenger car side over the past years. It can, therefore, be assumed that the architecture of OBD systems for heavy duty vehicles will be similar to those systems already in use in light vehicles.

In February 1999, a special OBD working group assigned by the EU-Commission met with ACEA to evaluate the possibilities for OBD application on heavy

Table 18. Emission regulations for heavy duty vehicles in Europe

Related Information	ECE-Regulation Emission Standards in [g/kWh]			EU-Directive Emission Standards in [g/kWh]		
	ECE-R49/00			-		
In force and applicable as of	April 15-1982			-		
Standards	HC: 3.5 CO: 14 NO_x: 18			-		
	ECE-R49/01			88/77/EEC		
In force as of	May 14-1990			February 9-1988		
applicable as of	May 14-1990			October 1-1990		
Standards	"Euro 0" HC: 2.4 CO: 11.2 NOx: 14.4			"Euro 0" HC: 2.4 CO: 11.2 NOx: 14.4		
	ECE-R49/02			91/542/EEC		
In force as of	December 30-1992			October 25-1991		
applicable as of	Step A: July 1-1992 ("Euro 1") Step B: October 1-1995 ("Euro 2")					
Standards	Step A HC: 1.1 CO: 4.5 NO_x: 8.0 PM: 0.36 Step B HC: 1.1 CO: 4.0 NO_x: 7.0 PM: 0.15					
	ECE-R49/03			1999/96/EC		
In force as of	December 27-2001			February 16-2000		
applicable as of	Step A : October 1-2000 ("Euro 3") Step B1: October 1-2005 ("Euro 4") Step B2: October 1-2008 ("Euro 5")					
Standards ESC&ELR: valid for conventional Diesel with or w/o	Valid as of	CO	HC	NO_x	PM	Opacity $[m^{-1}]$ ELR
Oxicat/with or w/o EGR	Oct. 2000 (Euro3)	2.1	0.66	5.0	0.10	0.8
	Oct. 2005 (Euro4)	1.5	0.46	3.5	0.02	0.5
	Oct. 2008 (Euro5)	1.5	0.46	2.0	0.02	0.5
Standards ETC: additionally valid for	Valid as of	CO	NMHC/CH_4	NOx	PM	n. a.
Diesel with $DeNO_x$ and/ or PM-Filter	Oct. 2000 (Euro3)	5.45	0.78/1.6	5.0	0.16	n. a.
	Oct. 2005 (Euro4)	4.0	0.55/1.1	3.5	0.03	n. a.
	Oct. 2008 (Euro5)	4.0	0.55/1.1	2.0	0.03	n. a.

CH_4-standards for gas engines.

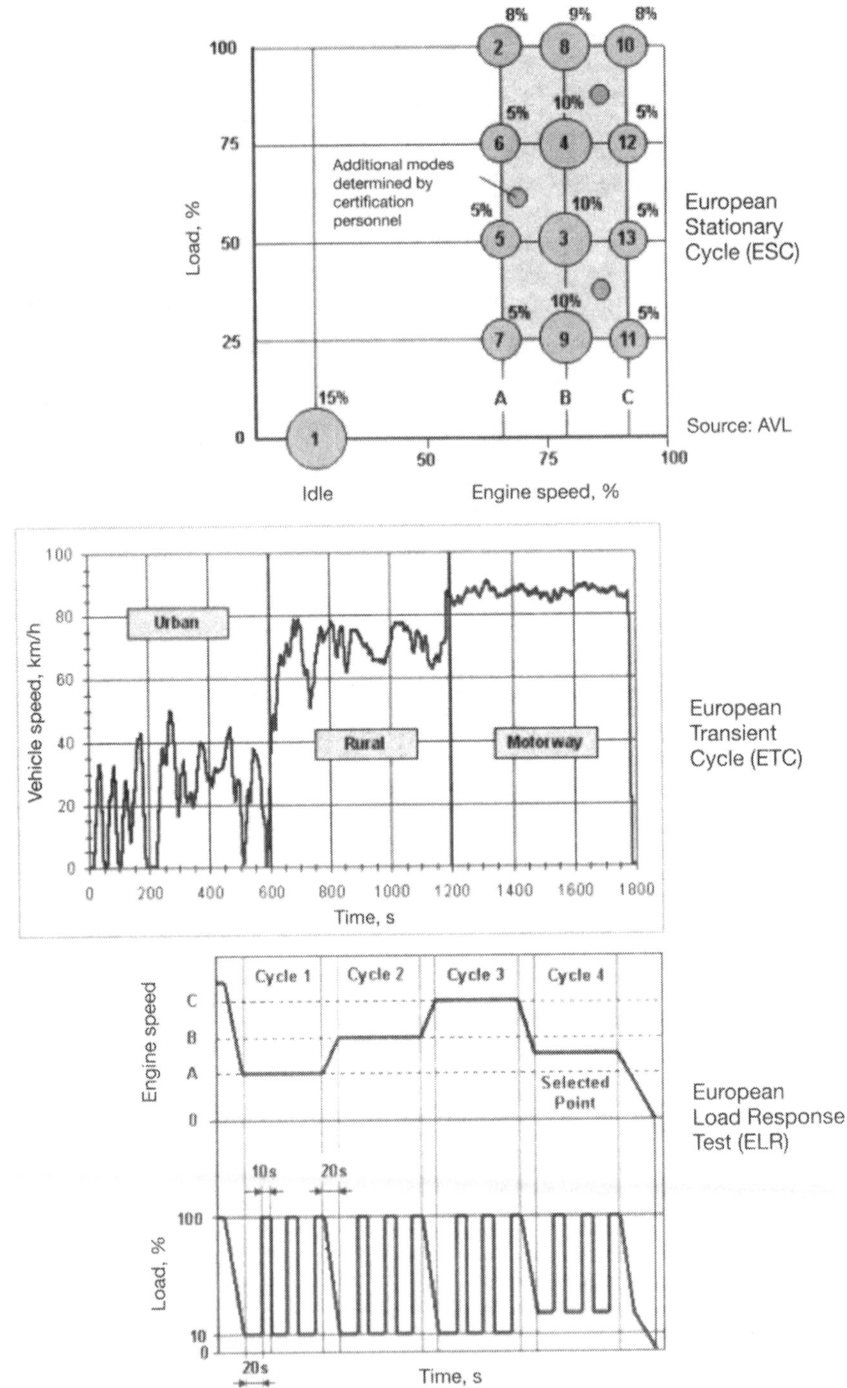

Fig. 23. The new driving cycles for heavy duty vehicle engines valid as of Euro 3 (year 2000)

duty (Diesel) vehicles. The background for this work is Article 4 of EU-Directive1999/96/EC which specifies that OBD shall apply to new types of heavy duty vehicles from 1st October 2005 and to all types from 1 October 2006. A Commission proposal due by 31st December 2000 will provide the requirements for heavy duty OBD.

At the 4th meeting of the Commission OBD working group (a sub-group of the Motor Vehicle Emissions Group – MVEG) ACEA – after consultation with the concerned industry – presented its vision of OBD for heavy duty vehicles. Consequently, ACEA were given the task of preparing a first proposal for heavy duty vehicle OBD. This proposal was to provide the framework for further discussion towards a complete and effective OBD concept to be applicable to new vehicles from October 2008, with an intermediate stage being applicable to new vehicles from October 2005.

The draft proposal had to take into account the OBD requirements already in force for passenger cars and light commercial vehicles equipped with Diesel engines, as required through directive 70/220/EEC and should consider aspects such as:

- an intermediate stage of OBD, mandatory for new types of compression-ignition engines and new types of vehicles with compression-ignition engines from 1st October 2005; mandatory for all types of compression-ignition engines and all types of vehicles with compression-ignition engines from 1st October 2006;
- a final stage of OBD, mandatory for new types of compression-ignition engines and new types of vehicles with compression-ignition engines from 1st October 2008; mandatory for all types of compression-ignition engines and all types of vehicles with compression-ignition engines from 1st October 2009;
- the feasibility of a manufacturer choosing to fit an OBD system to a vehicle before 2005 or 2008
- OBD threshold limits and out of range values together with an appropriate test cycle and test procedures for an OBD type-approval demonstration test that will be applicable from 1st October 2005 and 1st October 2008.

A first Commission proposal was due on heavy duty OBD at the end of 2000 and was further evaluated since then. A final version is expected from the Commission in early 2003.

Similar to the developments in Brussels, a heavy-duty vehicle OBD had been identified as a topic for development as a Global Technical Regulation (GTR) within the context of the 1998 UN-ECE Global Agreement. The first meeting of the heavy-duty vehicle OBD group (WWH-OBD) took place in Tokyo on 21 February 2002 and was attended by representatives from Japan and the EU.

Equipped with modern technology for meeting upcoming stringent standards for particulate matter and NO_x emissions in combination with OBD systems of the second generation, i.e. systems which do not only detect functionality of components but also monitor emission levels against defined thresholds, heavy duty vehicles will maintain optimal emission performance in field operation and will thereby consolidate their position in an environment-friendly transportation system.

6
Summary and Outlook

Within the previous sections an attempt was made to highlight important milestones of the evolution of automobile emission-related regulations in selected legislative scenarios from its beginning to date. Starting with the determination of base emission levels of uncontrolled automobiles and efforts towards the establishment of appropriate standards, measurement techniques and test procedures in California about four decades ago this development has become a unique chapter in the history of the automobile.

Although the described legislations have achieved to date that evaporative and exhaust emissions from new automobiles are no longer a matter of concern from the environmental point of view, they will continue to impact on the concerned industry by further strengthened or newly established requirements.

Countries like the Triad states described before, where stringent legislation has forced the application of latest state-of-the-art emission control technologies allowing vehicles to approach a zero emission level will shift their political emphasis from further reducing conventional emission standards towards new aspects like e.g.:

- consideration of so far unregulated emissions,
- new definition of the basic target with regard to the reduction of particulate matter emissions,
- revision of today's measurement technique for the determination particulate matter emissions to better address,
  - expanding the scope of OBD-requirements,
  - strengthening emission durability requirements,
  - emissions surveillance testing of vehicles in field operation,
- incentives for new emission control technologies and alternative engine/vehicle concepts with zero or close-to-zero emissions and/or very low fuel consumption (CO_2-emissions),
- requirements for the specification of oils and fuels necessary to allow introduction and safe operation of new emission control technologies.

Countries where local conditions do not yet allow the introduction of today's most advanced emission control requirements which, however, strive at further improvements of their environmental situation, will continue to work on the establishment of the necessary preconditions in a cost-efficient step-by-step approach.

In this respect it can be anticipated that important markets like, e.g., India and China, will further follow their already started way of taking over ECE regulations into their national auto emission legislative framework.

On this way they should be accompanied and supported by further advanced countries in order that they can benefit from already available experience.

Acknowledgement. The author thanks Mr. Jean-Pierre Pochic, DCAG Stuttgart, for his valuable contributions to this document especially with respect to developments in the political arena in Europe.

7 References

1. Patterson DJ, Henein NA (1974) Emissions from Combustion Engines and Their Control Ann Arbor, Michigan ISBN 0-250-97514-9
2. Haagen-Smit AJ (1952) Chemistry and Physiology of Los Angeles Smog, Ind. Eng. Chem. 44 (6), 134
3. US Department of Commerce (1977) Statistical Abstract of the United States, Bureau of the Census, 98th edition
4. State of California Department of Public Health (1960) Technical Report of California Standards for Ambient Air Quality and Motor Vehicle Exhaust, Section. III: The Technical Bases for Motor Vehicle Exhaust Standards, Chapter XVII, Data
5. Haas GC (1960) The California Motor Vehicle Emission Standards, SAE-Paper 210 A. In: SAE "Vehicle Emissions", part I. Progress in Technology, vol 6
6. State of California (1971) Air Pollution Control in California, Resources Agency, Air Resources Board, Annual Report, January 72
7. State of California, Air Resources Board (1969) Test Procedure for Assembly Line or Pre-Delivery Testing of Motor Vehicle Exhaust Emissions
8. Berg W (1982) Aufwand und Probleme für Gesetzgeber und Automobilindustrie bei der Kontrolle der Schadstoffemissionen von Personenkraftwagen mit Otto- und Diesel-Motoren. (Efforts and Problems for Legislator and Automobile Industry with the Control of Pollutant Emissions from Passenger Cars with Otto and Diesel-Engines); Doctoral Thesis, Technical University of Braunschweig
9. Title 13 – California Code of Regulations, Section 1968.1 as adopted September 14, 1989, Malfunction and Diagnostic System Requirements – 1994 and Subsequent Model Year Passenger Cars, Light Duty Trucks and Medium Duty Vehicles and Engines (OBD II), (OBD I valid as of model year 1988)
10. California Health and Safety Code, Section 1960. 1: California Exhaust Emission Standards for 1981 through 2003 model year passenger cars, light duty trucks and medium duty vehicles, (LEV I Regulations and non-LEV so-called "Tier I"- standards for 1995 through 2003 model years) as effective 12/1/1999
11. ibid Section 1961.1(LEV II Regulations)
12. California Air Resources Board, Board Hearing January 25, 2001
13. Liptak, B.G. (1974) Environmental Engineer's Handbook, Vol.2, Air Pollution, Radnor, Pennsylvania, ISBN 0-8019-5692-7
14. Air Pollution Control. Hearings before a Special Subcommittee on Air and Water Pollution of the Committee on Public Works, US Senate, 89th Congress, 1st session on S. 306 "A Bill to Amend the Clean Air Act to Require Standards for Controlling the Emission of Pollutants from Gasoline Powered or Diesel Powered Vehicles, to Establish a Federal Air Pollution Control Laboratory, and for Other Purposes", April 6/8/9, 1965, Washington D.C.; April 7, 1965, Detroit Michigan; US Gov. Printing Office Washington 1965
15. Progress in the Prevention and Control of Air Pollution; First Report of the Secretary of Health, Education and Welfare to the United States Congress (pursuant to P.L. 909-148 – The Air Quality Act of 1967), June 28, 1968; US Gov. Printing Office Washington 1968
16. Motor Vehicles, Air Pollution and Health; A Report of the Surgeon General to the US Congress in Compliance with P.L. 86-493 ("The Schenck Act"), US Department of Health, Education and Welfare, Public Health Service, Division of Air pollution, 87th Congress, 2nd Session, House Document No. 489, US Gov. Printing Office Washington 1962
17. Public Law 86-493 (The "Schenck Act" or the "Motor Vehicle Exhaust Act"), H.R.8238, 86th Congress, June 8, 1960, Sec 1&2
18. US Department of Health, Education and Welfare, Public Health Service, "The Clean Air Act" December 17, 1963, (P.L. 88-206), As Amended October 20-1965 (P.L. 89-272); October 15, 1966 (P.L. 89-675), "Title I", Sec.106 (b)

19. US Department of Health, Education and Welfare, Public Health Service, "The Clean Air Act" December 17, 1963, (P.L. 88–206), As Amended October 20, 1965 (P.L. 89–272); October 15, 1966 (P.L. 89–675), "Title I", "Title II" – Control of Air Pollution from Motor Vehicles, footnote: Title II added by Sec. 101 of P.L. 89–272, Oct. 20, 1965
20. Air Quality Act of 1967, P.L. 90–148, November 21, 1967, "Title I", Sec. 101 (a) (2)
21. US Department of Health, Education and Welfare, Public Health Service, "The Clean Air Act" December 17, 1963, (P.L. 88–206), As Amended October 20, 1965 (P.L. 89–272); October 15, 1966 (P.L. 89–675), "Title I", Sec. 208 (a) to (c)
22. Bonine, J.E. (1975) The Evolution of "Technology Forcing" in the Clean Air Act, in: "Environment Reporter", Monograph No. 21, The Bureau of National Affairs, Inc., Vol. 6, No.13, July 25, 1975
23. US Environmental Protection Agency (1970) The Clean Air Act (as amended) Washington D.C., December 1970, Sec. 104 (a) (1) D and E
24. Berg W (1990) Die neue Abgasgesetzgebung der USA ("The New Emission Legislation of the US"); Information about the 1990 Clean Air Act Amendments and the 1990 Health and Safety Code Amendments; Daimler-Benz AG, November 1990
25. US Environmental Protection Agency, 40 CFR Parts 80, 85 and 86, Control of Air Pollution from New Motor Vehicles: "Tier 2" Motor Vehicle Emission Standards and Gasoline Sulfur Control Requirements, Final Rule December 21, 1999
26. US Federal Register, 40 CFR, Part 9 et al. Vol. 64, No. 185, Final Rule, Compliance Assurance Program (CAP 2000) May 4, 1999
27. Environment Agency Japan (1975) Quality of the Environment in Japan
28. Agreement Concerning the Adoption of Uniform Conditions of Approval and Reciprocal Recognition of Approval for Motor Vehicle Equipment and Parts; ECE, Geneva, March 1958, Treaty Establishing the European Economic Community; Rome, March 25, 1957
29. Berg W (1985) Evolution of Motor Vehicle Emission Control in Europe-Leading to the Catalyst Car?; SAE-Paper 850384 and Presentation at SAE-Congress Detroit, February 25–March 1, 1985
30. Berg W (2001) Automobilemissionen von PKW und NFZ – Politik, Gesetze, Technik; (Passenger Car and Heavy Duty Vehicle Emissions – Policies, Legislation, Techniques); Lecture at the Technical University of Vienna since 1998/99
31. Council Directive 70/220/EEC of March 20,1970 on the Approximation of the Laws of the Member States relating to Measures to be taken against Air Pollution by Emissions from Motor Vehicles
32. Way G (1958) Field Survey of Exhaust Gas Composition; SAE Paper 11 a; January 1958; in: SAE Vehicle Emissions Part I, PM Vol. 6
33. Implementation of the Clean Air Act Amendments of 1970 – Part 3 (Title II): Hearings before the Subcommittee on Air and Water Pollution of the Committee on Public Works – US Senate – 92nd Congress, 2nd Session, US Gov. Printing Office, Doc. Serial No. 92-H31, S. 1540
34. Huls TA (1973) Evolution of Federal Light Duty Mass Emission Regulations; SAE Paper 730554
35. Barth DS (1980) Federal motor vehicle emission goals for CO, HC and NO_x based on desired air quality levels. Journal of Air Pollution Control Association (JAPCA), Vol. 20, No.8; August 1980
36. The Impact of Auto Emission Standards; Report of the Staff of the Subcommittee on Air and Water Pollution to the Committee on Public Works, US Senate, Serial No. 93–11, US Gov. Printing Office, October 1970
37. Motor Vehicles Department, Ministry of Transport (MOT) Tokyo, (September 1977) Automotive Type Approval System in Japan
38. Personal Communication with Mr. Takayama, Director Environmental Pollution Control Division, Motor Vehicles Department, Road Transport Bureau, Ministry of Transport (MOT), Tokyo, January 1980
39. Personal Communication with different Departments of the Japanese Environment Agency, Tokyo, January/February 1980

40. Automotive Pollution Control Division, Air Quality Bureau, Environment Agency, Tokyo (March 1977) Air Pollution and Motor Vehicle Emission Control in Japan
41. Commission of the European Communities, Brussels-Palais des Congrès (21st & 22nd September 1992) European Symposium "Auto Emissions 2000"
42. Berg W (1999) Politische Anforderungen an den Diesel-Motor – Ein Antriebskonzept zwischen Förderung und Kritik.; 20. Internationales Wiener Motoren-Symposium, Wien 1999 (Political Requirements for the Diesel Engine. A Propulsion System between Promotion and Critique; 20th International Vienna Engine Symposium, Vienna 1999
43. Berg W (2001) Schwerpunkte aus der Internationalen Abgasgesetzgebung für Kraftfahrzeuge (Highlights from International Emission Regulations for Automobiles), Uniti-Technical Congress Hohenheim 2001, University of Stuttgart-Hohenheim, March 20–22, 2001
44. Personal Communication with Mr. Shigeru Tsuda, Western Automobile Co.Ltd. Tokyo, Japan, 1980
45. Yoshizumi K (1976) Analysis of Traffic Flow in Urban Area, Tokyo 1976
46. Luther H, Schmidt U (1964) Die Entgiftung der Abgase von Verbrennungskraftmaschinen (Emission Control on Motor Vehicle Engines); Technical University Clausthal (Prof. Luther) and German Automobile Manufacturers Association (VDA) (Prof. Schmidt), Frankfurt 1964
47. Obländer K, Kräft D (1969) Abgasreinigung an Kraftfahrzeugen – Messverfahren und Testzyklen (Emission Control on Motor Vehicles – Measurement Techniques and Test Cycles); Automobiltechnische Zeitschrift (ATZ) 71, 1969 (4)
48. Legatski TW (1937) Fuel System of 1937 Automobiles; API Proceedings; Vol.18, Sec.3
49. Wentworth JT (1958) Carburetor Evaporation Losses; SAE Paper No. 123, January 1958, in SAE "Vehicle Emissions – Progress in Technology", Part II, 6:146–156
50. Muller HL, Kay RE, Wagner TO (1967) Determining the Amount and Composition of Evaporation Losses from Automotive Fuel Systems. Originally published in SAE Transactions, vol 75
51. Automobile Manufacturers Association; Fuel System Evaporation Losses. AMA Engineering Notes 616, September 1961
52. Coordinating Research Council, Inc.; 1965 CRC Motor Vehicle Evaporation Loss Tests; June 1966 (CRC Report No.391, CRC Project No. CM-58-65; January 1966 (revised: February 1966 and June 1966)
53. Coordinating Research Council, Inc.; 1966 CRC Motor Vehicle Evaporation Loss Technique Evaluation; CRC Report No.400, January 1967
54. Martens SW, Thurston KW (zzzz) Measurement of Total Vehicle Evaporation Emissions; SAE Paper 680125 in SAE "Vehicle Emissions – Progress in Technology", Part III, 14:191
55. 40 CFR Part 86, revised as of July 1, 2000; § 86.130–78: Test sequence, Fig. B78–10, p.471
56. 40 CFR Part 86, revised as of July 1, 2000; §86.096–8: Emission Standards for 1996 and later model year light duty vehicles, p. 310 etc.
57. 40 CFR Part 86, revised as of July 1, 2000; §86.098–8: Emission Standards for 1998 and later model year light duty vehicles, p. 337
58. California Environmental Protection Agency, Air Resources Board: "California Exhaust Emission Standards and Test Procedures for 2003 and subsequent Model Zero Emission Vehicles and 2001 and subsequent Model Hybrid Electric Vehicles in the Passenger Car, Light Duty Truck and Medium Duty Vehicle Classes", adopted August 5, 1999; Sec. C3: Baseline Partial ZEV Allowance
59. Lenz HP (2000) Technical University Vienna, Austria: Anteil des Strassenverkehrs an den CO_2-Emissionen (Share of Road Traffic in CO_2-Emissions), Presentation during Conference by " Haus der Technik" ("Fuel Economy Improvement – A Contribution of the Automobile to Saving Resources and Lowering CO_2-Emissions"), Munich, November 13/14
60. Lange B (2000) Automotive Expert for the European Parliament – Committee on Environment, Health and Consumer Protection: The Kyoto Agreements and the Objectives of the European Union, Presentation during Conference by " Haus der Technik" ("Fuel Economy Improvement – A Contribution of the Automobile to Saving Resources and Lowering CO_2-Emissions") in Munich, Nov. 13/14

61. Schindler KP (2000) The Contribution of the DI Diesel Engine to Reduce Fuel Consumption and CO_2-Emissions, Presentation during Conference by " Haus der Technik" ("Fuel Economy Improvement - A Contribution of the Automobile to Saving Resources and Lowering CO_2-Emissions") in Munich, Nov. 13/14
62. Berg W (1978) Vorschriften über den (Kraftstoff-)Verbrauch von Personenwagen; ("Regulations about the Fuel Consumption of Passenger Cars"), Automobil Revue No. 15
63. Personal communication with Ed Wall, US Department of Energy, Coordinator for the DOE-program "Partnership for a New Generation of Vehicles" (PNGV)
64. National Research Council - 7th Report: Review of the Research Program of the Partnership for a New Generation of Vehicles
65. Testimony of David K. Garman, Assistant Secretary, Energy Efficiency and Renewable Energy, US Department of Energy, November 1, 2001
66. Daisho Y (2000) Department of Mechanical Engineering, Waseda University Tokyo: Technical Measures for Reducing CO_2 Emission in the Transportation Sector in Japan, Presentation during Conference by "Haus der Technik" ("Fuel Economy Improvement - A Contribution of the Automobile to Saving Resources and Lowering CO_2-Emissions") in Munich, Nov. 13/14
67. California Air Resources Board (May 26, 1999), On-Board Diagnostics II (**OBD II**) Regulatory Review, Mail Out #MSC 99-1
68. Berg, W. In-Use Testing of Heavy Duty Diesel Vehicles - A Study [for ACEA] about Literature, Test Programs and Experience in different Countries. Brussels, 9-14-1998
69. The Central Environment Council: Future Policy for Motor Vehicle Exhaust Emission Reduction (Fifth Report); Tokyo April 16, 2002

The Handbook of Environmental Chemistry Vol. 3, Part T (2003): 255–288
DOI 10.1007/b11990

Fuels

Dušan Gruden

Dr. Ing. h.c. F. Porsche Aktiengesellschaft, Porschestrasse, 71287 Weissach, Germany
E-mail: dusan.gruden@porsche.de

The symbiotic relationship between internal combustion engines (Otto and Diesel) and petroleum-derived fuels (gasoline and Diesel oil) has characterized the entire road traffic scene for more than 100 years. To date, the increasingly stringent demands on vehicles and engines could be complied with by harmonizing the further development of engines and fuels. According to the current state of the art, the "conventional" fuels (gasoline and Diesel oil) are the most economic alternative for most transportation tasks. The polluting character of mineral oil is a growing handicap. It is up to the oil industry to improve the fuel quality and thus clearly lower the pollutant emissions.

The current engine generation and the engines and exhaust gas aftertreatment systems being developed for future applications place new and even more severe demands on fuel quality. The demands have been laid down in the World Wide Fuel Charter drawn up in a joint approach by the international automotive industries. The new demands on fuel quality are evolving parallel to the development of a new generation of engines. The focus is on the legal requirement of simultaneously reducing the exhaust pollutant and CO_2 emissions (fuel consumption).

The current priorities in the field of fuel development can be summarized as follows:

Improvement of the quality of conventional fuels, reduction of fuel consumption of Otto and Diesel engines and intensively deal with potential alternative fuels.

The fundamental characteristics of alternative fuels such as liquid gas, natural gas, methanol, ethanol and hydrogen are very similar to those of current fuels. To the engine designer this means that the combustion engine will continue to prevail as the main automotive power unit but that it will have to be adapted to potential new fuel characteristics.

From this situation, the oil industry should draw the conclusion that everything must be done to optimize the fuel qualities and to assist the automotive industry in its efforts to meet the requirements of modern society.

Keywords. Gasoline, Diesel fuel, Octane, Cetane number, Exhaust gas emission, WWFC, Additives, Sulfur, Alternative fuels, Alcohols, Bio fuels, Hydrogen, LPG, CNG

1 Introduction

The discovery of mineral oil in the middle of the 19th century paved the way for the invention of the internal-combustion engine. Mineral oil serves as the basis for various fuels (Fig. 1) with gasoline and Diesel oil being the most widely known ones.

The two main fuels derived from mineral oil have determined the configuration of the combustion engines for which they are used. The properties of the gasoline and Diesel engine clearly depend on the fuel qualities. The symbiosis be-

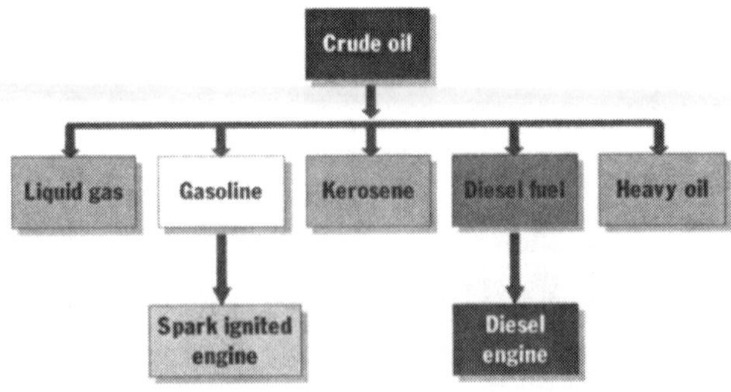

Fig. 1. Fuels and engines

tween a comfortable, low-priced fuel and an efficient combustion engine has allowed this type of drive unit to proliferate.

Despite various attempts, no other drive concept has been able to assert itself besides the gasoline and Diesel engines during more than 100 years of automotive history.

Both gasoline and Diesel engines are optimum energy-conversion units transforming the chemical energy of fuels into mechanical work with utmost efficiency.

2 Fuels and Engine Characteristics

2.1 Boiling Curve

No fuel – no combustion engine. This does not mean, however, that combustion engines tolerate any type of fuel. The mineral-oil boiling curve illustrates that Otto and Diesel engines respond in an extremely sensitive way to the fuel quality (Fig. 2).

According to current knowledge, gasolines can be burnt efficiently only in spark-ignited engines with their homogeneous air/fuel mixtures and externally supplied ignition. Diesel fuel is suited for heterogeneous mixtures ignited through self-ignition. To date, all attempts of using gasoline in Diesel engines and Diesel oil in Otto engines have been to no avail.

The dream of a multi-fuel engine capable of "swallowing anything" will not come true since engines have to comply with numerous requirements and meet more and more stringent exhaust emission standards.

This means that, in the foreseeable future, gasoline and Diesel engines will continue to prevail as the main machines for the transformation of the chemical fuel energy into mechanical work.

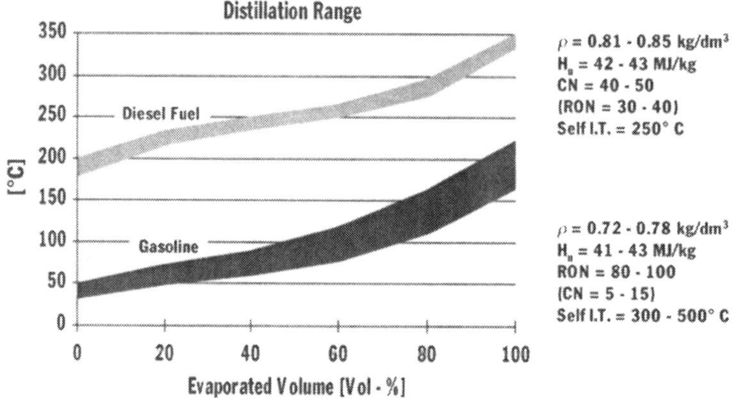

Fig. 2. Characteristics of fuels for Otto and Diesel engines

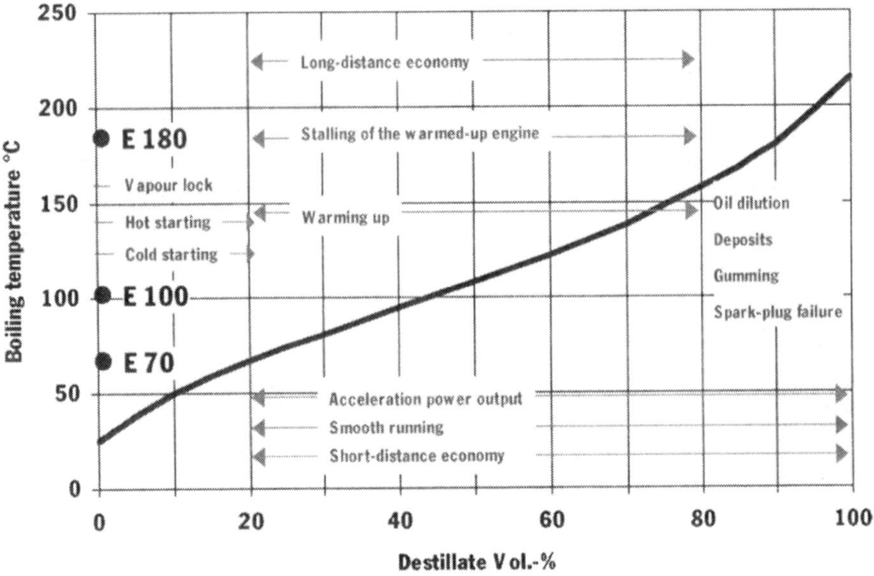

Fig. 3. Boiling curve of a gasoline

The boiling curve of the mineral oil is not only decisive for the amount of gasoline and Diesel oil consumed but also has a considerable influence on the operational characteristics of the engines.

Figure 3 shows those areas of the distillation curve which are of particular importance for the behavior of an gasoline engine. Apart from the characteristic items such as the initial and final boiling points the amounts of fuel evaporating at 70°C, 100°C and 180°C (points E70, E100, E180) are also of importance for the evaluation gasolines.

As far as Diesel fuel is concerned, the point T95 – at which 95% of the fuel volume evaporates – is of particular importance. Apart from the boiling curve, there are several other fuel properties which influence the engine characteristics.

2.2 Octane and Cetane Numbers

The layout of an Otto engine vitally depends on the octane number of the gasoline. The octane number informs about the knock resistance of the fuel. Likewise, the cetane number of Diesel oil is important for the operational behavior of the Diesel engine. It describes the ignition performance of the Diesel fuel. There is a close correlation between both numbers (Fig. 4): The higher the octane number the lower the cetane number and vice versa.

When trying to lower the fuel consumption and CO_2 exhaust emissions of modern Otto engines consideration must be given above all to the compression ratio.

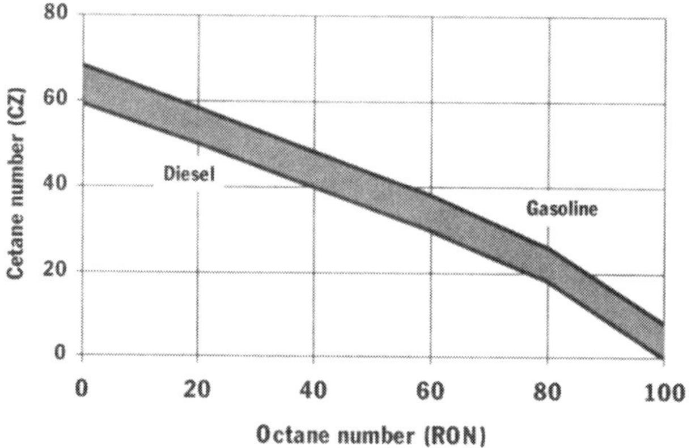

Fig. 4. Correlation between octane and cetane numbers

To achieve low fuel consumptions the compression ratios must be relatively high ranging between ε=10 and 12. The compression ratio, in its turn, strongly depends on the octane rating of the gasoline: The higher the octane number the higher the compression ratio which can be chosen for the respective engine (Fig. 5).

As various pertinent tests and collected data show, the octane number has a clearly positive influence on the specific work (power output) and fuel consumption of the gasoline engine (Fig. 6).

The actual octane requirement of an Otto engine is not constant, however, but varies strongly with the operating condition (Fig. 7).

At low loads no combustion knock occurs even with low octane number of fuels. The octane requirement of the engine exclusively depends on the full-load demands. Theoretically, the best solution would be an engine with either a variable compression ratio or the possibility of varying the fuel quality (on-board

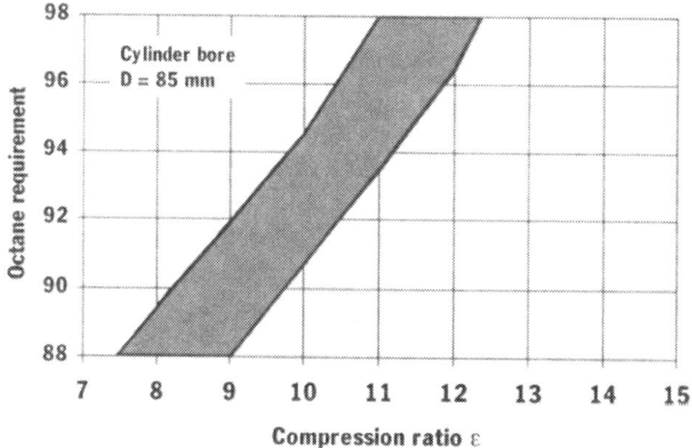

Fig. 5. Compression ratios and octane requirements of modern Otto engines

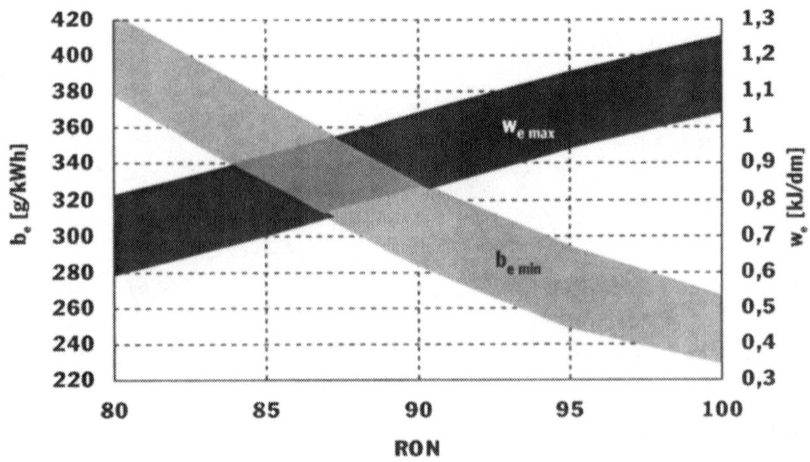

Fig. 6. Octane number, min. spec. fuel consumption and max. spec. work

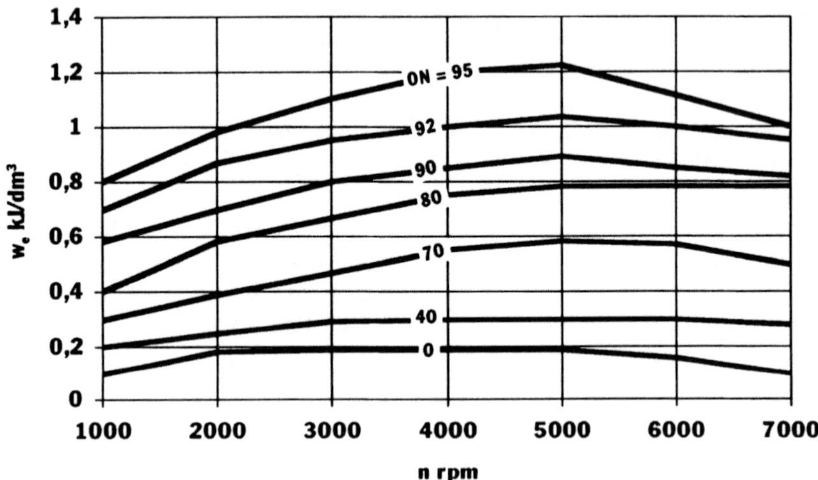

Fig. 7. Octane requirements of a spark ignited engine

mixing of fuels with high and low octane ratings). So far, it has not been possible to realize an engine of that kind.

As far as Diesel fuels are concerned, the cetane number is the decisive criterion for the ignition performance and the ignition delay, that is the interval between the fuel injection into the combustion chamber and the beginning of combustion. The higher the cetane number the shorter the ignition delay (Fig. 8) and the better the control of the combustion process in the Diesel engine.

As it proved to be difficult to determine the cetane number by way of testing, a mathematical value – the so-called cetane index – was defined which correlates clearly with the cetane number.

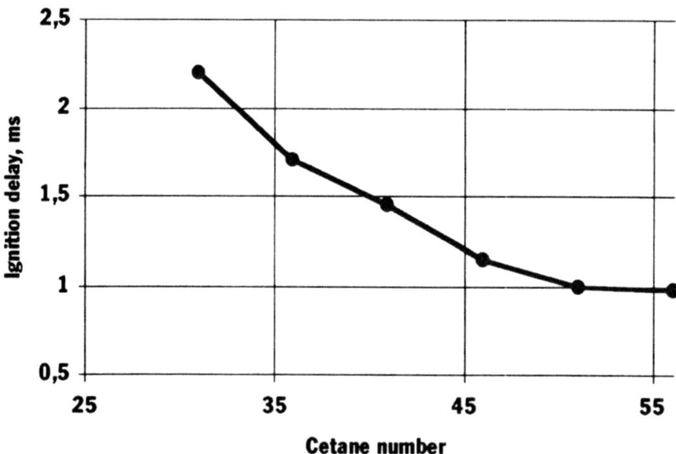

Fig. 8. Ignition delay versus cetane number [2]

The chosen examples of the close interdependence between the fuel quality and the engine characteristics indicate that the fuel is as important an engine part as other components such as the pistons, crankshaft, valves etc. The engine manufacturers and their suppliers are used to do their engine developments in close co-operation with each other and to jointly submit the mechanical parts to comprehensive and most severe tests. Unfortunately, in the field of fuel development it has not been possible yet to establish a similar close co-operation with the oil industry even though corresponding efforts have been made for many years.

Today, there still is no real harmonization between fuel producers and engine manufacturers.

3
Fuels and Exhaust Gas Emissions

Even though fuels are known to have an influence on exhaust emissions and therefore clearly defined fuel qualities are to be used for emission testing, their inherent potential for pollutant reduction through variation of the fuel composition has not been fully exploited.

With the exception of nitrogen oxides and oxygen, all the compounds found in automotive exhaust gases have their direct origin in the fuel composition.

Fuels for gasoline and Diesel engines are mixtures of unsaturated and saturated hydrocarbons (paraffins, olefins, aromatic compounds). As the structural formulae of these hydrocarbons show, their combustion products must be of highly different composition (Fig. 9).

n-Hexane C_6H_{14}

CnH_{2n+2}

Isooctane C_8H_{18}

CnH_{2n-6}

Benzene C_6H_6

CnH_{2n}

1-Hexene C_6H_{12}

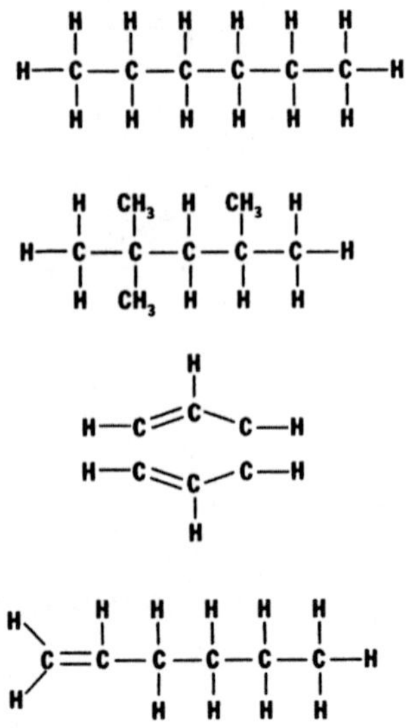

Fig. 9. Saturated and unsaturated fuel components

3.1 Regulated Exhaust Gas Constituents

There is a world-wide consensus that efforts to clean up vehicle emissions must utilize engine and fuel technology. However, the battle has recently focused on one key issue: which sector should incur most of the burden?

With regards to fuel quality, among the most important regulatory development are:

- the Clean Air Act Amendments (CAAA) and the resulting U.S. Reformulated Gasoline (RFG) Program, and
- EU Auto-Oil Program and the resulting Fuel Directive 98/70/EEC.

The EU-Auto-Oil Program was the first program of its kind in Europe to bring together the resources and expertise of the automotive sector through the European Association of Automobile Manufacturers (ACEA) and the oil refiners through the European Petroleum Industry Association (EUROPIA) in collaboration with the services of the European Commission.

The automotive manufacturers, the oil industry and the European Union carried out a joint automobile/oil program aimed at examining the effects which the fuel composition has on the noxious exhaust gas constituents and CO_2 emissions of vehicles with three-way catalyst and oxygen sensor. It was repeatedly found that nearly all of the fuel characteristics have a considerable influence on exhaust composition.

By varying the boiling curve and increasing the boiling volume at E 100 from 35 to 65%, the CO and HC emissions can be reduced by 17% and 10 to 30%, respectively. At the same time, fuels containing a high percentage of aromatic com-

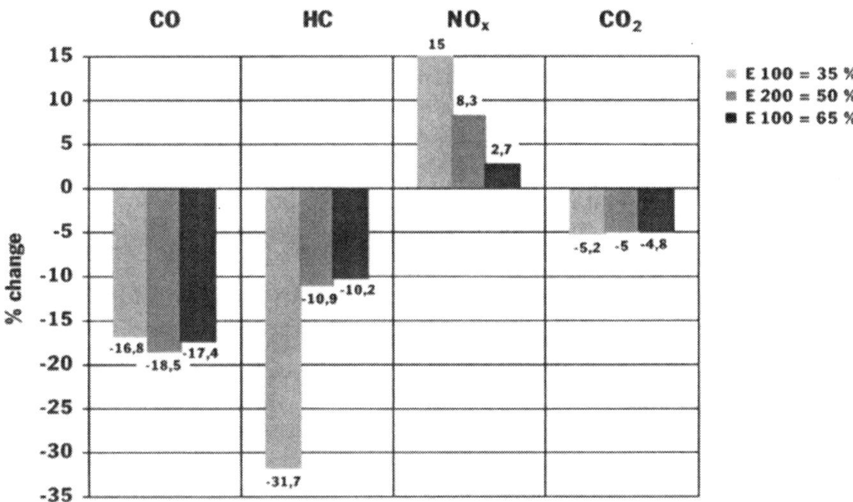

Fig. 10. Relative effects of reducing fuel aromatic content from 50 to 20% on composite emissions over the gasoline EPEFE fleet [11]

pounds result in higher NO_x emissions unless the mixture formation system provides for fuel-specific corrections (Fig. 10).

Lowering the percentage of aromatic compounds is an efficient means of reducing the CO and HC emissions (Fig. 10). In engines with a three-way catalyst for exhaust aftertreatment, varying the A/F ratio with changed fuel composition results in higher NO_x emissions. If the λ-window is chosen correctly, the reduced aromatic compounds help to lower the NO_x emissions.

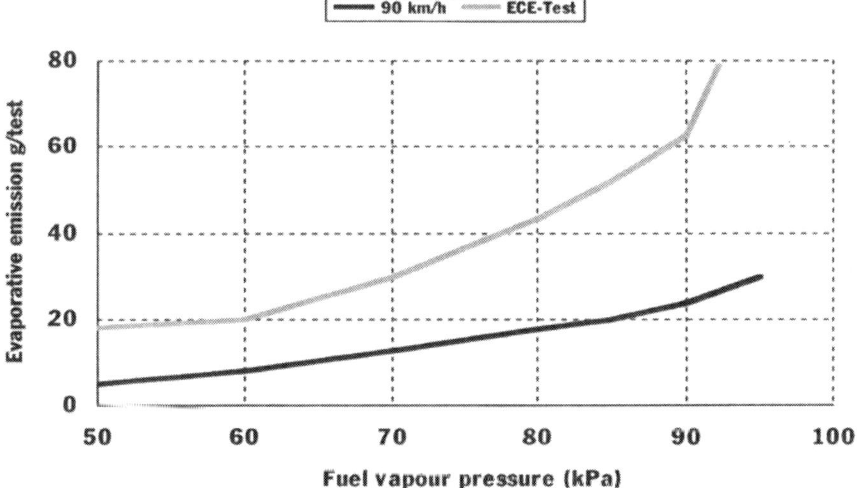

Fig. 11. Effect of vapor pressure on vehicle evaporative emission

Lowering the fuel vapor pressure, for example, is an efficient means of reducing fuel evaporation (Fig. 11). Vapor pressure differences play a decisive role during cold starting and engine warming up.

A study performed by Porsche in co-operation with Shell AG has shown that the exhaust gas emissions of an Otto engine can be clearly lowered by varying the composition of the fuel used: HC by 10 to 30%; NO_x by 15 to 30% and overall aromatic compound emission by between 20 and 80% (Fig. 12).

The fuel composition, too, has a non-negligible influence on exhaust gas emission of a Diesel engine.

High cetane numbers improve self-ignition and combustion and also have a beneficial influence on pollutant emissions in general and on particle emission in particular (Fig. 13).

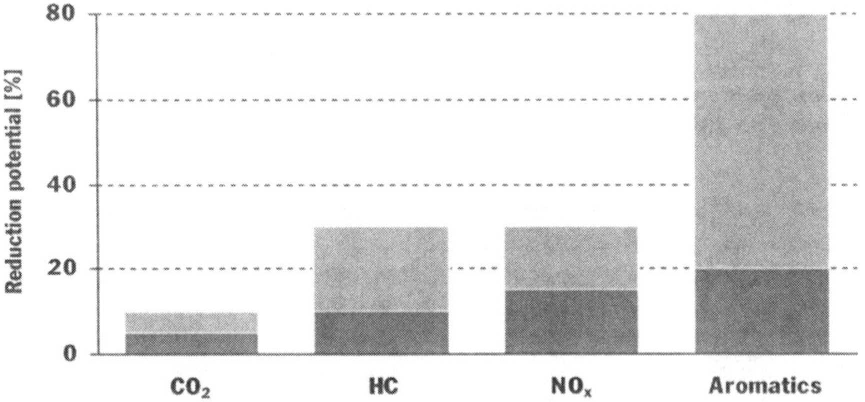

Fig. 12. By changing the fuel composition, many pollutant components can be reduced

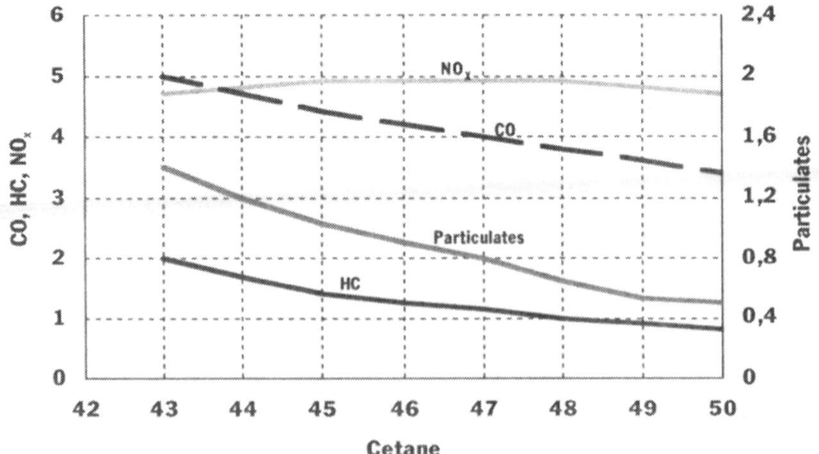

Fig. 13. How cetane number affects emissions

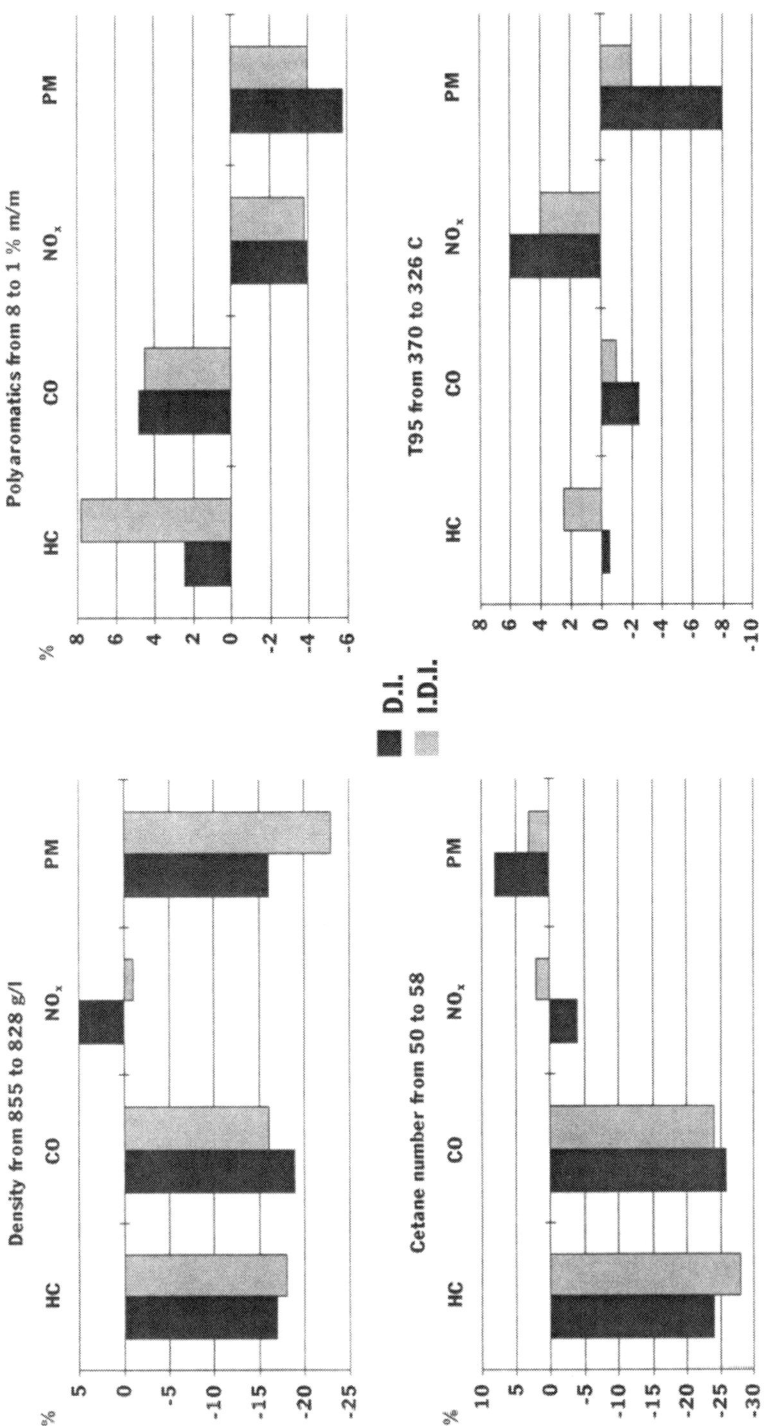

Fig. 14. Effect of fuel composition changes on emissions – DI compared to IDI

Within the scope of the EU/Auto/Oil program, the influence of Diesel fuels on the emission behavior of both direct injection and prechamber Diesel engines was examined (Fig. 14). Besides the cetane number Diesel engine emissions also significantly depend on the fuel density, the percentage of polyaromatic compounds and the boiling characteristics.

The US-Auto/Oil program carried out in the United States also confirmed the influence of the fuel composition on exhaust gas pollutant emissions.

3.2 Unregulated Exhaust Gas Emissions

Besides the "traditional" pollutants for which exhaust emission limits have existed for many decades, legislators are showing more and more interest also in so-called unregulated exhaust-gas constituents such as benzene, polycyclic aromatic hydrocarbons (PAH), formaldehyde, methane etc.

In the USA, cancer-causing exhaust-gas constituents such as benzene, formaldehyde, polycyclic aromatic compounds etc. are known under the designation "Air-Toxic Components".

All these exhaust-gas components as well as the so-called ozone-formation potential are influenced by the fuel composition.

Figure 15 shows the influence of the aromatic-compound content on benzene emissions. The discussion about the reduction of the automotive benzene emissions is therefore closely linked with the discussion about fuel composition. According to earlier tests the presence of benzene in the exhaust gas is directly related with the benzene contained in the fuel.

3.3 CO_2 Emission

In the public discussions on non-limited exhaust-gas constituents the exhaust component number one is carbon dioxide (CO_2) – a complete-combustion product which is held responsible for its potential influence on the greenhouse effect and possible climatic changes. In the struggle against the possible effects of CO_2 emission, the primary task for the automotive manufacturers is to reduce the fuel consumption of their cars. With fossil fuels, the CO_2 emissions are directly proportional to fuel consumption.

For gasoline-fueled engines the following applies:

$$CO_2 \text{ g/km} = 24 \times B \tag{a}$$

For Diesel engines and Diesel fuels, the following conversion factor is used:

$$CO_2 \text{ g/km} = 27 \times B \tag{b}$$

where B=fuel consumption (L/100 km).

At present, comprehensive developments are under way which are aimed at reducing automotive fuel consumption. The measures to be taken are not limited to the engine but include the complete car, the vehicle weight, air drag and rolling resistance, the drive train etc.

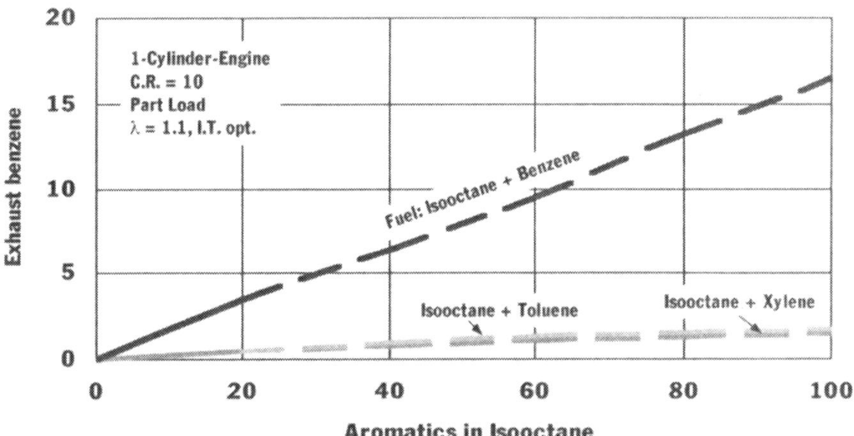

Fig. 15. Influence of aromatics in fuel on exhaust benzene

The automotive industry in Europe, organized under the roof of ACEA, committed itself to lower the CO_2 emissions of its products by 25% between 1995 and 2008, thus reducing the average value from 186 g CO_2/km to 140 g CO_2/km. This corresponds to a fuel consumption reduction from an average of 7.8 to 5.9 L/100 km.

The German automotive industry has additionally committed itself to lower the average fuel consumption and resulting CO_2 emissions by 25% between 1990 and 2005.

The European automotive industry will have to meet the EU3 (since 2000) as well as the even more stringent EU4 (2005) exhaust emission standards while also keeping their promise of further cutting down the CO_2 emission levels of their car fleets. This is an enormous challenge which cannot be managed with the currently available technologies only.

Therefore, when looking for possibilities to lower the CO_2 emissions, the potential contributions of the fuels should not be left out of account (Fig. 16).

Lowering the percentage of aromatic compounds and increasing the content of saturated hydrocarbons helps to reduce the CO_2 emission. Modifications to the composition of currently available fuels are expected to lower the CO_2 emission by 4 to 8%.

The clearly perceptible influence of the fuel quality on the exhaust-gas composition in both the Otto and Diesel engines justifies the demand of automotive industry for tighter fuel specifications and standardization. The pollutants for which emission standards exist (CO, HC, NO_x and particulate matter) can be cut down by up to 30% while the non-limited exhaust constituents such as benzene, sulfur compounds, formaldehyde, PAH etc. can be reduced to varying degrees ranging from few percent to full elimination. By varying the fuel composition, the CO_2 emissions, too, can be significantly lowered (4 to 8%).

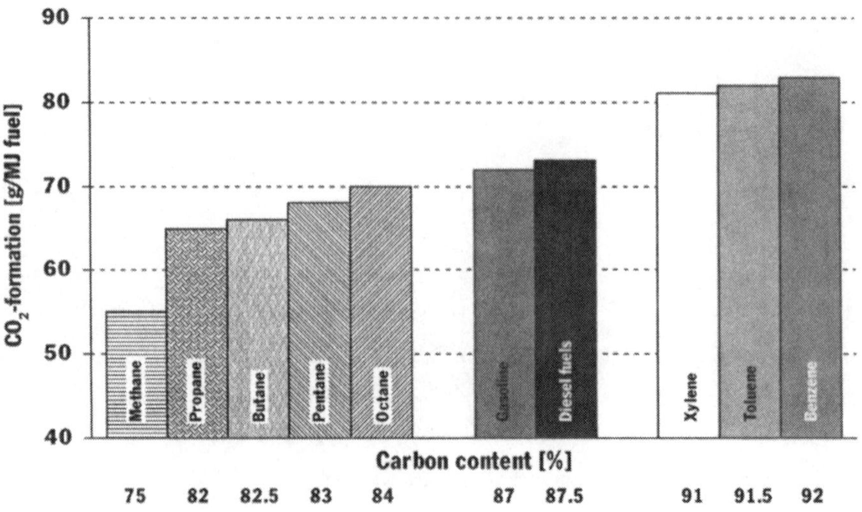

Fig. 16. With an increasing carbon content in the fuel the CO_2 emission increases during the combustion

4 Requirements for Fuels

As stated above, the design of a combustion engine primarily depends on the properties of the fuel used. Engines, on the other hand, have certain demands on the fuel quality to enable them to function properly.

In the early 1990s already, the American Environmental Protection Agency (EPA) carried out a study concerning the inherent potential for further reducing the pollutant emissions of the registered vehicles in traffic. According to the results of this investigation, the pollutant emissions could be lowered by 30% through regular maintenance and technical control of the cars. A 15% reduction could be achieved by using better, cleaner and recomposed fuels whereas more severe emissions limits for new cars resulted in pollutant decreases of as little as 2% only.

The advantages offered by the newly developed engine and exhaust aftertreatment technologies can only be put to good use if the entire system is optimized accordingly. Fuels with their vast range of inherent properties must be included in this optimization. Fuel quality and automotive technology are inseparably linked; they influence each other and must be carefully harmonized if efficient results are to be obtained. The automotive manufacturers and the oil industry will have to co-operate closely in order to satisfy the demands of the customers and meet the ever increasing legal and environmental requirements.

For an entire century, it was the carburetor engine which prevailed as the main propulsion system for passenger cars. It used especially developed so-called "carburetor fuels" whose characteristics had been optimized to suit the carburetor concept. These fuel characteristics are specified in DIN EN 228.

In the course of the last 10 years, carburetor engines have almost entirely disappeared. Modern engines use manifold injection or so-called Multi-Point Injection (MPI) systems for mixture formation. It is interesting to note that the demands of these systems on the fuel quality have not yet been precisely defined, even though many of the boundary conditions known from the carburetor variants have changed such as, for example, fuel pressure and temperature, intake manifold vacuum, dwell time in the intake manifold etc.

In the World Wide Fuel Charter (WWFC) [5], the major automotive manufacturers in the world, represented by their respective associations (ACEA, Alliance, EMO and JAMA), have defined minimum demands to be fulfilled by fuel quality world-wide. These demands take into account the respective regional differences and technical development levels.

Automotive industry asks for the WWFC demands to be complied with since it is through the combined effect of automotive technology and fuel quality that the pollutant emissions of the current vehicle population can be improved.

The European Commission has published new proposals for the amendment of Directive 98/70/EC on the quality of gasolines and Diesel fuels [4].

These amendments primarily concern the amount of aromatic compounds, benzene and sulfur contained in gasolines as well as the cetane number, density and percentages of polyaromatic compounds and sulfur of Diesel fuels (Table 1).

In the course of the years to come, the demands the new engine generation will place on fuel quality will be more and more clearly specified as engine development progresses.

The success of direct-injection gasoline engines will strongly depend on the availability of suitable fuels and lubricants. At the time being, currently available fuels are used to develop these engines but there is no guarantee that they will be able to fully comply with the special demands of direct injection systems.

From today's point of view the following fuel characteristics will continue to play a decisive role in the development of future engine generations:

- The octane number of gasolines,
- The cetane number of Diesel fuels,
- The oil industry and the automotive manufacturers should concert their efforts to find out which octane number is the most efficient one from an eco-

Table 1. Fuel characteristics according to EU Directive 98/70/EC

Gasoline	Year 2000	Year 2005
Aromatic compounds vol.%	42	35
Benzene vol.%	1	<1
S_2 ppm	150	<50 (10)
Diesel oil		
Cetane number	Min. 51	
Density g/cm^3	845	
Polyaromatic compounds vol. %	11	
S_2 ppm	350	<50 (10)

nomic point of view on the one hand and with regard to its emission-reducing potential on the other. The recommendation of RON = 95 as the best octane rating dates back to the late 1970s and should therefore be checked in the light of today's boundary conditions.

- The **evaporative characteristic** (distillation curve) plays an outstanding role mainly during cold starting and warming up and has a substantial influence on driveability and on the fuel behavior during injection and ignition. It might be necessary to develop new test methods for further optimization.
- The **thermal stability and formation of deposits** in the injection system, combustion chamber, pistons, spark plugs and EGR valves.
  - It is not clear yet what must be done to keep the intake valves of direct injection engines clean. The same is true for the combustion chamber where deposits are formed.
- In addition, comprehensive consideration must be given to the required **aromatic compound, benzene and olefin contents** of gasolines as well as to the **density, content of polyaromatic compounds, final boiling point** and **lubricity** of Diesel fuels such as described in the WWFC.
  - The tighter specifications for future gasolines and Diesel fuels will result in far more exacting tolerances at the refineries and in a higher amount of saturated paraffinous molecules or, in other words, in a higher H/C ratio of the fuel.

4.1 Additives

Despite design modifications of combustion systems and new performance parameters carried out by car manufacturers, the most desired and widely applicable approach would be to produce clean fuels, that reduce regulated and unregulated emissions. To achieve this objective, the most convenient and versatile, approach would be to use fuel additives that improve performance in the combustion chamber and exhaust gas system.

The importance of these additives, i.e., chemical substances added to fuels and lubricants, for the maintenance of the required fuel characteristics is constantly growing. To further improve the fuel quality it will therefore be necessary to optimize both their basic composition and the additives they contain.

As new challenges face vehicle and engine builders, they look to the suppliers of fuels and additives for help. A good understanding of motor industry needs is essential if the additives industry is to plan and prioritize its research effort effectively.

Optimized operating agents/additives help to improve the following phenomena:

- Gasoline Engine
  - **Knock resistance** – octane improver will experience the best gains.
  - **Cleanliness** of the intake system. Recently, detergent and dispersant additives have been developed to reduce deposit build-up within the gasoline engine's inlet system. Use of fuel additives can considerably reduce the mass of inlet valve deposits and thus eliminate the potential performance and emission problems.

Injectors require gasoline with detergent additives if they are to stay clean and give optimum fuel economy and emission performance. Fuel injector fouling indicated considerably cleaner injectors for additive containing gasolines.
  - **Deposits in the combustion chamber.** Under the influence of market pressures mainly in Europe and legislative pressure, starting in the USA, there is a growing interest in detergent additives to lower deposit formation in the combustion chamber and thereby maintain emission closer to design values and in combustion improvers in general.
  - **Exhaust-gas aftertreatment.** Noble metal catalyst systems and oxygen sensors are sensitive to certain elements and this has to be allowed for.
- Diesel Engine
  Additives in Diesel fuel have been used for a long time for various purposes, e.g., improved ignition quality and cold fuel flow properties. Some use has also been made of soot oxidation promoters to lower black smoke.
  - **Cetane improvers** for easier start in cold weather and overall increases in engine efficiency.
  - **Combustion promoters.** These combustion additives in Diesel fuels will continue to be the prime and cost-effective approach for efficient Diesel combustion and auto emission control.
  - **Smoke suppressant.** Smoke and particulate emissions of modern Diesel engines, although very significantly reduced, have not been eliminated and much additive research is focused on this point.
  - Cold flow improvers for Diesel fuels.

Besides the need for improved cold start performance, the combination of combustion additives, ignition accelerators, cetane improvers and deposit control additives as a multi-functional Diesel additive package is necessary for formulating premium clean Diesel fuels. Overall it is concluded that average reduction of emissions show clear advantages for the additive fuels, especially on particulate and hydrocarbon emission as well as fuel consumption.

Fuel additives end their life as combustion products, but the contribution they make to regulated emissions of toxic substances is normally very small and is outweighed many times over by the major vehicle emissions reductions, which could not have been made over the past few years without their development.

In certain cases, inappropriate use of fuel additive formulation can lead to unforeseen problems. These developments will certainly require more sophisticated measurement, testing and analytic capabilities if additives are to be cost-effectively applied. All new fuel additives must be designed with exhaust-gas aftertreatment system compatibility on mind.

It is clear from reading the World-Wide Fuels Charter (WWFC) that metallic-based additives can no longer be part of tomorrow's additives. Growing evidence exists of their deleterious effect on exhaust oxygen sensor and increased particulate emissions. Non-metallic ashless compounds will therefore replace metallic additives.

4.2 Sulfur Content

As of January 1, 2005, all the new cars sold in the EU must comply with the Euro 4 emission standards. Parallel to these activities, the automotive industry must intensify the development of new technologies for the further reduction of the CO_2 emissions. All newly developed solutions must meet most stringent exhaust-emission standards which can be achieved only by providing for corresponding new exhaust-gas aftertreatment systems.

The technologies for the reduction of the CO_2 emission require fuels with modified qualities and, above all, with a lower sulfur (S_2) content.

The most promising concepts are Otto engines with direct fuel injection while direct-injection Diesel engines are the current state of the art already.

For both engine types, new efficient systems for NO_x reduction in lean-burn mode (air excess) are needed (see chapters on Otto and Diesel engines).

The following solutions have been investigated:

- NO_x storage catalyst,
- Lean De-NO_x catalyst,
- Diesel oxidation catalyst, and
- Particulate filter for Diesel engines only.

The efficiency of all these exhaust-gas aftertreatment systems declines with increasing sulfur content of the fuel (Fig. 17).

The alkaline and alkaline earth elements of the NO_x storage catalysts are noted not only for their thermally resistant nitrates but also for their strong propensity

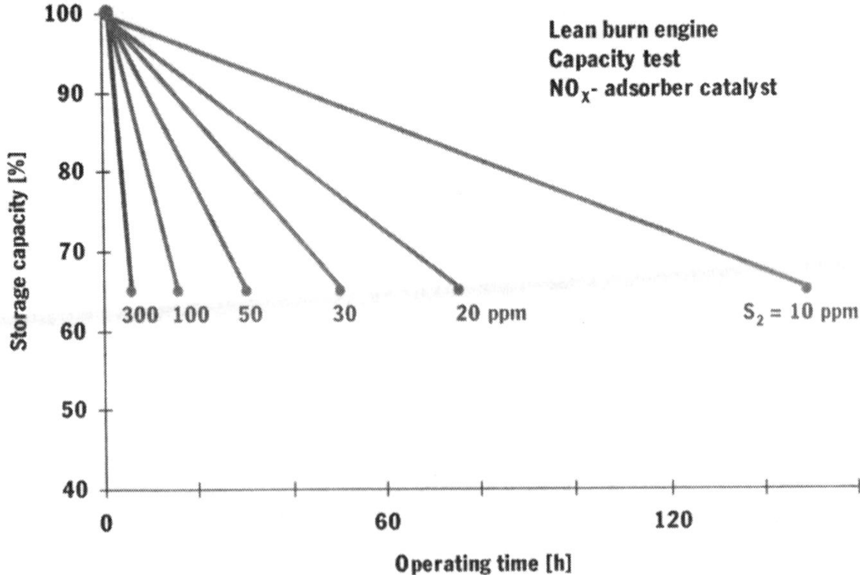

Fig. 17. Influence of sulfur on the storage capacity of the NO_x catalyst

to form sulfates resulting in an irreversible poisoning by the sulfur contained in the fuel.

The thermal decomposition of the sulfates starts at temperatures of more than 1000°C provided that there is sufficient oxygen in the exhaust gas. These are exhaust gas temperatures which are not reached under realistic driving conditions.

The sulfur content also decisively influences the particulate emissions of Diesel engines (Fig. 18).

S_2 oxidizes in the particle filter forming sulfates which can be measured in the exhaust gas in the form of particulates.

According to pertinent results, sulfur also has an enormous bearing on the catalyst conversion rate in modern engines. High levels of sulfur in gasoline adversely affect the performance of catalytic converters used to reduce vehicle tailpipe emissions. Although sulfur in gasoline does not poison the catalyst as severely as leaded gasoline, its effect is significant enough that the automotive industry and engine manufacturers would prefer to see sulfur-free fuel.

A high sulfur content deteriorates the efficiency of a 3-way catalyst mainly in the warming-up phase and this effect is the more critical the lower the emission levels which a car is expected to comply with. Sulfur results in irreversible deteriorations of the catalyst wash-coat and of certain metallic catalyst components [21]. At the time being, no S_2-resistant catalysts are available yet.

The sulfur content also affects the ability of OBD (On Board Diagnosis) to detect failures of the catalyst conversion system. The oxygen sensor is sensitive to S_2, and sulfate deposits (SO_4) on the sensor falsify the signal. In addition, high sulfur levels in the fuel strongly impair the oxygen-absorption capacity and HC conversion rate of the catalyst causing the OBD system to be supplied with false information.

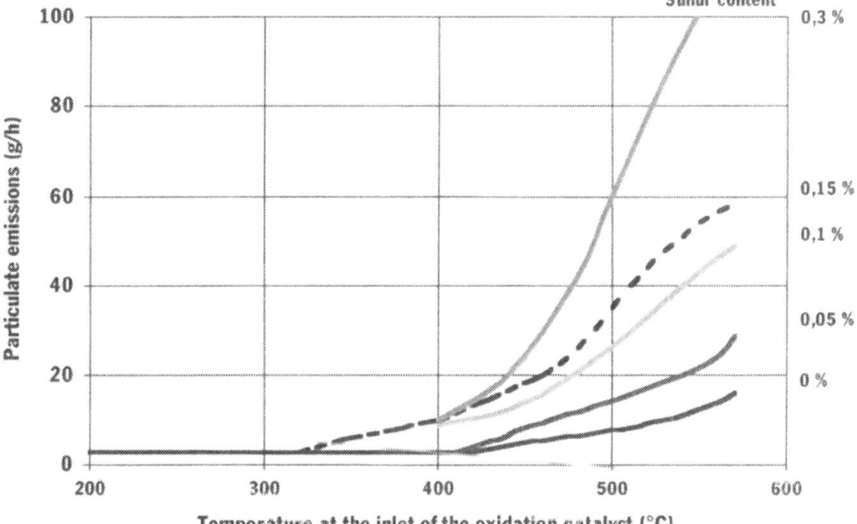

Fig. 18. Influence of the sulfur content on the particulate emission of a Diesel engine

Fuel injectors require low-sulfur gasoline with detergent additives if they are to stay clean and give optimum fuel economy and emission performance.

The sulfur content of automotive fuels clearly affects the pollutant emissions of modern Diesel engines as well. The sulfur-free so-called City Diesel (Sweden Class 1) available in Sweden yields 5–50% less CO, 10–15% less HC and NO_x and 5–20% less PM than today's EU Diesel fuel (350 ppm S_2). Even if the S_2 content is lowered from 50 ppm to 10 ppm the PM emissions will drop by as little as 5% only.

In general, a reduction in sulfur and aromatic content and an increase in cetane number result in improved fuel quality for diesel engines.

Summarizing it can be said that the engines and emission aftertreatment systems which exist already or are currently under development require fuels with extremely low sulfur contents, in order to ensure satisfactory efficiency over the stipulated lifetime.

That is why the demand of the automotive industry for sulfur-free fuels is fully justified. Sulfur is a basic constituent of the crude oil and participates in the combustion process unless removed from the fuel during refining. The elimination of sulfur from the fuel and improving of fuel quality would have immediate positive effects:

- Improvement for all the currently registered vehicles;
- Extended efficiency of exhaust-gas aftertreatment systems;
- Reduction of
  - CO by 10 to 50%,
  - HC by 10 to 30%,
  - NO_x by 15 to 30%,
  - Particulates by 5 to 20% (Diesel engines);
- Simultaneous reduction of the CO_2 emissions by 4 to 8%.

The latest proposal in the EU [4] prescribes the production and introduction of sulfur-free fuels (10 ppm max. S_2) in the EU as of 2005. As of 2011, these fuels will be compulsory in the entire EU.

As of January 1, 2004, US refineries will have to supply gasoline fuels having an S_2 content of less than 30 ppm. In addition, EPA in the USA is asking for 15 ppm S_2 as of June 1, 2006 versus 500 ppm S_2 today. And discussions are underway about a future sulfur level of as low as 5 ppm. The technologies for producing "clean" Diesel fuel that also require high cetane numbers, low aromatics as well as lower T_{95} and density are not necessarily the same as the sulfur reduction technologies. Additional reactions have to be carried out to improve Diesel cetane quality or to reduce the level of aromatics.

4.3
A Life Cycle Comparison – Well to Wheel Analysis

The reduction or elimination of sulfur at refinery level requires additional energy which comes along with higher CO_2 emissions during refining. It must therefore be made sure that lowering the sulfur content does not affect the other fuel characteristics in general and the octane and cetane number in particular.

The additional energy needed to produce sulfur-free fuels is expected to generate about 0.5 to 78 kt of additional CO_2 per million tons of fuel at refinery level. According to a CONCAWE study [10], the emissions of CO_2 in the EU caused by the production of S_2-free fuels will increase by 4.6 million tons or, in other words, by 0.5 to 0.7% per year.

The fuel cycle for petroleum-based fuels consists of four stages: crude oil extraction and recovery; crude oil transportation and storage; crude refining in a petroleum refinery to produce different individual fuels; and transportation, storage and distribution of fuels.

In the overall CO_2 balance of a fuel throughout its entire life cycle – from crude oil exploration to refining, transportation, distribution and combustion in an automotive engine – the refining processes account for about 4 to 7% of the entire CO_2 emissions (Fig. 19). More than 85% of the overall CO_2 emissions are caused by the combustion process. By introducing sulfur-free fuels, the fuel consumption of the vehicle population and thus also the CO_2 emissions would be lowered by about 5%. This means that the higher CO_2 levels during the production of cleaner fuels in the refinery are more than compensated by lower CO_2 emissions in current automotive traffic.

In many of the existing refineries, the production of sulfur-free fuels will require additional capital investments to be made and result in a certain increase of the operating expenses which will depend on the condition of the respective refinery and the sulfur content of the crude oil. The fuel prices to be paid by the consumers are expected to rise by

- 0.1 to 0.43 € cent/L for gasoline and
- 0.2 to 0.67 € cent/L for Diesel fuel.

The oil industry frequently argues that the contribution of fuel quality improvements to the reduction of exhaust emissions is comparatively small [11]. Without the help of correspondingly adapted fuels, however, it will not be possible to implement the new technologies needed to guarantee the prescribed low pollutant emissions over the life time of car (100,000 miles=160,000 km).

The better the fuel quality the easier the task of building good engines capable of complying with the various requirements. To meet the multiple demands of society in a practical way the complete system from the refinery to the recycling of used cars must be comprehensively optimized. Anyway the motto is that "clean engines need clean fuels".

	CO_2-Emission [%]
Exploration and transport of crude oil	2 – 5,7
Refining	4 – 7
Distribution	< 1
Use in car	85 - 94

Fig. 19. CO_2 life cycle evaluation

5 Alternative Fuels

More than 125 years ago, a solid symbiosis between fossil fuels and the internal-combustion engine came into being. This symbiosis will continue to exist as long as mineral oil will maintain its role as the primary energy source of mankind.

With today's vehicle population running almost exclusively on "classical" mineral-oil-based fuels, a smoothly functioning oil supply is vital to guarantee man's mobility and satisfy his transportation requirements.

The forecasts concerning the remaining crude oil reserves have to be corrected at regular intervals (Fig. 20).

According to the forecasts made in the early 1980s, there would be just enough oil for next 30 years to come. Recent forecasts predict sufficient reserves for the next 4 to 7 decades – that is until far into the 21st century – depending on which boundary conditions are taken into account [12].

When discussing future energy requirements, however, we must give due consideration also to the people living in this world: the energy requirements of the ever increasing world population and the upcoming economic regions is constantly growing. The people living in those developing countries are striving to reach the same standard of living as the industrialized nations. Consequently, the demand for crude oil and other energy sources will grow enormously.

No matter how long the oil reserves of our earth will last in the end – they are definitely limited.

The search for alternatives to replace gasolines and Diesel oil as automotive fuels is not a new approach at all. The constant driving force behind this search for new alternative energy sources has been the awareness that the crude oil reserves will not last for ever and that the political situation in most of the oil-producing

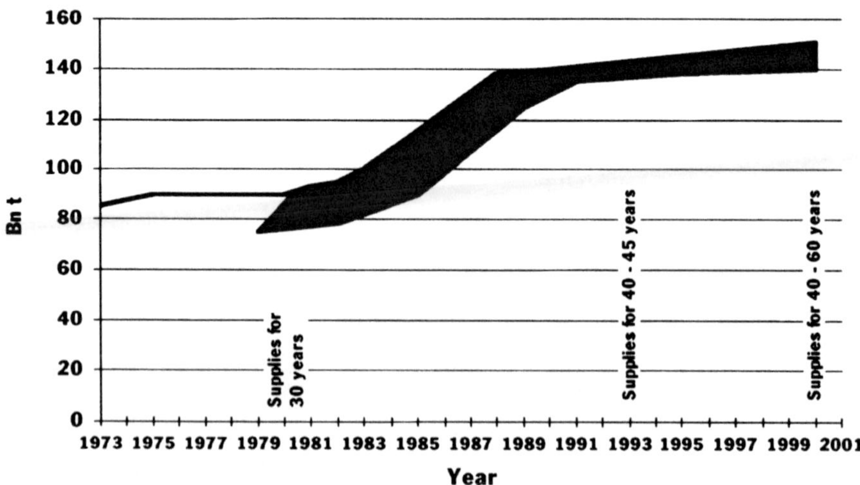

Fig. 20. Forecasts about "remaining" crude oil reserves

Fossile	Non-Fossile
Mineral oil	Biomass
Natural gas	Water
Coal	Wind
Oil shale/oil sand	Solar energy
	Tidal power
	Geothermal Energy
	Nuclear Power

Fig. 21. Energy sources

regions of this world is unstable. Mainly during the oil crises of 1973 and 1981, alternative possibilities had been examined in order to reduce the dependence on oil supplied by the OPEC countries.

During the last 30 years or so of the 20th century, this problem was further enhanced by concerns about environmental pollution through the combustion products from millions of motor cars. One of the most urgent targets must be to develop clean alternative fuels to complement the existing mineral oil products in a first approach and to completely replace them later on.

Figure 21 lists potential energy sources of fossil and non-fossil origin which will be intensively checked for their inherent potential as alternative fuels in the foreseeable future.

Until non-fossil energy carriers can be fully exploited, the main task of automotive manufacturers will be to further reduce the fuel consumption in order to make the existing crude oil reserves last and thus gain precious additional time for the development of new, seemingly inefficient or completely unknown energy sources today.

5.1 Alternative Fuels of Fossil Origin

20 years ago, the introduction of alternative fuels – mainly won from oil shale, oil sand and coal – was predicted for the end of the 20th century. The fossil and non-fossil fuels considered as alternative fuels today (Fig. 22) are the same even if their future prospects are viewed quite differently.

Currently, the discussions about alternative fuels are mainly focusing on environmental issues with emphasis on their potential of reducing the CO_2 emissions. Among such fuel sources number natural gas (methane and ethane) and liquid gas (propane and butane), hydrogen, methanol (from natural gas, biomass or coal) and fuels obtained from biomass (ethanol and vegetable oils). Electric vehicles, too, number among the "environmentally friendly" automobiles.

Alternative fuels are suited for short to mean-term application only if they are tolerated by the Otto and Diesel engines of current vehicle populations and can be distributed via the existing infrastructure system.

Energy Source	Forecasts in 1980	Current Status	Forecasts in 2000
Mineral Oil	- Limited resources - Politically unstable sources - Prices will increase dramatically	⟷	- identical - identical - less expensive than in 1980
Natural gas	- Limited resources	↑	- confirmed reserves for 30 to 60 years – Main fuel for electric energy production and transportation
Coal	- Most economic alternative for synthetic fuels & electric current	↓	- Concerns about greenhouse effect might hamper the exploitation
Nuclear Power	- Will be the main source of electricity	↓	- No capacity extension planned in the USA
Oil shale and oil sands	- Enormous potential as a source of synthetic fuels	↓	- Currently of no importance; might be of interest in the future
Solar energy	- Solar energy and biomass will play a predominant role in 2000	↓	- Is currently in the F&E phase No alternative of importance for the next years to come
Nuclear Fusion	- Ultimate dream	⟷	- Identical

Fig. 22. Primary energy sources

5.1.1
Natural Gas and Liquefied Petroleum Gas

Natural gas (LNG, CNG) and liquefied petroleum gas (LPG) are the only alternatives currently used besides the mineral-oil-derived fuels. Natural gas (methane and ethane) and liquefied petroleum gas (propane and butane) are by-products of the production of fossil fuels out of mineral oil and are widely available. The existing reserves are estimated to last for more than 60 years.

In compressed, liquefied or absorbed forms (CNG, LNG and ANG, respectively) they are used as an alternative to gasoline for fueling internal-combustion engines.

Liquefied petroleum gas, or LPG (commonly called propane), is a liquid mixture (at least 90% propane, 2.5% butane and higher hydrocarbons, and the balance ethane and propylene). It is a by-product of natural gas processing or petroleum refining.

Propane is the most accessible of the liquid and gaseous alternative fuels. Publicly accessible fueling stations exist. Propane has been used as a transportation fuel around the world for more than 60 years.

Range on LPG is somewhat less than that of comparable gasoline-powered vehicles. Power, acceleration, payload, and cruise speed are comparable with those obtained with an equivalent internal-combustion engine.

Thanks to the improved mixture formation and resulting more efficient combustion of gas-powered engines, the carbon monoxide (CO) and unburnt hydrocarbon (HC) emissions are lower than those measured with liquid fuels. The composition of natural gas (mainly methane CH_4) and liquid gas (mainly propane C_3H_8 and butane C_4H_{10}) with their favorable C/H ratio results in lower CO_2 and NO_x emissions.

Gaseous fuels contribute to the reduction of limited and non-limited exhaust-gas constituents and of the particulate emission. With natural gas (CNG), the non-methane HC emission is very low.

It is difficult for gaseous fuels, however, to undercut the low emissions reached with modern gasoline engines fitted with three-way catalyst and oxygen sensor.

All the gaseous fuels have one particular feature in common: due to their low energy density when compared with gasoline and Diesel fuel they require essentially more sophisticated, bigger, heavier and more expensive fuel tanks. Although utility companies have gas pipelines in place there are few stations designed for refueling passenger cars. Refueling natural gas vehicles is a slow process.

The service-proven technology for the operation of internal-combustion engines on either natural gas (CNG) or liquid gas (LPG) has been known for many years. Quite frequently, engines are laid out as bi-fuel engines because there is no distribution infrastructure yet for gaseous fuels.

In the future, it will also be possible to use gaseous fuels as a basic material for the production of liquid fuels by means of the GTL-Gas-to-Liquid Procedure. This method is economically justifiable at a mineral oil price of 20–22 US $/barrel [12] and produces Diesel fuels of very good quality.

5.1.2 *Methanol*

Methanol (methyl alcohol or wood alcohol; CH_3OH) can be made from natural gas, coal or wood. Methanol is less flammable than gasoline, and can be cheaply and easily made from natural gas. Methanol engines have 30–50% fewer toxic and organic emissions than gasoline engines (except for formaldehyde) and are more efficient. Methanol has none of gasoline's carcinogenic ingredients. It is easy to make vehicles that run on gasoline, methanol, or a combination of both, making for a simple, large-scale transition from gasoline to methanol.

A methanol vehicle has only 60% of a gasoline vehicle's range. Methanol itself is also toxic, although safe product designs can limit the danger. Methanol engines emit toxic formaldehyde (although catalytic converters limit this). Pure methanol (M100) burns invisibly, and has trouble starting an engine in cold weather. M85, which is 85% methanol and 15% gasoline, avoids these problems.

Tests with methanol-fueled Diesel engines have shown that no soot occurs since alcohol combustion is soot-free.

5.1.3 *Dimethyl Ether (DME)*

Lately, dimethyl ether (DME) has been proposed as an alternative fuel of fossil origin for Diesel engines (Fig. 23).

Dimethyl ether (CH_3-O-CH_3) can be gained from natural gas, coal and biomass. It has a high oxygen content of 35% and burns soot-free.

	DME	Diesel oil	Methanol	Ethanol	CNG (Methan)
Chemical formula	CH_3-O-CH_3		CH_3-OH	CH_3-CH_2-OH	CH_4
Lower calorific value (MJ/kg)	27.6	42.5	19.5	25.0	50.0
Density (g/ml)	0.66	0.84	0.79	0.81	
Cetane number (-)	>> 55	40 ÷ 55	5	8	
Self-ignition temperature (°C)	235	250	450	420	650
Octane number (-)	-	-	111	108	130
Stoichiometric A/F ratio (-)	9.0	14.6	6.5	9.0	17.2
Boiling point (°C)	-25	180÷370	65	78	-162
Evaporation heat (kJ/kg)	460 at −20° 410 at 20°	250	1110	904	-
Ignition limits (% of gas in the air)	3.4÷18	0.6÷6.5	5.5÷26	3.5÷15	5÷15
Weight-% of carbon	52.2	86.0	37.5	52.2	75.0
Weight-% of hydrogen	13.0	14.0	12.5	13.0	25.0
Weight-% of oxygen	34.8	0	50.0	34.8	0

Fig. 23. Combustion-relevant characteristics of DME versus Diesel oil an other alternative fuels

From an ecological point of view, DME is beneficial since it results in clear reductions of all pollutant emissions.

The future of this alternative fuel will depend on whether it will be possible to produce the required great amounts at economically justifiable costs.

5.2 Fuels from Renewable Resources

When looking for fuels to replace the traditional mineral-oil-derived products the renewable energies derived from the sun, biomass, water and wind seem to be particularly attractive from an ecological point of view.

Every year, the surface of the earth receives solar energy corresponding to about 10 times the energy stored in the entire world-wide reserves of fossil fuels and uranium or to 15,000 times the yearly world-wide energy requirement. To date, a tiny fraction only of this enormous energy potential is being made use of. The problem of solar energy is in its very low density.

The live plants on our earth store as much energy as all the confirmed coal, oil and natural gas reserves put together.

Nature produces 170 billion tons of biomass each year; 6 billion tons or 3.5% of which are used by man as food. A tiny part only serves as raw material for petrochemistry. Thanks to appropriate biological procedures, it has been possible, however, to cover 5% of the world-wide energy consumption with biomass-derived energy.

Biological fuels have accompanied the development of the internal-combustion engine from its very beginning. In 1912, Rudolf Diesel stated: "The amount of vegetable oils used as automotive fuels might be negligible today. But over the

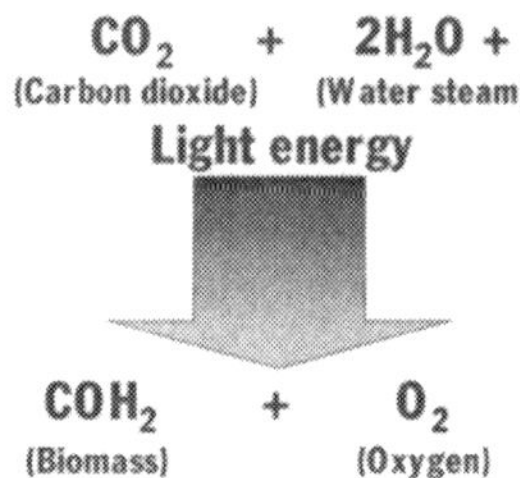

Fig. 24. Simplified formula of photosynthesis

years, such oils might become as important as petroleum and the currently used coal/tar products."

Accordingly, the fuel shortages of mid 1970s brought a renewed interest in bio fuels produced from renewable feedstocks, using vegetable oils including used frying oil, waste animal fats and even landfill gases.

"Biomass" is a collective noun which designates all vegetable materials and vegetable waste from woods and cultivations as well as all animal waste and organic matter from private households.

Despite the inherent enormous energy potential of biomass, the production of vegetable energy has hardly been considered to date.

In our age of discussions about recycling and closed circuits, however, fuels made from renewable resources are an attractive solution.

The photosynthesis (Fig. 24) is one of nature's most excellent inventions. Theoretically, it also is an ideal way of recycling fuels. Under the influence of solar energy, the carbon dioxide and water contained in the atmosphere combine to produce biomass – an excellent raw material for high-quality engine fuels such as alcohols and vegetable oils. The combustion products of these fuels – that is carbon dioxide and water steam – are the very raw materials from which biomass is produced, thus closing the natural carbon and hydrogen circuit of nature. There is no further increase of the carbon dioxide concentration in the atmosphere. The wood waste from 1 ha of forest, for example, can replace 200 to 300 kg of heating oil per year.

In the future, the contribution of biogenic fuels will continue to increase and allow the demand for energy to be covered in an ecologically and climatically more compatible way. According to current knowledge, however, the process of exploiting fuels from biomass is not yet economically efficient enough.

When producing bio fuels, only 20% of the energy input is recovered at the end of the conversion chain. And renewable raw materials still have some other drawbacks: they are not readily available over the year in sufficient quantities and at acceptable qualities. Their compositions vary and the prices are unstable.

5.2.1 *Ethanol*

Ethanol – also called ethyl alcohol or gain alcohol (C_2H_5OH) – is made by fermenting wheat, fruit, sugar cane and sugar beets. Ethanol produces fewer organic and toxic emissions than gasoline, and ethanol-burning engines are more efficient than those run on gasoline. It can be produced from a variety of renewable resources. Researchers are investigating how to make ethanol from the wood and plant cellulose found in biomass, which could make ethanol economically viable as well as ecologically sound. Current vehicles can be modified to run as flexible-fuel vehicles that take either gasoline or ethanol. Ethanol is a very promising reproductive resource which is also attractive for its neutral CO_2 balance.

Ethanol is quite expensive. Producing ethanol in large volumes would drive up the prices of both the fuel and its source crops. Dedicating huge amounts of cropland to fuel production could have negative ecological effects, including the production of greenhouse gases. Ethanol vehicles also have only 75 to 90% of the range of gasoline vehicles with the same size fuel tank. In the U.S., ethanol is currently produced with the help of a tax credit from the government, and is sold as E85 (85% ethanol, 15% gasoline). Ethanol has a high octane number and could also provide the oxygen currently required in reformulated gasoline.

Fuels containing oxygenous constituents are mostly used as additives for conventional fuels and less as fuels in their pure form. It is relatively easy to adapt internal-combustion engines to the requirements of such fuels. The higher octane ratings of alcohols and ethers can be made use of to improve the efficiency of the engine.

5.2.2 *Vegetable Oils*

Diesel engines tolerate a wide variety of vegetable oils. They have high cetane ratings, no sulfur and good lubricity. However, the manufacturers of vegetable oils must be provided with clear specifications for the production and supply of bio fuels in form of a fuel specifications. The most widely known bio fuel besides ethanol is rapeseed methyl ester.

Biodiesel advocates cite several advantages to biodiesel. Because it can be used in existing Diesel engines without costly modifications or retrofits, it represents the last costly way to comply with different energy and environmental requirements. It can also be stored in regular Diesel tanks and pumped with regular equipment.

On the environmental side, since it is made from feedstocks that take carbon dioxide out of the atmosphere, using biodiesel adds no net carbon dioxide to the atmosphere which is important to reducing greenhouse gases.

When used in a conventional Diesel engine, combustion of biodiesel which is an oxygenated fuel results in a more complete combustion and improved emission. This results in a reduction of unburnt hydrocarbons, carbon monoxide and particulate matter.

Biodiesel is non-toxic, biodegradable and free of sulfur. Because there is no sulfur in biodiesel, it works well with catalysts, particulate traps and exhaust gas recirculation systems technologies that will have increasing use in the next years. Currently, the main drawback of biodiesel is the significantly higher costs compared to petroleum diesel.

The use of alternative fuels such as alcohols and bio fuels having very different properties compared to conventional fuels brings on the need for special new additive treatments.

5.2.3
Hydrogen (H_2)

Hydrogen is a particularly interesting alternative fuel for internal-combustion engines. It is the true primary substance which fuels the universe and man's main great hope with regard to the power supply of the future.

For many years, this fuel has been successfully used in research combustion engines. Combustion engines run on hydrogen emit nothing more than water vapor and nitrogen oxides though in clearly smaller quantities than conventionally-fueled engines.

Hydrogen also is optimally suited for the most promising alternative automotive power plant concept for the future: the fuel cell.

Hydrogen, however, is not easy to produce, transport, refuel and store. It can be stored either in liquid form at –253 °C or as condensed gas at 300 bar.

Today, more than 96% of the hydrogen is derived from mineral oil, natural gas and hard coal releasing great amounts of CO_2 in doing so. The appeal of hydrogen will increase only if we succeed in producing it electrolytically from water with the help of solar energy. The corresponding technological prerequisites will probably be available by end of the 20th century. To date, mass producing hydrogen from water has been one of the unfulfilled dreams of mankind. Currently, the costs of deriving hydrogen from water by way of electrolysis are 60 times as high as the production costs of gasoline or Diesel fuels.

In Germany, hydrogen is considered to be the fuel of the future [18] whereas in the USA it does not seem to have any chances of introduction [19].

Corresponding studies performed by DOE (Department of Energy) and CARB (Californian Air Resource Board) emphasize the difficulties expected in handling hydrogen. The low energy density of hydrogen makes it very expensive to transport and store. Further, there is no appropriate distribution infrastructure available. And it would take 100 billion US $ to produce and distribute the amount of hydrogen required to cover just 10% of the energy needed for US road traffic. According to the CARB study, "hydrogen is suited neither technically nor economically for automotive application and will not be so in the foreseeable future". A potential solution is liquid hydrogenous fuels: An on-board reformer decomposes these fuels into H_2 and CO_2. There are quite a number of fuels which might serve as H_2 sources such as gasoline, methanol, ethanol, crude oil and bio fuels. Each of these has its particular advantages and drawbacks and must be made to comply with the new requirements of fuel cell application.

What would be particularly desirable are saturated hydrocarbons (paraffins) with their high H_2 content. And it is also important that fuel-cell fuels be completely sulfur-free. Sulfur is sheer poison for both the reactor and the fuel cell itself. The octane number is no longer of importance for fuel cell application and there is no need either for oxygenous components.

6 Outlook on the Future

The symbiotic relationship between internal combustion engines (Otto and Diesel) and petroleum-derived fuels (gasoline and Diesel oil) has characterized the entire road traffic scene for more than 100 years. To date, the increasingly stringent demands on vehicles and engines could be complied with by harmonizing the further development of engines and fuels.

The automobile will continue to be the most important means of transportation far into the 21st century. Many of the future solutions with regard to new environmentally friendly vehicles will directly affect also the fuels used to power those vehicles.

The current engine generation and the engines and exhaust gas aftertreatment systems being developed for future applications place new and even more severe demands on fuel quality.

The demands have been laid down in the World Wide Fuel Charter drawn up in a joint approach by the international automotive industries. The new demands on fuel quality are evolving parallel to the development of a new generation of engines. The focus is on the legal requirement of simultaneously reducing the exhaust pollutant and CO_2 emissions (fuel consumption).

Modern fuels for Otto and Diesel engine applications must comply with octane and cetane requirements, be free of sulfur and meet a series of additional fuel specifications which are defined in the WWFC.

According to the current state of the art, the "conventional" fuels (gasoline and Diesel oil) are the most economic alternative for most transportation tasks. The polluting character of mineral oil is a growing handicap. It is up to the oil industry to improve the fuel quality and thus clearly lower the pollutant emissions.

The current priorities in the field of fuel development can be summarized as follows:

1. Improvement of the quality of conventional fuels. Engines with good performance characteristics require high-quality fuels to make full use of their inherent potential. The better the fuel quality the easier the building of good, high-efficient engines with low pollutant emissions. As early as in 1919, Charles F. Kettering wrote that "the fuel problem has an automotive and a petrochemical aspect and can be solved only through close co-operation between both industries".
2. Reduction of the fuel consumption of Otto and Diesel engines. By reducing the fuel consumption the existing mineral oil reserves can be made to last longer allowing precious time to be gained for the development of new energy sources which, at the time being, may appear to be uneconomic or still be completely unknown.
3. It appears to be useful and necessary now to intensively deal with potential alternative fuels. So far, except for some local applications, alternative fuels have gained no major importance yet on a world wide level.

Major reasons for the failure of alternative fuels to assert themselves are:

- High production costs. The relatively low oil prices of the last few years have clearly delayed the development of alternative fuels. This is also true for the most promising ones. According to forecasts in the late eighties, more than 10% of the primary energy needed in 2000 would be covered by reproductive energy sources. Today, the world-wide share is less than 2%.
- **Complicated handling** as well as loss of comfort and/or restricted vehicle utilization. Figure 25 compares the energy densities of various energy carriers.
- The range of a vehicle with a constant tank volume is proportional to the energy density.

When considering fuels under an ecological point of view, it is essential to also take the CO_2 emission into account. Figure 26 compares the overall CO_2 emissions of various fuels giving consideration not only to the exhaust emissions during vehicle operation but also during fuel supply (production, transportation).

According to previous experience, it takes fundamental scientific discoveries several decades to be commercially applied. In the field of automotive fuels, there still is a lack of such fundamental developments which might revolutionize the fuel market in the next 10 to 15 years to come.

The development of automotive engines depends on the available fuels. Any modification to the fuel type necessitates a new engine concept. Since it takes 4 to 6 years to redesign an engine the automotive manufacturers have great interest in future-oriented fuel developments.

The fundamental characteristics of alternative fuels such as liquid gas, natural gas, methanol, ethanol and hydrogen are very similar to those of current fuels. To the engine designer this means that the combustion engine will continue to pre-

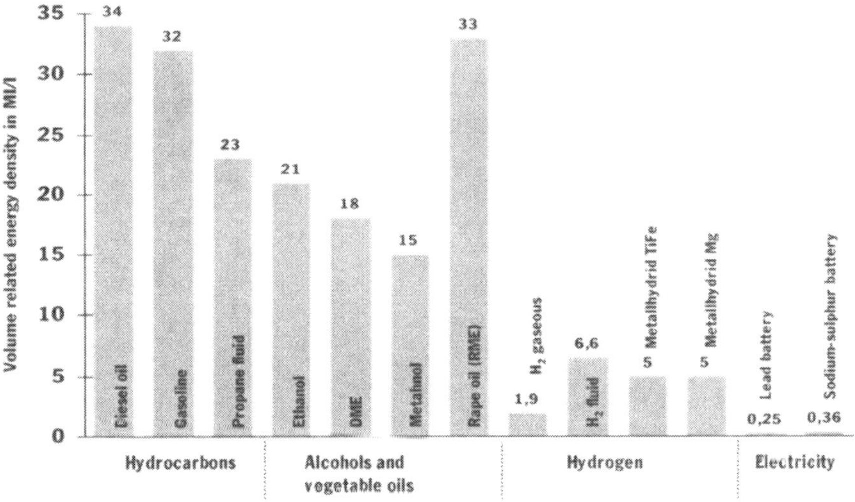

Fig. 25. Energy densities of various energy carriers

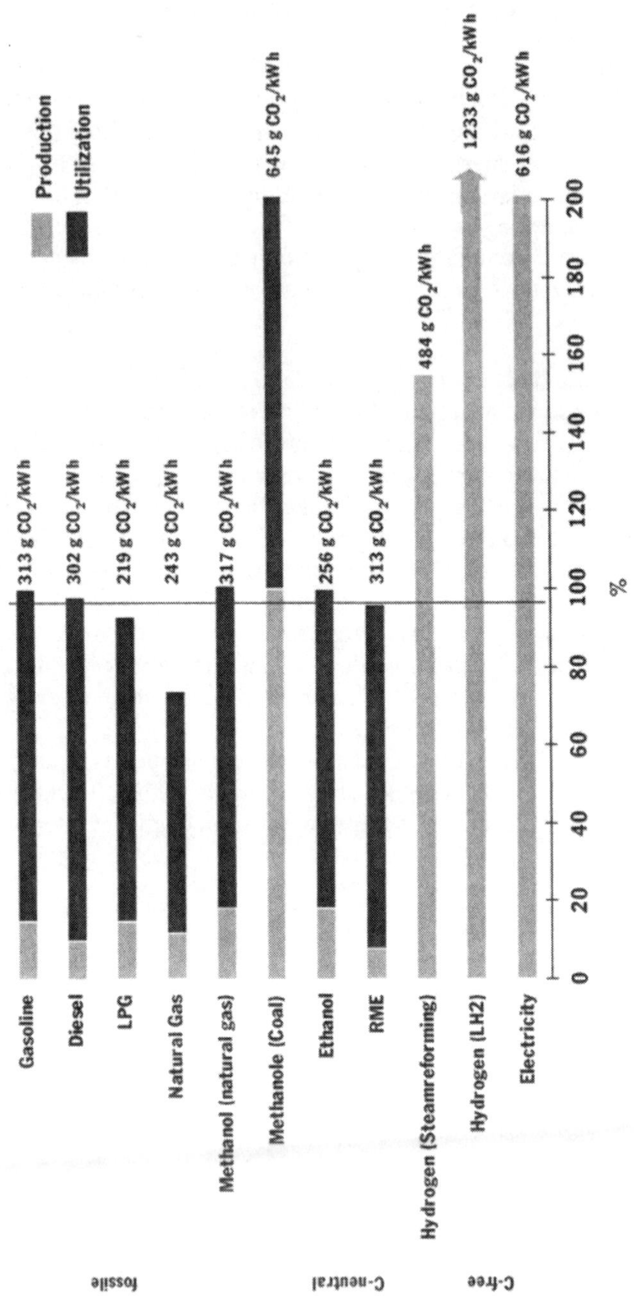

Fig. 26. CO_2 emission of different fuels

vail as the main automotive power unit but that it will have to be adapted to potential new fuel characteristics.

From this situation, the oil industry should draw the conclusion that everything must be done to optimize the fuel qualities and to assist the automotive industry in its efforts to meet the requirements of modern society.

Of all the potential alternative propulsion systems for the future, the only alternative intensively examined at the time being is the fuel cell concept. Hydrogen, the most efficient fuel for this type of propulsion system, will probably have to be ruled out for the foreseeable future. Other fuels under development are liquid hydrocarbons such as methanol and gasoline. Some of the requirements on fuels for fuel-cell operation are in compliance with the demands on engines: high H/C ratio, absence of sulfur. Others are quite different such as the octane or cetane numbers and oxygen contents. The solutions to be found in terms of fuel production, transportation, replenishing and storage for fuel cell applications will be beneficial for internal-combustion engines as well. All the fuels suited for fuel cells can also be used in internal-combustion engines.

The future will show which requirements the fuels for future-oriented propulsion systems will have to comply with. What is known and generally acknowledged today is that "clean engines need clean fuels".

7 References

1. Guibet JC (1997) Fuels and Engines. Edition Technics Paris
2. Stocky TP, Henly TJ, et al (2001) Opportunities in Distillate Blending. World Refining April
3. World Wide Fuel Charter (2000) ACEA-Brüssel, April
4. Directive of the European Parliament and of the Council on the quality of petrol and Diesel fuels and amending directive 98/70/EC. Commission of the European Communities, Brüssel 2001
5. Armstrong AP (1999) Alternative transport fuels. Horses for courses. I. Mech. E. Paper S 607/002/99
6. Data of the sulphur effect on advanced emission-control technologies. ACEA-Report, Brüssel 2000
7. Göbel U, Kreuzer T, et al (1998) Moderne NO_x-Adsorber-Technologien. VDA-Technischer Kongreß
8. Diesel-Emission Control – Sulphur Effects. Automotive Engineering International, December 2000
9. Refining Industry divided: Implementing Diesel-Sulphur Rule. World Refining Nov. 2000
10. EU-Consultant Issues. Summary Report on Sulphur in Fuel. Walsh-Car Lines. Dec. 2000
11. Fuel quality, vehicle Technology and their interactions. Concawe Report No. 99/55. Brüssel 1999
12. Giere H-H, Nierhauve B (2001) Kraftstoffe aus Sicht eines Mineralölunternehmens. 3rd Colloquium Fuels. TA-Esslingen
13. Advanced Technology Primer. Alliance of Automobile Manufacturer. Washington D.C. 2000
14. Uddin N (2000) There's No Fuel like an Old Fuel: Powertrain International
15. Thrän D, Rösch C, et al (2000) Normen für biogene Festbrennstoffe. BWK 52, 12
16. Krumm H, Lange WW, et al (1996) Geringere Umweltbelastung durch verbesserte Fahrzeugbetriebsstoffe. Shell Technischer Dienst
17. AGELFI – 12th European Automotive Symposium. Straßbourg, 1996
18. Nierhauve B (2000) Herausforderungen an die Kraftstoffbereitstellung. Mineralöl-Anwendungstechnik 45

19. Berlonitz PJ, Darnell CP (2000)Fuel Choices for Fuel Cell Powered Vehicles. SAE-Paper 01–003
20. Graskow BR, Ahmadi MR, et al (2001) The influence of Fuel Additives on ultrafine Particulate Emissions from Spark Ignited Engines. Int. Conference Fuels. TA-Esslingen
21. Scott HW (1976) Energy – Past, Present and Future. SAE-Paper 760 402
22. Meurer JS (1986) Auch Automobile bedürfen neuer unerschöpflicher Energiequellen. Fisita-Kongreß. Belgrad
23. Kane ED (1980) The future of automotive fuels and lubricants. Automotive Engineering 88:10
24. Pester W (1987) Wasserstoff: der Kraftstoff für Autos der Zukunft. VDI-Nachrichten 16.01.1987.
25. Waldeyer H, Quadflieg H, et al (1987) Chancen und Risiken für Kraftstoffe auf der Basis nachwachsender Rohstoffe. Forschung und neue Technologien im Verkehr. Hamburg
26. Fabri J, Dabelstein WEA, et al (1990) Chancen alternativer Kraftstoffe unter dem besonderen Aspekt der Umweltverträglichkeit. Shell Technischer Dienst. 28. März
27. Graeser U, Keim W, et al (1995) Perspektiven der Petrochemie. Erdöl, Erdgas, Kohle 11:5.
28. Kraftstoffe von Morgen. Kollektion Forschung & Entwicklung. Renault 1993.
29. Weniger Energie auch im Verkehr. Esso Energieprognose 93
30. Fuel Additives, Deposit Formation and Emissions. European Commission, ACEA, EUROPIA, Brüssel, 1995.
31. Hattiangadi U, Spoor M, et al (2000) Clean fuels: a strategy for today's refiners. Petroleum Technology Quarterly. Winter 2000/01
32. Joshi S, Lave L, et al (2000) A Life Cycle Comparison of Alternative Transportation Fuels. SAE-Paper 01–1516
33. Lee CK, McGovern S (2001) Comparison of clean diesel Production technologies. Petroleum Technology Quarterly (PTQ), Winter 2001/02
34. Siuru B (2001) Bio diesel Remains an Option in Alternative Fuel Mix. Diesel Progress, June

Subject Index

Printing: Saladruck Berlin
Binding: Lüderitz&Bauer, Berlin

Lightning Source UK Ltd.
Milton Keynes UK
19 May 2010

154413UK00007B/34/P

9 783540 000501